Environmental Biotechnology

BIMAL C. BHATTACHARYYA

Professor, Department of Food Technology,
Techno India, Kolkata

RINTU BANERJEE

Professor, Department of Agricultural and Food Engineering,
Indian Institute of Technology Kharagpur

OXFORD

UNIVERSITY PRESS

OXFORD
UNIVERSITY PRESS

Oxford University Press is a department of the University of Oxford.
It furthers the University's objective of excellence in research, scholarship,
and education by publishing worldwide. Oxford is a registered trademark of
Oxford University Press in the UK and in certain other countries

Published in India by
Oxford University Press
22 Workspace, 2nd Floor, 1/22 Asaf Ali Road, New Delhi 110 002

First published 2007
10th impression 2023

ISBN-13: 978-0-19-568782-8
ISBN-10: 0-19-568782-5

Typeset in Palatino
by Planman Technologies, India
Printed in India by Repro India Limited, Haryana

For product information and current price, please visit www.india.oup.com

To the loving memory of

Late Gopal Chandra Banerjee
Late Mahim Chandra Bhattacharyya
Late Sukhamoyee Devi

for their divine blessings

Preface

Environmental pollution is one of the three major crises being faced by the world today. The other two are energy scarcity and food shortage. Environmental pollution is causing ecological imbalance, which is posing a threat to the existence of the living system; the rapid depletion of conventional energy sources such as fossil fuels is creating a grim situation for the furtherance of the present day civilization; food shortage is causing millions around the world to suffer from malnutrition. All these problems can be effectively tackled directly or indirectly by appropriate application of the knowledge of aerobic and anaerobic waste treatment processes.

Scientists and technologists are attempting to identify alternative eco-friendly energy sources, such as solar, wind, biogas, and tidal, as substitutes for fossil fuels. The indiscriminate use of chemical fertilizers instead of the vastly beneficial and harmless organic manure not only hampers land fertility, but also causes land and water pollution. Most of the causes of pollution can be successfully eliminated through the use of biological treatment methods.

The main functionaries of the biological treatment of contaminated solid and liquid wastes are micro-organisms. For their growth or survival one group of micro-organisms (aerobes) needs air or oxygen, while another group (anaerobes) cannot tolerate the presence of dissolved oxygen in their environment. The aerobic system reduces the concentration of organic pollutants in wastes, while the anaerobic system, along with the reduction of pollutants, yields biofuel (CH_4 gas) and a humus-like residue enriched with nitrogen, phosphorus, and potassium. Waste thus becomes useful!

The knowledge obtained from the authors' earlier works forms the nucleus of this text. This book presents both aerobic and anaerobic pollution control methods.

Though a large number of articles on this subject appear regularly in journals, and several conferences and symposia are held frequently, there are no textbooks that tackle this subject appropriately or serve the needs of students, especially in India. This book has been written with a view to cater to the need of the various sections of the society interested in this subject. This book is obviously not the first to deal with the subject, but could well be the first to incorporate a fundamental but consolidated treatment of the application of anaerobic systems for converting waste into wealth.

Though this textbook has been prepared primarily for undergraduate and postgraduate students of engineering colleges and universities, it will also serve as a useful reference for scientists and technologists who are actively involved in the application of biological waste treatment processes.

Content and Coverage

The text has been broadly divided into eight chapters.

Chapter 1, the introductory chapter, starts with the definition of environmental pollution. It explains the nature of pollutants and the microbiology and biochemistry of aerobic and anaerobic treatment methods, including bioreactors. The chapter concludes with information on the minimum national standards for waste disposal. The main objective of this chapter is to introduce readers to this subject. It will serve the purpose of a detailed introduction to those who do not want to delve deep into the technicalities of the subject matter.

Chapter 2 discusses in detail the nature of liquid and solid wastes, the physico-chemical characteristics of wastes depending on their sources and availability, and the nature of effluents from specific industries.

Chapter 3 discusses the basic scientific aspects of anaerobic treatment processes for biofuel generation from waste. It includes the microbiological and biochemical aspects of waste treatment, strain improvement methods, and environmental aspects influencing the anaerobic digestion system. Biotechnologists or environmental engineers as well as bioscientists and the rest of the scientific community will find this chapter very useful. The information covered in this chapter is essentially required for the design of bioreactors.

Chapter 4 extensively covers the analytical methods employed to determine the levels of pollutants in waste before and after treatment. It includes all the relevant techniques.

Chapter 5 elaborates on engineering basics such as material and energy balance as well as reaction kinetics, which constitutes the starting step of bioreactor design.

Chapter 6 discusses the design procedures for bioreactors or fermenters for the treatment of liquid waste. It explains the design steps systematically, so that students can use this information to solve problems related to liquid waste treatment.

Chapter 7 tackles the management of solid wastes. It covers non-biological treatment processes such as incineration and pyrolysis as well as biological treatment processes including landfill bioreactor design and vermicomposting.

Chapter 8 discusses the socio-economic aspects of biological treatment of organic wastes. It includes two case studies—(i) Design and Economic Evaluation of an Integrated Biogas Plant for a Dairy Farm and (ii) Design Analysis of a Community Biogas Plant.

Acknowledgements

We would like to acknowledge the support received from the members of the Oxford University Press, India. Though this book is mainly the outcome of our vast experience in teaching and research as well as some years of extension service in rural areas, the topics covered in different short-term courses, seminars, and workshops held in India have been immensely useful. We acknowledge these and other references (listed in the chapters), which have been consulted from time to time to give the book its present shape.

We also acknowledge with great appreciation the assistance rendered to us by our students, scholars, and colleagues during the preparation of the manuscript. Though it is difficult to mention the names of all of them, we would like to expressly thank Sukanta Bhattacharya, Mithu Das, B. Sanyasi Rao, Sandipan Karmakar, Prabhat Kumar, Amiya Bhowmik, and Lakshmishri Roy, without whose devoted help this book would not have been completed in time.

Finally we express our gratitude to Smt. Arati Banerjee and Brahmachari Vivekananda for their encouragement, appreciation, and blessings.

BIMAL C. BHATTACHARYYA
RINTU BANERJEE

Contents

Introduction

Introduction

Environmental biotechnology, an important branch of biotechnology, deals with the detection of environmental contaminants contained in industrial, agricultural, and domestic wastes and the remediation of the pollution caused by this contamination. The study of this discipline primarily spans two main areas: (i) *environment science* and (ii) *biotechnology*. Environmental biotechnology involves applying the knowledge of biotechnology to solve environmental problems. This chapter presents an overview of the causes of pollution, the nature of pollutants, and the application of *green technology* to remediate the damage caused by pollution, and presents the minimum national standards for waste disposal into the environment.

Environment

Environment has been defined as the 'sum total of all the conditions and influences that affect the development and life of organisms'. This is a comprehensive definition as it stresses the totality of the environment and encompasses every living organism, from the lowest level to the highest, including human beings who have their own environment. For human beings environment consists of air, water, and land. Human existence on the earth is, therefore, dependent upon the maintenance of symbiotic equilibrium between human beings and their environment. In recent centuries, this relationship has changed drastically because of the rapid industrialization taking place without any concern for its impact on the environment. To meet our ever-increasing needs we have not left any part of the biosphere untouched and thus have put our own survival at stake.

Our environment performs three functions.

1. It provides living space.
2. It provides resources such as water, air, minerals, and soil.
3. It acts as a sink by assimilating the wastes produced by humans.

However, it is to be understood that the capacity of the environment to perform the aforementioned functions is limited, which is a matter of grave concern. It is important to ensure that the amount of stress imposed on the environment due to the rapid exploitation of its resources does not out do nature's capacity to replenish them. The amount of chemical and industrial effluents must not exceed the assimilative capacity of the environment. Fortunately, today all sections of our society are aware of the problem of environmental pollution. Various laws have been introduced in almost all countries to maintain air and water quality. To make this relationship between human beings and the environment more amenable and sustainable, scientists are depending more and more upon the knowledge of biotechnology.

Biotechnology

Biotechnology has been referred to as one of the greatest technological achievements of the twentieth century. It is also considered one of the most important technologies of the present century. The recent spectacular developments in biotechnology and genetic engineering emphasize their multifunctional role in achieving social development in the face of many challenging problems such as growing population, diminishing non-renewable resources, and ecological hazards of development and growth.

Biotechnology can be defined as *the application of scientific and engineering principles to the processing of materials by biological agents and services.* This shows that in its development the *frontiers in biosciences, physical science, and engineering techniques are coalescing for achieving techno-economic multiphase applications.* It is interesting to note that the term 'biotechnology' is derived from the fusion of the terms 'biology' and 'technology'. In other words, biotechnology is concerned with the utilization of biological components for generating useful products. According to the US Science Foundation, biotechnology involves the controlled use of biological agents such as micro-organisms or cellular components for beneficial use. The European Federation of Biotechnology defines it as the integrated use of biochemistry, microbiology, and engineering sciences in order to achieve technological application of the capabilities of micro-organisms, cultured tissues, or cells and parts thereof. According to British biotechnologists the application of biological organisms, systems, or processes constitutes biotechnology.

From these definitions and statements, it is clear that an enhanced understanding at the cellular and molecular levels of biological systems coupled with the development of processes based on new biological techniques have led to an increase in agricultural productivity and the development of vaccines, various diagnostic tools, drugs, improved animal husbandry techniques, hazardous wastewater treatment technologies, etc. It has been realized that a closer interaction and collaboration between bioscientists and engineers is necessary to achieve the techno-economic viability of biotechnology-based processes.

A common misconception among people is that biotechnology includes only DNA and genetic engineering. Genetic engineering is not a discipline, but a tool or technique that can be applied to all aspects of biotechnology such as microbes, plants, and animals; it can even be used for the remediation of environmental pollution. Biotechnology consists in a wide spectrum of sophisticated techniques that range from recombinant DNA technology and hybridoma technology to enzyme technology and enzyme engineering. The major branches of biotechnology are *microbial/industrial biotechnology, plant/agricultural biotechnology, animal/healthcare biotechnology, and environmental biotechnology.*

Though the term biotechnology is of recent origin, the discipline itself is very old. Microbes have been employed for making wine, vinegar, curd, bread, etc. since 5000 BC. Some of these processes are so common and have become such an integral part of our daily lives that we may hesitate to associate them with biotechnology. These processes are based on the natural capabilities of micro-organisms and are referred to as *old biotechnology.* A major area of biotechnological innovation has been in food production. Scientists have hybridized animals and plants to create a greater genetic variety. The offspring from the crosses were repeatedly bred to produce the greatest number of desirable traits, which eventually led to the production of many of the food products in use today.

At the beginning of the twentieth century biotechnology began to bring industry and agriculture together. During World War I, fermentation processes were developed that produced paint solvents and acetone from starch for the rapidly growing automobile industry. The outbreak of World War II and the consequent sufferings led to the manufacture of penicillin. Biotechnological focus then shifted to pharmaceutics. The cold war years were dominated by work on micro-organisms, meant for the preparation of biological warfare as well as antibiotics and fermentation processes. Present day biotechnology has its roots in chemistry, physics, and biology.

Table 1.1 Applications of biotechnology

Microbial biotechnology	Plant biotechnology	Immunotechnology	Environmental biotechnology
Enzymes for food processing, fine chemicals, biosensors, diagnostic kits Fuels: alcohol, hydrogen, methane Single-cell proteins (SCPs) Antibiotics Biofertilizers Biopesticides	Plant cell culture for fine chemicals such as steroids, alkaloids Tissue culture for transgenic plants and hybrid plants, which can resist adverse environmental conditions, and for artificial seeds	Vaccines Enzymes Fine chemicals such as hormones, antibodies, and blood factors by mammalian cell culture Diagnostic kits	Waste treatment for eco-friendly environment Waste utilization for value-added products, to make waste treatment process more economical

Today, biotechnology is a significant part of our lives, as it contributes to many areas such as agriculture, bioremediation, energy production, food production, waste disposal, mining, and medicine. After successfully cloning other species, scientists are now working on human clones. DNA fingerprinting is becoming a common practice in forensics. Insulin and other medicines are being produced through the cloning of vectors that now carry the chosen gene. Immunoassays are being used not only in medicine for drug-level and pregnancy testing but also by farmers as an aid to detect unsafe levels of pesticides, herbicides, and toxins in crops and animal products. These assays also act as rapid field tests that help to evaluate the amount of industrial chemicals in groundwater sediments and soil. In agriculture, genetic engineering is being used to produce plants that are resistant to insects, weeds, and diseases. An example is the *Flavr–Savr* tomato developed by the use of recombinant DNA technology. This was the first food item created by such a technique to be sold commercially. Thus, new biotechnological techniques have permitted scientists to manipulate and bring together the desired traits.

Table 1.1 lists the applications of different branches of biotechnology.

Environmental Biotechnology

Environmental biotechnology is one of the several areas of biotechnology, as mentioned in the earlier section, which applies the principles of microbiology to various environmental issues such as treatment of industrial and

Secret industrial waste dumping site discovered on ocean bottom

municipal wastewater; improvement in the quality of drinking water; restoration of commercial, archaeological, and other sites which are being destroyed by hazardous materials; protection of rivers, lakes, and coastal waters from environmental contaminants; prevention of the spreading of pathogens through water or air; production of environmentally benign chemicals such as ethanol, methanol, methane, SCP, etc.; and reduction of industrial residuals in order to check the production of pollutants requiring disposal.

Today the world is facing three major crises. These are environmental pollution, energy deficit, and shortage of food. The most alarming among these is environmental (air, water, and land) pollution, which is threatening to destroy the ecological balance. This has occasioned the need for increased awareness among people and governments regarding the consequences of this imbalance, propelling environmental biotechnology into prominence. A good example of the application of environmental biotechnology is the biogas plant, which recycles different types of wastes to produce useful end products such as biomanure and biogas.

Figure 1.1 shows how the process of recycling maintains ecological balance. The sun is the original source of energy for all living beings and it maintains ecological balance by the natural process of recycling. From Fig. 1.1 it can be seen that in the natural process the greenhouse effect is controlled by the

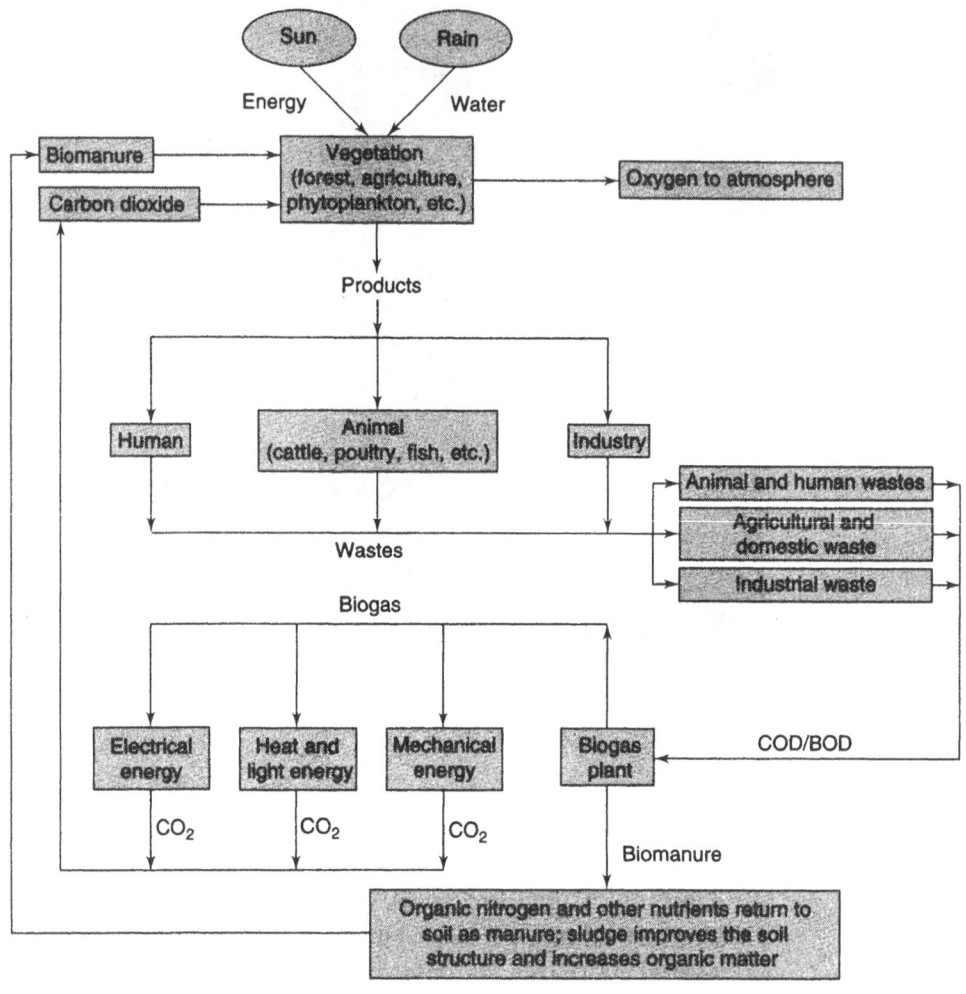

Fig. 1.1 Integral recycling process for combating pollution (COD denotes chemical oxygen demand)

conversion of carbon dioxide into biomass and oxygen. For sun rays to pass, clear skies are essential. The presence of smoke and industrial fumes hinders the process of the sun's energization of photochemical reactions. Indiscriminate cutting of plants and trees and disposal of refuse from various sources such as animals, households, and industry pollute the land and water by increasing the BOD (biological oxygen demand) level, thus affecting the growth of plants on land and algae, plankton, etc. under water, which are responsible for the reduction of CO_2 in the atmosphere and the supply of O_2. As 75% of the earth is covered by water bodies, water plants are mainly responsible for the production of O_2. If their growth is hampered by pollution, the O_2 generation rate will reduce and the CO_2 accumulation in the atmosphere will increase.

Types of Environmental Pollution

There are three forms of environmental pollutants, namely, gases, liquids, and solids. The pollutants emitted through industrial chimneys and exhausts, mainly oxides of nitrogen, hydrocarbons, CO_2, H_2S, SO_2, Cl_2, etc., cause *air pollution*. There are no direct biological means by which these pollutants can be arrested. For that reason air pollution does not come within the purview of this book.

Waste discharge from various sources pollute water, which leads to land pollution too, in four ways. Accordingly there are four types of water pollution.

1. Physical pollution
2. Chemical pollution
3. Physiological pollution
4. Biological pollution

Physical Pollution

Physical pollution of water is caused by the solid constituents of industrial effluents and sewage water. The nature of these solids varies depending on the type of industry. For example, in tannery effluents, the total solid content, including suspended solids, may vary from 20,000 ppm (parts per million) to 150,000 ppm, of which a major portion may be $CaCO_3$, hair, flesh, etc. from the lime yard and sludge from the vegetable tanning yard. On the other hand, in paper and sugar industries, the solids in the effluents are mostly lignocellulosic in nature. Most of the solids can be separated from the liquid by some physical process such as settling tank and filtration.

Chemical Pollution

Solid and colloidal chemicals that cannot be separated easily by any physical method, even by ultra-filtration, cause *chemical pollution*. These may be either organic or inorganic in nature. For example, an appropriate *p*H of water is important for the healthy living of aquatic species. For fish to survive, the desirable *p*H range is 5–9. However, due to the discharge of effluents containing inorganic soluble chemicals such as acids, salts, and alkalies, the *p*H of a natural water source may change, which is not desirable. The same thing is true in the case of land pollution also. Heavy metals, such as Cr, Ba, and As are also responsible for the toxicity of water when effluents containing such contaminants are discharged into water bodies.

Many water-soluble organic chemicals are also responsible for the toxicity of the receiving water. Some examples are phenols, chlorinated phenols, weak acids, etc. When highly oxidizable chemicals, such as condensed types of

Types of drums in which hazardous waste is stored for transportation and treatment

tanning and reducing types of bleaching agents, are discharged, occluded oxygen deficiency occurs in aquatic life, causing respiratiory problems. It is, therefore, essential that all such oxidizable chemicals present in wastewater be peroxidized either by adding oxidizing agents, such as H_2SO_4 and potassium chlorate, or by blowing air into the wastewater before discharging them into the receiving water. This will prevent the depletion of occluded oxygen by the effluent chemicals in the receiving water. To effect this, the amount of oxygen needed to stabilize the chemical oxygen demand (COD) of the wastewater must be determined.

Physiological Pollution

Physiological pollution, though different from chemical pollution, is also caused by the soluble chemicals and colloidal substances present in wastewater. As mentioned earlier, chemical pollution occurs when the presence of suspended solids, soluble chemicals, and colloidal substances causes the COD of the effluent and the toxicity of some heavy metals to exceed their threshold limit. After the removal of the suspended particles and stabilization of the chemicals, the effluent may become harmless in a general sense. However, such a harmless effluent may still not be acceptable to the public due to the following reasons.

(a) The effluent may impart colour, produce odour, or even change the taste of the receiving water and the fish living in it. For example, H_2S is harmless at a concentration lower than 15 ppm but can be smelled even at a concentration of 0.0011 ppm. Similarly, though chlorophenol at a concentration of 0.001 ppm is not harmful, it can be tasted even at such a low concentration. Astonishingly, trace amounts of phenol in the receiving water can make the fish obtained from this water taste bad, which is not acceptable.

(b) Colloidal solids, which do not settle easily, cause the receiving water to become turbid. This reduces the penetration depth of the sunlight and lowers the rate of the natural process of purification by photochemical reaction.

There may be some other factors responsible for physiological pollution which cannot be properly tackled by chemical treatment or physical filtration. Therefore biological methods of wastewater treatment are essential for mitigating the causes of physiological pollution.

Biological Pollution

Biological pollution is caused by the organic compounds present in wastewater or solid wastes. The various types of micro-organisms present in air, water, and soil decompose these polymeric complex compounds into simpler ones and finally convert them into CO_2 and H_2O by consuming large quantities of dissolved oxygen, thereby rendering the water or the surroundings oxygen deficient. The following correlations show how oxygen is consumed during the biological degradation processes:

$$\text{Organic complex compound} + O_2 \longrightarrow \text{smaller intermediate}$$
$$\text{compound} + CO_2 \text{ (respiratory)}$$
$$\text{Intermediate compound} + O_2 \longrightarrow CO_2 + H_2O$$

If proteinaceous materials are present in the waste, then micro-organisms convert the nitrogen into nitric acid by oxidation–reduction reactions. CO_2, H_2O, and HNO_3 are the end products of biological degradation leading to the stabilization of biodegradable organic compounds.

Under ideal conditions non-polluted river water becomes saturated with oxygen at a concentration of 9.2 ppm. However, for various reasons, no natural water body can actually achieve this saturation value. According to Henry's law, a sparingly soluble gas such as O_2 reaches its maximum concentration at a given temperature, which is proportional to the partial pressure of O_2 in the atmosphere. Due to rapid industrialization and deforestation, the atmosphere is getting polluted and this, in turn, is affecting the concentration of O_2 in air. Hence the factors that affect the rate of dissolution of atmospheric O_2 into the receiving water are (a) air/water contact area, (b) natural turbulence of receiving water, (c) surface condition of receiving water, etc. The surface condition of the receiving water is affected by surface-active agents, detergents, fats, oils, etc., resulting in the dissolution of a lower quantity of atmospheric O_2 into the receiving water.

Due to the aforementioned reasons, for all practical purposes, the occluded oxygen content of unpolluted, shallow, turbulent river water may be taken as 7.0 ppm. For freshwater biota, the minimum dissolved O_2 requirement is 5.0 ppm. This requirement may increase during cold weather or spawning. From the point of view of calculations, it may therefore be said that 2.0 ppm of the dissolved O_2 is usually available to stabilize the effluent, both chemically and biologically. It should be noted that the quantity indicated is the maximum value and the remaining oxygen (5.0 ppm) is just sufficient for the survival of fish and hardly adequate during spawning.

Measurement of Pollution Level

The measurement of the pollution level is essential for the design of biological treatment plants used for waste stabilization. There are two ways of expressing pollution intensity.

1. *By measuring the concentration of the toxic components present in the effluents*   Heavy metals, which are carcinogenic to human physiology, should not be present in the environment, especially in receiving water, beyond a certain concentration. These toxic elements may be consumed by humans through fish, fruit, and vegetables. The same is true in the case of chemicals such as cyanides, sulphides, and phenols. Concentration is expressed in mg/L, mg/kg, or ppm. The US Environmental Protection Agency has listed some 129 heavy metals and chemical compounds as priority pollutants and identified their threshold values. The Indian Standards Institution also has similar publications for Indian conditions.

2. *By measuring how much O_2 is required to stabilize one litre of an effluent*   Such measurement is expressed in terms of the BOD or COD. The total organic carbon (TOC) present in the waste can also be used as a measure of the pollution level. The COD is actually a measure of the amount of O_2 required for stabilizing a particular amount of waste by chemical oxidation. For example, consider the conversion of sugar into CO_2 and water:

$$C_6H_{12}O_6 + 6O_2 \longrightarrow 6CO_2 + 6H_2O$$

On the other hand, the BOD measures the O_2 utilized by different bacteria for stabilizing the waste. In biological treatment processes the BOD plays a very important role. For example, a high BOD value of an effluent indicates that it contains a large quantity of biodegradable organic substances and that the receiving water, water into which this effluent is discharged, will be highly polluted. The pretreatment of such an effluent is a must to reduce the BOD level. The BOD value also helps us to decide which treatment process (aerobic or anaerobic) will be more economical.

Microbiology and Biochemistry of Pollution Abatement

Microbiology and biochemistry are disciplines of biotechnology. To understand the principles behind biological treatment processes, the relevant characteristics of micro-organisms and their biochemical functions should be known. A biological process cannot be performed without micro-organisms and therefore it is necessary to know the conditions under which they function better. Some of the questions to be asked are the following: Do they need atmospheric O_2 for their proliferation or is it detrimental to their growth? What should the C/N (carbon/nitrogen) ratio of waste materials be so that micro-organisms can grow properly? What biochemical pathways do they follow to convert an organic compound, responsible for causing pollution, into suitable end products to minimize the pollution problem?

Biological Treatment Process to Abate Pollution

As mentioned earlier, wastes are of two types, namely, solid and liquid. The overall treatment process is divided into two stages:

1. Primary treatment
2. Secondary treatment

In the case of solid wastes, primary treatment involves aggregation and separation of inorganic substances from organic ones. Metal, glass, building material, etc. belong to the category of inorganic wastes and are non-biodegradable. The separation of such contaminants is necessary to reduce the volume of the reactor required for secondary treatment. Organic solid wastes that are free from inorganic components can either be composted or used as landfill.

As liquid wastes are physically different from solid wastes, primary treatment of liquid wastes means separation of suspended and insoluble materials from liquid streams. *Screening, settling or lagooning, filtration*, etc., which are physical treatment processes, are primary treatment methods for liquid wastes. Primary treatment can eliminate more than 80% of the suspended solids and nearly 80% of the BOD and COD of the effluents.

After primary treatment the effluent contains some suspended solids and soluble organic compounds including colour and toxic materials. Secondary treatment brings down the toxicity and BOD as well as COD of the final discharge within the tolerance limits specified by the World Health Organisation (WHO) and the ISI.

Biological treatment methods have many advantages over other conventional waste treatment processes. Some of them are mentioned here.

(a) While physico-chemical treatment can remove specifically a single or small group of pollutants from a wide range, biological treatment removes a wide variety of pollutants in a single step. This makes the latter process more cost-effective.

(b) As mild temperature and ordinary pressure conditions are favourable, biological treatment processes are less energy intensive.

(c) Risks of explosion, corrosion, toxic gas leakages, etc. are minimal in biological treatments.

(d) As less complex equipment is required to maintain the desired conditions for a biological treatment process, the capital investment is low.

One major disadvantage of biological treatment processes is that they require large space or land area. The processing time is also high.

Biodegradation Methods

The main objective of allowing biodegradation is to stabilize the organic wastes. Depending upon the nature of the micro-organisms used, biodegradation methods are divided into two categories. These are aerobic and anaerobic. There may be some combined aerobic and anaerobic methods also. Normally, when the BOD and the colour intensity of the effluent are low (less than 2500 ppm), the pure aerobic method is recommended. On the other hand, for wastes with high BOD, the anaerobic process is preferred. However, the exact demarcation of BOD values for these two methods is difficult to set. The following section compares aerobic and anaerobic processes.

Comparison Between Aerobic and Anaerobic Processes

Source of oxygen

The basic difference between aerobic and anaerobic methods is that in the case of the former, supply of gaseous oxygen is required. However, in the anaerobic method, the micro-organisms utilize the oxygen already present in their food, i.e., in organic wastes, and gaseous oxygen is not only *not* required but also proves to be harmful.

Micro-organisms

Aerobic and anaerobic micro-organisms are different in nature and the end products of the degradation of organic compounds by these two classes of micro-organisms are also not the same. The end products of aerobic degradation are CO_2, H_2O, and NO_3 along with the formation of some colloidal or suspended residues of non-degradable nature. On the contrary, anaerobic digestion results in the production of methane gas (CH_4), CO_2, and a humus-like residue. Interestingly, depending on the nature of wastes and the degradation conditions, the ratio of CH_4 and CO_2 may vary from 1:1 to 3:1. Since CH_4 is a valuable and non-polluting fuel, the anaerobic waste treatment method proves to be more profitable, provided the methane gas obtained is free from H_2S. If wastes contain sulphur compounds, keratin, hair, etc., then the methane gas thus produced will always be contaminated with H_2S gas. For large-scale applications, methane must be free from H_2S to protect the combustion system from severe corrosion. Unfortunately the cost involved in the separation of CH_4 and H_2S is so high that the process of purification of methane gas does not prove to be economical.

Micro-organisms to food (degradable wastes) ratio

This is a very important controlling parameter in both the aerobic and anaerobic systems. For the smooth functioning of the process of fermentation, the micro-organisms to food ratio should be low so that the micro-organisms have sufficient food for metabolism. However, as the digestion process nears completion, this ratio becomes high and a new condition called *endogenous respiration* is set up, in which the micro-organisms eat each other and get destroyed completely. This ultimately leads to the end of the essential process of reproduction of micro-organisms as they are also converted into the end products.

Carbon balance

Under aerobic conditions, around 50% of the organic carbon is converted into biomass and the rest into carbon dioxide. However, in anaerobic conditions, almost 95% of the organic carbon gets decomposed into biogas and only about 5% is incorporated into biomass. This shows that the aerobic process generates a large quantity of sludge (cell mass) as well as CO_2. The anaerobic process, on the other hand, produces a low volume of cell mass and a large volume of biogas. Due to the high cell mass concentration the aerobic process is faster than the anaerobic process.

Energy balance

When wastes are treated aerobically about 60% of the energy gets stored in the large quantity of newly formed cells and 40% is lost as process heat. However, under anaerobic conditions, practically 90% of the energy stored in organic wastes can be recovered in methane gas, after 5% to 7% is used for the growth of new cells and 3% to 5% is wasted as heat. This clearly indicates that the anaerobic process has an advantage over the aerobic system.

BOD/COD reduction

The biological treatment of solid and liquid wastes effectively reduces the BOD/COD of wastes, resulting in the stabilization of biodegradable refuse. As mentioned earlier, BOD reduction is possible by both the aerobic and the anaerobic method. In the aerobic process, however, the presence of atmospheric O_2 (air) is unavoidable and this leads to reasonably high consumption of energy for the compression of the required amount of air. Alternatively, to save compression energy, a huge land area is required for spreading the wastes and allowing for the absorption of the atmospheric O_2 by surface contact. In the present day situation the second proposition is not practical and hence the forced circulation of air through the effluent in a confined space is a more acceptable option for aerobic waste treatment. The volume of air to be supplied for aerobic treatment depends on a few factors. These are the rate at which occluded O_2 is consumed by the micro-organisms involved in the process, the rate at which atmospheric O_2 dissolves in the effluent, the solubility of O_2 in the effluent, and the air pressure. As all the aforementioned factors have limitations, the volume requirement becomes high. For this reason the aerobic process is preferred only for the treatment of effluents with low BOD. For the anaerobic BOD reduction process, atmospheric O_2 is not required and the cost involved is comparatively low. That is why the anaerobic process is recommended for the treatment of wastes with high BOD. To reduce the processing time in the case of wastes with high BOD, anaerobic treatment followed by aerobic reduction is suggested.

Comparison between aerobic and anaerobic treatment processes

Aerobic Micro-organisms utilize biodegradable organic material for their growth and metabolism. This results in the generation of a huge quantity of cell mass and gaseous CO_2, mainly due to respiratory action. This process is faster and therefore the required reactor volume is less. It is an energy consumer and not an energy producer process. For example, to destroy 1 t of BOD, the electricity consumption is approximately 1100 kWh and cell mass generated is 400–600 kg.

Anaerobic While the aerobic process is more suitable for the treatment of liquid effluents, the anaerobic process can be utilized for the treatment of both solid and liquid wastes. This process is capable of treating high-strength organic wastes (BOD value more than 40,000 ppm). This is not an energy consumer process; it instead generates a huge quantity of gaseous energy in the form of methane. If 1 t of BOD is anaerobically consumed, the energy produced in the form of CH_4 gas will be equivalent to 1.16×10^7 kJ or 2.77×10^6 kcal along with net cell mass production of only 50–150 kg. Furthermore, the digested material can be directly utilized as a biofertilizer due to the presence of some essential nutrients, such as free NH_3, in it. The activated sludge produced in the aerobic process can also be utilized in the anaerobic process for further stabilization.

Environmental factors

For any biological process micro-organisms can act on organic matter only if the environmental conditions are favourable. Some of the factors to be considered are *p*H, acidity, alkalinity, concentration, food to micro-organism ratio, temperature, and toxic metallic ion concentration. For efficient operation of the biological system, the environmental conditions of the system should be maintained at their optimum level.

Aerobic Treatment Methods

Though there are several aerobic treatment methods, the most popular among them are the following.

(a) Activated sludge process
(b) Trickling filter method

Activated sludge process

This is a type of aerobic fermentation in which a high concentration of micro-organisms is maintained by recycling the active sludge. The system is depicted in Fig. 1.2. This process is further explained in Chapter 6.

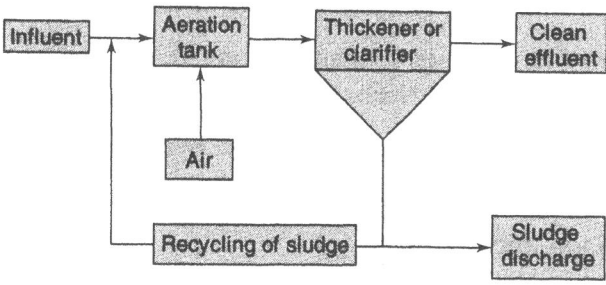

Fig. 1.2 Activated sludge system

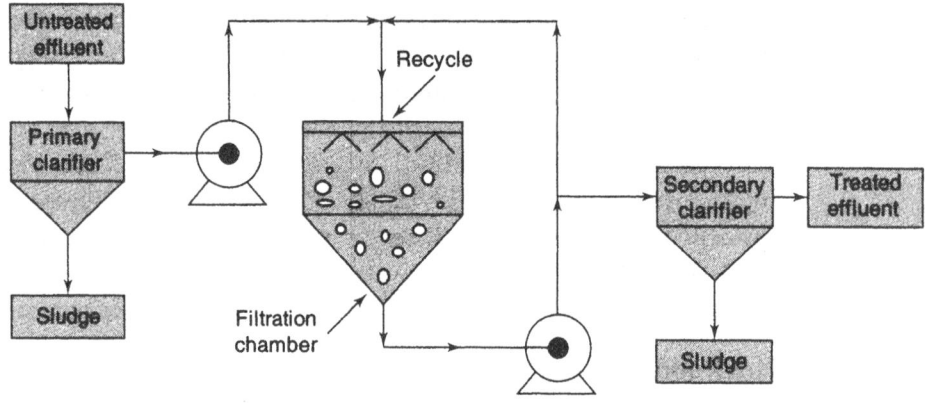

Fig. 1.3   Trickling filter

Trickling filter

The system of trickling filtration, which has been schematically shown in Fig. 1.3, consists mainly of a primary clarifier, a filtration chamber, and a secondary clarifier. The primary and secondary clarifiers separate the suspended solids and sludge from the untreated and filtered effluent, respectively. The filtration chamber consists of a large shallow, circular, horizontal-bed filter. The bed is prepared with porous sandstones, charcoal, or gravel, which are first soaked in a solution containing the required varieties and concentrations of micro-organisms and nutrients. The soaking process can be performed either inside or outside the chamber. The filter chamber is normally top-open. A sprinkler, which is a horizontally rotating, perforated pipe closed at both the ends, is fitted above the filter bed for sprinkling the effluent from the primary clarifier onto the stones through a pump. The sprinkled effluent trickles down through the stone bed, which contains the desired micro-organisms, and pollution-free, stabilized water comes out from the bottom of the filter unit along with certain biological end products.

Non-conventional Aerobic Treatment Methods

Some non-conventional aerobic treatment methods are also being used. Two systems among others seem to be quite promising. These are the following.

(a) Fluidized bed aerobic system developed by Dorr–Oliver
(b) Rotating biological contactor (RBC) system

Fluidized bed aerobic system

The fluidized bed aerobic system is a highly efficient and compact aerobic system suitable for BOD removal, nitrification, and denitrification of

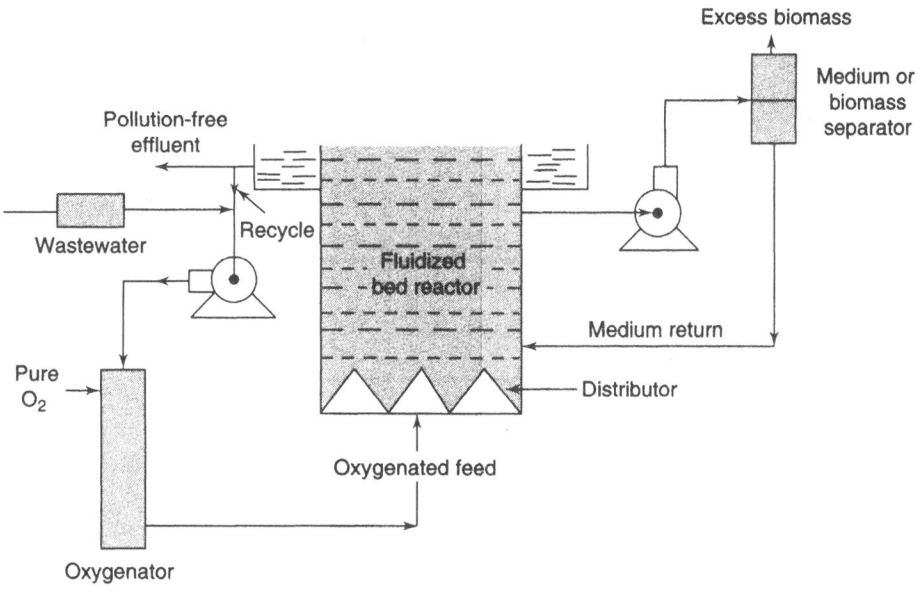

Fig. 1.4 Fluidized bed aerobic system

municipal and industrial wastewater. Organic chemicals, pulp and paper, petroleum refining, and textiles are a few industries best served by this system.

The system has been schematically shown in Fig. 1.4. To support the cells in fluidized condition, the medium selected for the bed is specially graded sand, which provides a vast area for biological growth. An interesting feature is that the fixed film biomass concentration per litre reactor volume is as high as 10,000–30,000 ppm, which results in the extremely high volumetric efficiency of the system. This also requires minimum reactor volume. In this system there is no provision for sludge recycling, as it is not necessary; as a result, the cell mass concentration is automatically maintained in the bed. As the system comprises a specially designed oxygenator in which pure O_2 is dissolved in the effluent stream prior to entering the reactor, it allows for the oxidation of at least 50 ppm of oxygen-demanding effluent per cycle through the reactor. Another important aspect is the positive control of biofilm thickness achieved by introducing a suitable medium or biomass separation system.

Rotating biological contactors

This is a non-conventional aerobic system which is beginning to attract the attention of environmental engineers. The RBC system has been found very effective for treating low-strength organic wastes. The system consists of a

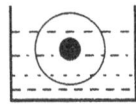

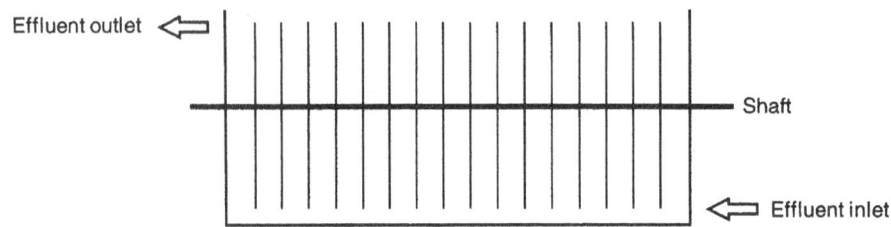

Effluent outlet

Shaft

Effluent inlet

Fig. 1.5   Rotating biological contactors

semi-cylindrical trough fitted with a large number of circular discs and a horizontal shaft maintaining a small gap between them (Fig. 1.5). The shaft along with the discs is rotated by a motor at low speed (about 10 rpm). The dimensions of the trough and discs are so chosen that, while treating the effluent, not more than 60% of the disc area is immersed in the effluent at any given time. The basic working principle of this system is to immobilize the microbial cells on the discs. While these circular discs rotate at low speeds, parts of the discs alternately remain exposed to air and immersed in the effluent. When a disc is exposed to air, O_2 is absorbed into its liquid film; when it passes through the liquid, the O_2 present in the wet surface of the disc is transferred into the bulk of the effluent contained in the trough. In this way, oxygen supply for microbial metabolism is maintained and BOD is reduced. In India this process has been successful. Although O_2 from the atmosphere is transferred by surface diffusion, the amount transferred is still considerably high because of very large surface area available for the system.

Anaerobic Treatment of Solid and Liquid Wastes

While the aerobic system is mostly suitable for the treatment of liquid wastes, the anaerobic process can successfully handle both solid and liquid wastes. As mentioned earlier, when the colour intensity and BOD of the effluent are very high anaerobic degradation methods are recommended. In such cases the aerobic treatment method has been found unsuitable due to economic reasons.

As the name implies, there is no O_2 supply in the anaerobic system. Now the question is: How do the anaerobes satisfy their oxygen requirement? These micro-organisms are capable of degrading organic matter using the oxygen present in the substrate and multiplying normally. The anaerobic process is without doubt economically attractive because the system requires neither atmospheric O_2 nor mechanical agitation. However, the system fails to provide intimate contact between the micro-organisms and the organic matter. For this reason, the rate of degradation of organic matter is much slower in the anaerobic process compared to aerobic fermentation. For example, if it takes 10 to 12 days for the stabilization of a particular waste with the anaerobic process, then it will take hardly 5 to 6 days for the stabilization of the same waste with the aerobic process. Therefore, for treating the same volume of effluent, the initial investment in an anaerobic system will be higher than that for aerobic treatment. This is also true in the case of the bioreactor and the area of land required.

Further, due to the difference in the fermentation processes, the end products produced in the two systems are also different. An anaerobically digested effluent produces foul-smelling, dark coloured residues, which cannot be discharged without first being treated aerobically in oxidation ponds for several hours to remove the bad smell and dark colour of the residual liquid.

For the disposal of solid wastes, especially municipal solid wastes (MSW), the landfill system is recommended. The normal time period for the stabilization of MSW in a landfill is 10 years. This period can be drastically reduced if the operational aspects of the landfill system are modified.

Mixed municipal waste

Several bioreactor systems are available for the anaerobic treatment of industrial effluents and municipal sewage water. Some of the most popular processes are discussed here.

(a) Stirred tank reactor (STR)
(b) Anaerobic contact process
(c) Fixed film reactor/anaerobic filter/packed bed reactor
(d) Expanded bed reactor
(e) Fluidized bed reactor
(f) Semi-fluidized bed reactor
(g) Upflow anaerobic sludge blanket (UASB) reactor
(h) Two-stage biomethanation system

Stirred Tank Reactor

The most popular and versatile among the available bioreactors is the STR. Its operating conditions are considered to be ideal. Perfect mixing can be achieved in this system and thus the hydraulic retention time becomes equal to the solid retention time. With this advantage, the main drawback of the system lies in the leakage of the microbial cells through the outlet along with the fermented broth. This disturbs the cell to food ratio, leading to inconsistency in the performance of the STR. Figure 1.6 gives a simple outline of an anaerobically operated CSTR (continuous STR). It may be mentioned here that CSTR is recommended for treating effluents with low solid content (2%–10% dry matter). Among its other advantages, uniform distribution of heat, if applied, and the ability to prevent accumulation of scum are worth

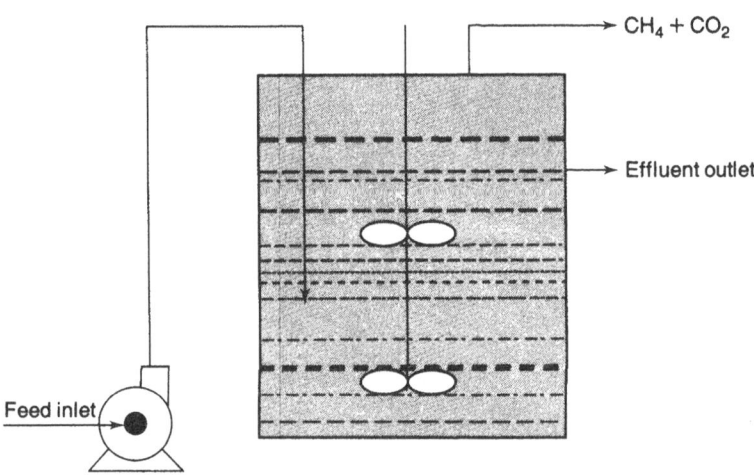

$CH_4 + CO_2$

Effluent outlet

Feed inlet

Fig. 1.6 Stirred tank reactor (continuous)

noting. The CSTR is the most common type of reactor used for the treatment of sewage or manure. The digesters currently being used for sewage treatment vary widely in size; for example, from 12,000 m³ to 100,000 m³.

Anaerobic Contact Process

The anaerobic contact process (ACP) is a modified version of the CSTR process and operates on the principle of the activated sludge process (aerobic treatment method), but without any O_2 supply. In the ACP, mixing is achieved either by using a mechanical agitator or by recycling the biogas produced in the system, instead of passing air as in the activated sludge process. A simple outline of the system is shown in Fig. 1.7.

In improving the performance efficiency of a bioreactor, the microbial population density plays an important role. Therefore, in the ACP a separate quiescent area or tank is incorporated to allow the micro-organisms to settle and subsequently return to the digester. This increases the biomass retention time in comparison to the hydraulic retention time. This is advantageous, as it leads to a higher micro-organisms to BOD/COD ratio, which in turn results in a more rapid and efficient removal of BOD/COD from the effluent stream.

In practice, the separation of anaerobic sludge from liquid media tends to be difficult because of reasons such as continuous gas production in the settlers' system. To tackle this problem, a company called Biomechanics Ltd developed a 'bioenergy' process and claimed improved performance by recycling cells. They applied the cold shocking technique to the liquor emerging from the ACP

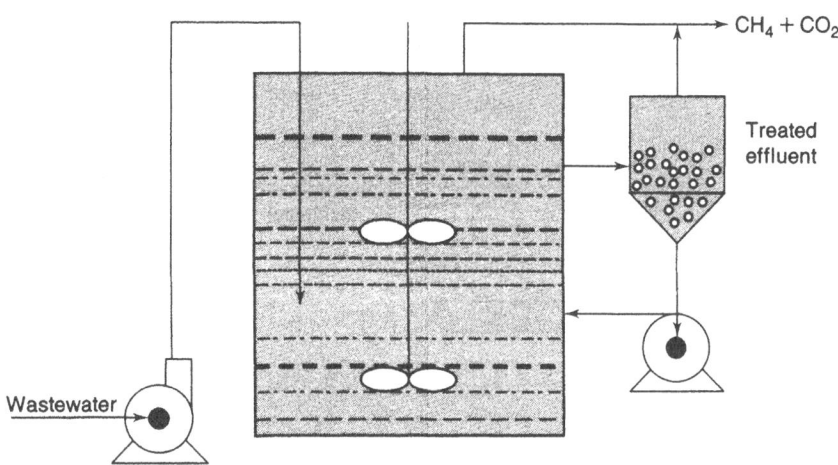

Fig. 1.7 Anaerobic contact process

to prevent gas production. This is, however, a process involving loss of energy. Other possible methods for separation of cells include floatation, centrifugation, gravity thickening, or flocculation followed by sedimentation. Full-scale contact process reactors are widely used for industrial wastewater treatment. For example, a 3000-m^3 reactor with a 300-m^3 settler is reportedly being used for treating pectin waste and is designed for a loading rate of 5 kg COD per m^3/day at a biomechanics plant. Similar units are also in operation in countries such as Germany and France.

Fixed Film Reactor/Anaerobic Filter/Packed Bed Reactor

A fixed film reactor (FFR) or an anaerobic filter (AF) or a packed bed reactor (PBR) is actually an immobilized whole cell reactor. Depending upon the entry of wastewater, these reactors may be termed either *upflow* or *downflow* reactors. Figures 1.8 and 1.9 represent these two reactors, respectively. In both types, a solid matrix of inert support material is used to provide the surface for microbial attachment. A variety of solid supports are used, such as gravel, commercially available glass beads, baked clay, needle punched polyester, and activated carbon beads. Bacterial flock can also be inserted in the voids between the inert media. The high specific biomass concentration achieved due to whole cell immobilization and flock retention makes it possible to attain a low hydraulic retention time (HRT) with a reasonably high solid retention time (SRT). The system is, however, not suitable for

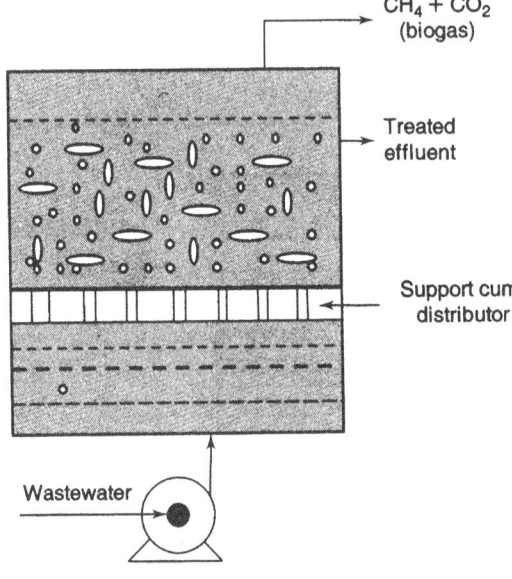

| Fig. 1.8 | Upflow fixed film or anaerobic filter bioreactor |

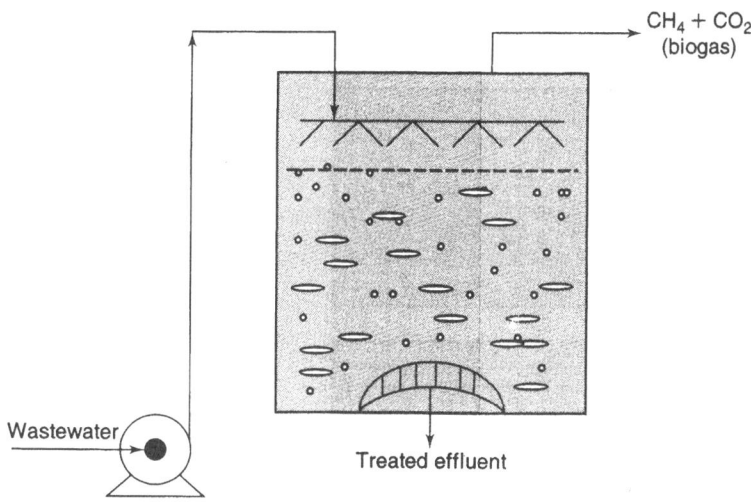

$CH_4 + CO_2$
(biogas)

Wastewater

Treated effluent

Fig. 1.9   Downflow fixed film or anaerobic filter bioreactor

treating wastewater containing large amounts of calcium salts, as these salts may clog the bed and channel the flow. Clogging and channelling may also occur if the effluent has a high concentration of suspended solids. In such a situation, the problem can be partially avoided if the system works alternately in the upflow and downflow modes. One more aspect is that during startup the microbial attachment may be slow and as a result the nature and arrangement of the support material may also affect the process to a large extent.

The field application of FFR/AF/PBR systems includes the treatment of different types of effluents containing 1%–10% dry matter; for example, vegetable processing wastes, animal wastes, wheat starch wastes, waste sulphite liquors, molasses wastes, and food processing wastes. A commercially developed upflow fixed film reactor is being used in Texas, USA. Elsewhere, a pilot plant (with a capacity of 10 m^3) based on the principle of downflow FFR is in operation to treat confectionary wastes. A similar 13,000-m^3 reactor is also in use to treat wastes from a rum-producing distillery.

Expanded Bed Reactor

This system is shown in Fig. 1.10 and can be roughly compared with the upflow fixed film reactor except in the case of bed-bulk density. In this system also the micro-organisms form a biofilm on the surface of small (0.2–0.3 mm) inert particles, preferably of low density, to minimize the energy requirement for expansion. Although ideal support media are difficult to find, usually sand, PVC particles, carbon granules, gravel, or ground glass are used.

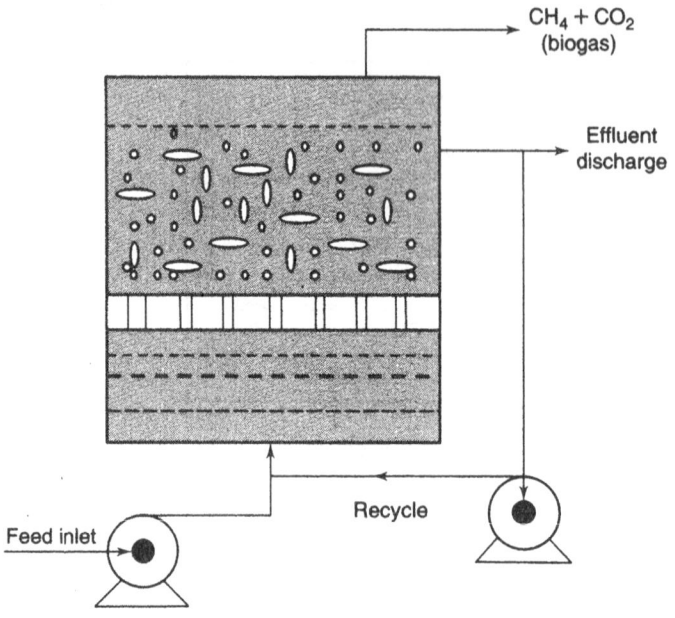

Fig. 1.10 Expanded bed reactor

Sufficient wastewater flowing up through the reactor expands the bed of particles to which the bacteria are attached. Depending upon the density of the particles, the superficial fluid velocity may vary between 2 and 10 m/h. If the volume of the feed liquid is not sufficient to accomplish such a velocity, then the same is attained by recirculating the outlet fluid. The recycle ratio may vary from 2 to 100. The expansion of the bed is restricted in such a manner that the relative positions of the particles within the bed can be retained, though the expansion volume might be higher by 20%–40%.

The expanded bed reactor (EBR) is quite useful for treating particulate wastes. However, there are no reports yet of the commercial use of the EBR system, though in a pilot plant study this system has been used to anaerobically treat whey, paper mill wastes, and wastes from molasses fermentation.

Fluidized Bed Reactor

The fluidized bed reactor (FBR) system is similar to the EBR in all respects except that the superficial fluid velocity is much higher in FBR, which results in the complete fluidization of granules. Typical velocities of 6–20 m/h are used. The particle size tends to be around 0.2–1 mm and the recycle ratio varies from

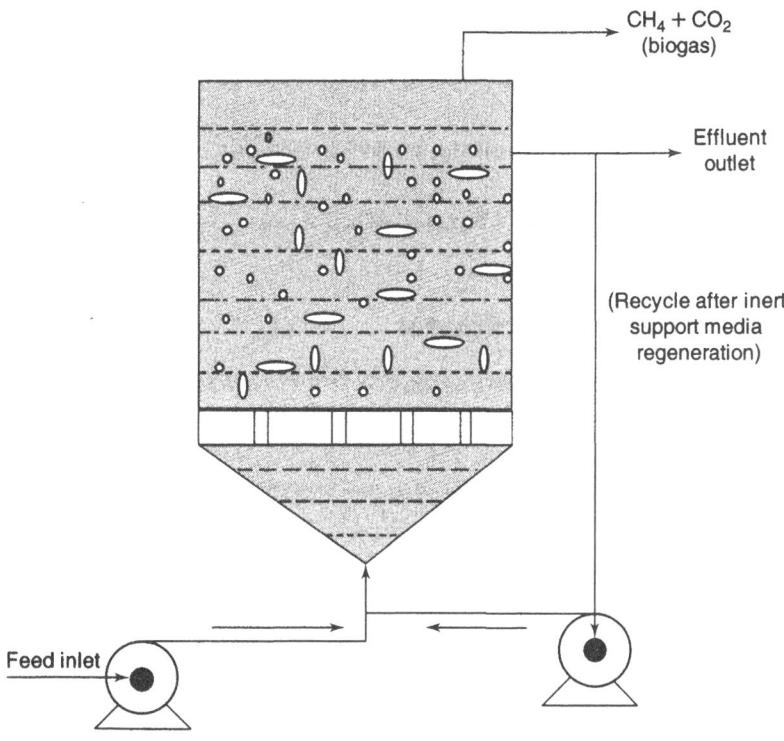

CH$_4$ + CO$_2$ (biogas)

Effluent outlet

(Recycle after inert support media regeneration)

Feed inlet

Fig. 1.11   Fluidized bed reactor

5 to 500. The expansion of the bed is also much greater (30%–100%). This system is shown in Fig. 1.11.

There are three major application areas in which the biofluidization system works successfully. These are the following.

1. Enzymes immobilized on a solid matrix.
2. Pure cultures of whole cells immobilized on a solid matrix.
3. A wide variety of wastewater treatment processes.

The FBR system for wastewater treatment has distinct advantages. In this system the suspended particles are constantly in motion and thus channelling or clogging can be prevented, resulting in the achievement of very efficient substrate degradation. Particulate wastes can be easily treated using the fluidized bed reactor. The FBR system has the following advantages.

1. Uniform solid distribution.
2. High concentration of active biomass per unit reactor volume.
3. Availability of large surface area for biological growth and contact with effluents, leading to a high rate of biodegradation.

The following are the disadvantages of the FBR system.

1. High energy input due to high recycle ratio as well as high density of bed particles.
2. Removal of supporting particles (usually sand) from the reactor is problematic.
3. Cleaning particles from the biomass and recycling them into the reactor is not that easy.

Semi-fluidized Bed Reactor

In a semi-fluidized bed reactor, both fluidization and inverted fluidization occur simultaneously. To achieve the condition of inverted fluidization, the upward velocity of the fluid must exceed the free-settling velocity of the particles. If a suitable porous support plate is placed at the top of the

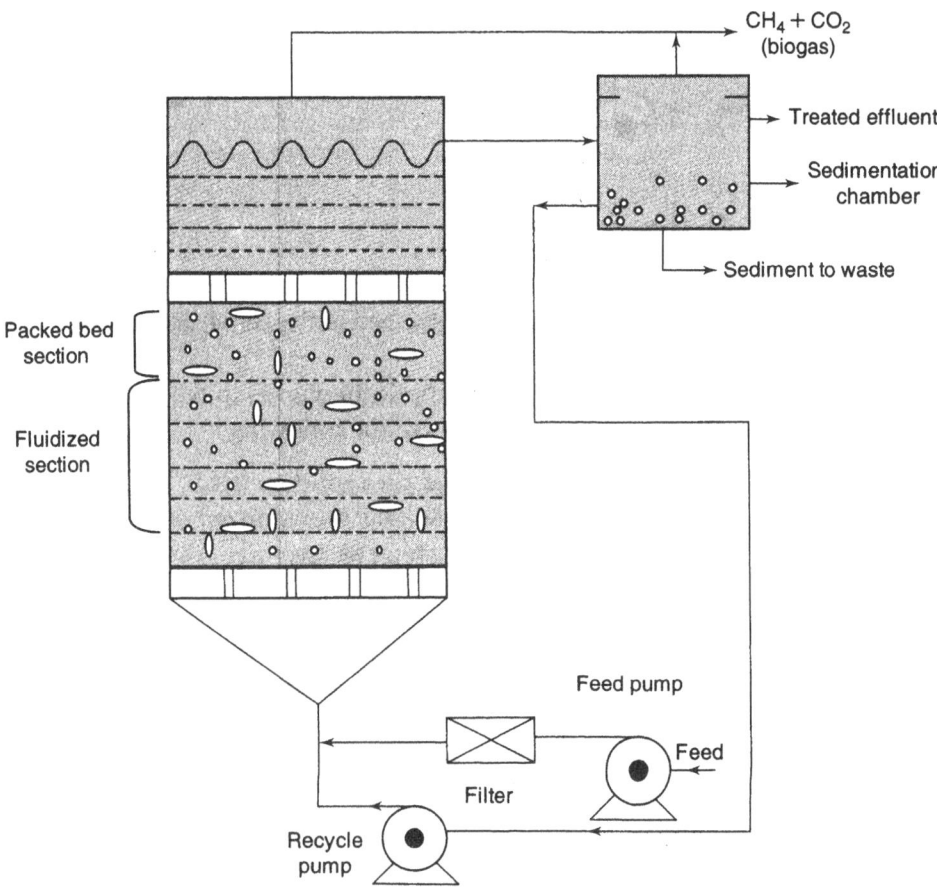

Fig. 1.12 Semi-fluidized bed reactor

column and the velocity is increased, then all the bed particles will collect at the top of the column forming a packed bed. This is the situation of inverted fluidi-zation. Now to attain the semi-fluidized condition, the fully developed fluidized bed is compressed from above using a perforated plate which nicely slides down inside the column. The pore size of the plate should be less than the particle size of the bed so that the particles are retained and only fluid is allowed to pass. As a result of the compression, a packed section is formed on top of the fluidized section, thus producing a semi-fluidized bed.

Successful application of the semi-fluidized bed reactor to treat wastewater shows that the effectiveness of the micro-organisms attached to the inert particles is drastically improved with regard to the reduction of TOC, BOD, TSS (total suspended solids), the degradation of hazardous organic contaminants and nitrogenous compounds, and the conversion of volatile solids into methane gas. In addition to the intrinsic advantages of a fluidized bed/packed bed system in series over the fluidized bed and packed bed reactors separately, the semi-fluidized bed reactor has other advantages such as the elimination of elutria-tion of bacteria-coated bed particles and unstable bed expansion, commonly encountered in the fluidized bed bioreactor. It also eliminates or reduces the clogging of the bed by suspended solids as experienced in fixed bed operation. Figure 1.12 shows an anaerobically operated semi-fluidized bed reactor system.

Upflow Anaerobic Sludge Blanket Reactor

The UASB reactor, developed in Holland, is particularly suited for the treatment of wastewater having a low concentration of soluble waste (about 1% dry matter). The design of the UASB reactor, shown in Fig. 1.13, has the following important features.

Wastewater enters from the bottom and gets evenly distributed throughout the bottom cross section through a properly designed liquid distributor. At the bottom of the reactor, above the feed distributor, a thick 'sludge blanket' consisting of active micro-organisms is formed; due to the settling characteristics of the microbial flock, a sludge blanket with a very high biomass concentration can be formed. The upward velocity of the liquid plays an important role in maintaining the sludge blanket undisturbed. It should be kept within the laminar flow zone and no turbulence should be created. The effluent to be treated is fed via the distributor into the base of the sludge blanket. Under appropriate conditions the flocks of bacteria are transformed into dense granules of 1–5 mm diameter having very good settling characteristics. The top portion of the reactor is provided with a larger cross section, so that the treated wastewater emerging from the blanket

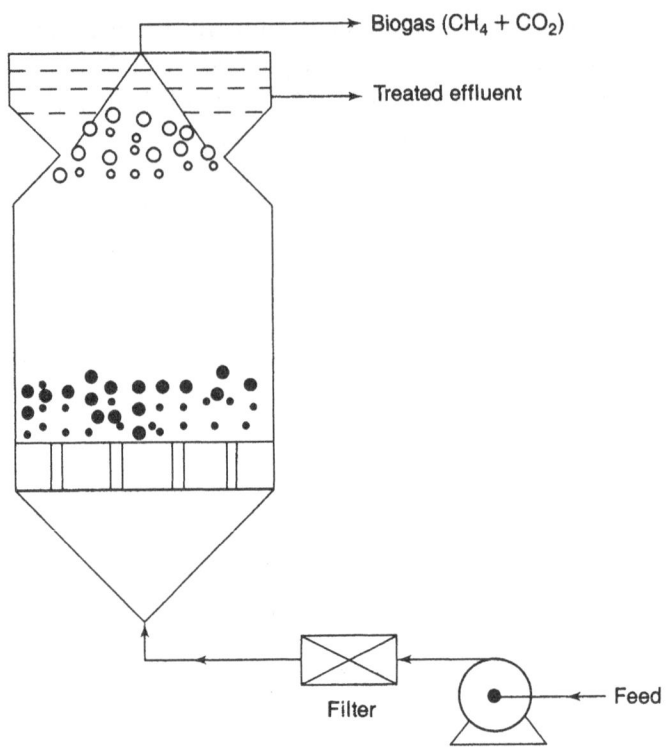

Biogas ($CH_4 + CO_2$)

Treated effluent

Filter

Feed

Fig. 1.13   UASB reactor

passes into a quiescent zone free of gas bubbles, where the bacteria that have detached from the blanket settle. The gas collection system is also attached at the top.

There are certain inherent limitations in the UASB design. The sludge blanket may get disturbed if the influent flow is too fast or if the gas production is too vigorous. Apart from this the digester does not treat particulate waste effectively, since the particles appear to interfere with flocculation and may also accumulate in the bed. This reduces the effectiveness of the bed per unit volume. Inefficient operation may also occur if the influent forces channels through the sludge blanket instead of passing uniformly through the whole volume of the bed.

UASB reactors are being used to treat a variety of food processing and dairy wastes, cane molasses distillery waste, and chemical industry effluents. In Holland and other countries this system has already been used on a commercial scale. Several such plants are working with a capacity of over 100 m^3. Suppose an application of UASB could be developed in which a high-rate system digests the effluent of a particular process industry to remove 70% of

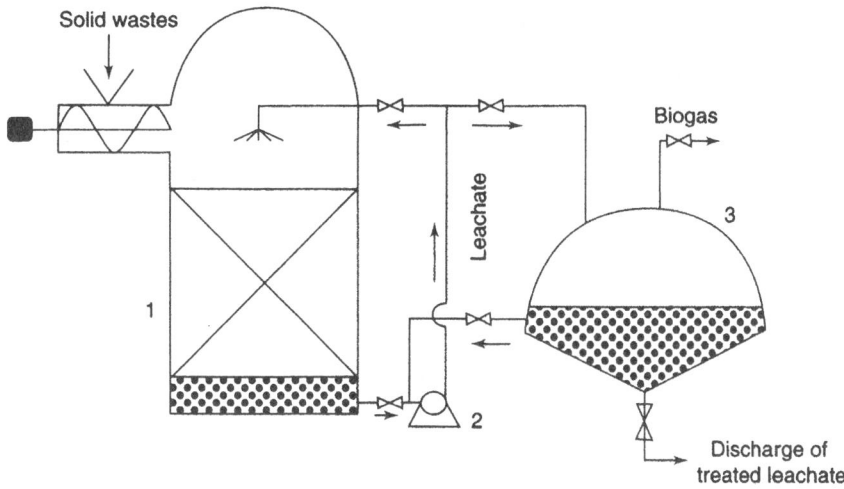

Fig. 1.14   Schematic diagram of two-stage biomethanation system
(1—leachate generator, 2—pump, 3—biogas generator)

the COD, with a biogas productivity of 0.7 m^3/kg of the COD destroyed, then
the potential gas production can be estimated as follows:

COD input	27,000 kg per day
COD destroyed	18,900 kg per day
Biogas produced	13,250 m^3 per day

If the system works for 300 days per annum, then the total biogas production
would be 3.97×10^6 m^3 per annum. If the COD loading rate is 15 kg/m^3 per day,
then the volume of the reactor would be 1800 m^3.

Two-stage Biomethanation System

In the two-stage biomethanation process (Fig. 1.14), the first stage comprises
of solid-state fermentation. Here insoluble complex polymeric compounds
are first hydrolysed into soluble and easily assimilable compounds such as
acetic acid and glucose. These compounds then act as the substrate for cell
growth and product formation. In the first stage the products are fatty acids
such as acetic acid and propionic acid. In the second stage, the substrate is a
mixture of volatile fatty acids and the product is biogas ($CH_4 + CO_2$), and
the reaction is carried out in the liquid phase. In both the stages the inoculum
is a mixed culture.

Relative Performance of Various Reactors   Table 1.2 shows the relative
performance of the reactors discussed here and Table 1.3 compares the performance

Table 1.2 Relative performance of various reactors

Reactor type	COD loading rate (kg/m³ per day)	COD removal (%)
Contact process	1–6	80–95
Upflow type	1–10	80–95
Downflow type	5–15	75–88
Fluidized and expanded	1–20	80–87
UASB	5–30	85–95

Table 1.3 Comparison of the performance of anaerobic reactors with reference to various parameters (data from different sources)

Parameter (units)	CSTR	Contact process	Anaerobic filter	EBR and FBR	UASB
Typical COD loading rate (kg/m³/day)	1.5–2.5	5	Up to 25	19	15
Typical COD reduction (%)	70	70	> 70	> 70	> 75
Typical CH₄ productivity (m³/kg of COD destroyed)	0.4	0.4	0.35	0.32	0.25–0.5
Hydraulic retention time (HRT) (h)	340–560	167	4.5–96	34	4–16
Solid retention time (SRT) (h)	340–560	Very high	Very high	~360	Very high
Typical biomass concentration (kg/m³)	5	20	15	25	15
Recycle ratio	-	1–2	-	5–500	-

of the CSTR, anaerobic contact process, fixed film reactor or packed bed process, expanded and fluidized bed systems, and the UASB process.

Indian Standards for Inland Surface Waters

Environmental legislation provides a legal tool using which activities that are detrimental to the environment and public health can be regulated. Table 1.4(a) presents the ISI specifications for the maximum permissible limits of the various constituents in industrial effluents that are being discharged into surface waters, into public sewers, and onto land used for irrigation. Table 1.4(b) lists the maximum permissible limits (specified by ISI) of the various contaminants in raw water used for public supply, fish culture, and irrigation.

Table 1.4 Indian standards for inland surface waters subject to pollution
(a) Pollution due to discharge into surface waters, into public sewers,
and onto land

Characteristics	Tolerance limits for industrial effluents discharged		
	Into inland surface water (IS:2490–1974)	Into public sewers (IS:3360–1974)	Onto land for irrigation (IS:3307–1965)
BOD (ppm)	30	500	500
COD (ppm)	250	-	-
*p*H	5.5–9.0	5.5–9.0	5.5–9.0
Suspended solids (ppm)	100	600	-
Total dissolved inorganic solids (ppm)	-	2100	2100
Temperature (°C)	40	45	-
Oil and grease (ppm)	10	100	30
Phenolic compounds	1	5	-
Cyanides (ppm)	0.2	2.5	-
Sulphides (ppm)	2	-	-
Fluorides (ppm)	2	-	-
Total residual chlorine (ppm)	1	-	-
Insecticides (ppm)	Absent	-	-
Arsenic (ppm)	0.2	-	-
Cadmium (ppm)	2	-	-
Hexavalent chromium (ppm)	0.1	2	-
Copper (ppm)	3	3	-
Lead (ppm)	0.1	1	-
Mercury (ppm)	0.01	-	-
Nickel (ppm)	3	2	-
Selenium (ppm)	0.05	-	-
Zinc (ppm)	5	15	-
Chlorides (ppm)	-	600	600
Boron (ppm)	-	2	2
Sulphates (ppm)	-	1000	1000
Sodium (percentage)	-	60	60
Ammoniacal nitrogen (ppm)	50	50	-
Radioactive materials			
1. Alpha emitters (μc/ml)	0.0000001	0.0000001	0.000000001
2. Beta emitters (μc/ml)	0.000001	0.000001	0.00000001

(contd)

Table 1.4 (*contd*)

(b) Pollution due to contaminants in raw water used for public supply, fish culture, and irrigation

Characteristics	Tolerance limits for inland surface waters		
	Used for public water supply	Used for fish culture	Used for irrigation
BOD (ppm)	3	-	-
Dissolved oxygen (% saturation)	40	40	-
pH	6–9	6–9	-
Total dissolved inorganic solids (ppm)	-	-	2100
Phenolic compounds as C_6H_5OH (ppm)	0.01	-	-
Cyanides as CN (ppm)	0.01	-	-
Fluorides as F (ppm)	1.5	-	-
Arsenic as As (ppm)	0.2	-	-
Chromium as Cr (ppm)	0.05	-	-
Free CO_2 (ppm)	-	6	-
Ammoniacal nitrogen as N (ppm)	-	1.2	-
Boron as B (ppm)	-	-	-
Sulphates as SO_4 (ppm)	-	-	2
Electrical conductance at 250°C	-	0.0001 mhos	0.003 mhos
Sodium (%)	-	-	60
Coliform organisms (monthly average maximum permissible number/100 ml)	Not more than 5000 with less than 5% of the samples with value > 20,000 and less than 20% of the samples with value > 5000	-	-
Radioactive materials			
1. Alpha emitters (μc/ml)	0.000000001	0.000000001	0.000000001
2. Beta emitters (μc/ml)	0.00000001	0.00000001	0.00000001

Minimum National Standards for Waste Disposal

The Central Pollution Control Board of India has specified certain minimum effluent standards at the national level known as the minimum national standards (MINAS). According to MINAS, any effluent or wastewater to be discharged from an industry must undergo pretreatment for the removal of pathogens, toxic materials, dissolved organic solids, and suspended materials, and proper pH control. Tables 1.5 to 1.9 represent MINAS specifications for some of the major industries causing water pollution.

Table 1.5   MINAS for synthetic fibre industries and breweries

Parameter	Maximum permissible concentration in effluent (in ppm, except pH)	
	Synthetic fibre industries	Breweries
BOD	30	30
TSS	100	100
pH	6.5–8	-

Table 1.6   MINAS for oil refineries

Parameter	Maximum permissible concentration in effluent (in ppm, except pH)	Maximum permissible quantum in kg/1000 t of crude oil processed
Oil and grease	10	7
Phenol	1	0.7
Sulphides	0.5	0.35
BOD	15	10.5
TSS	20	14
pH	6–8.5	14

Table 1.7   MINAS for sugar industries

Parameter	Maximum permissible concentration in effluent (in ppm, except pH)	
	Effluents disposed on land	Effluents disposed into water bodies or sewers
BOD	100	30
TSS	100	30
Oil and grease	-	10
pH	6.5–8	6.5–8

Table 1.8 MINAS for woolen industries

Parameter	Maximum permissible concentration in effluent (in ppm)
BOD	100 (disposal on land)
	30 (disposal into water bodies)
TSS	100
Oil and grease	10
Sulphides	2
Chromium	2
Phenolic compounds	5 (discharged into sewers)
	1 (discharged into water bodies)
Bioassay test	96% survival of test animals after 96 h

Table 1.9 MINAS for fertilizer industries (quantities in ppm, except pH)

Parameter	Nitrogen fertilizer plants	Complex fertilizer plants
TSS	–	100
Ammoniacal nitrogen	50	50
Total Kjeldahl nitrogen	100	100
Free ammonia	4	4
Nitrogen as nitrates	10	10
Vanadium	0.2	0.2
Arsenic	0.2	0.2
Chromium	2	2
Hexavalent chromium	0.1	0.1
pH	6.5–8	6.0–8.5

Summary

The various aspects of environmental biotechnology introduced in this chapter will now be discussed elaborately in the subsequent chapters. As environmental biotechnology deals with the remediation of environmental pollution problems, the primary need is to identify the nature of the pollutants and their sources. The next chapter, therefore, deals with the classification and characterization of pollutants based on their source.

Review Questions

1. Define the term environment and correlate it with your house environment.
2. What do you understand by the term biotechnology? What is the role of biosciences in biotechnology?

3. List the four major branches of biotechnology and their applications.
4. Define the term environmental biotechnology with reference to the environment and bio-technology.
5. How does the greenhouse effect cause global warming? How can it be overcome?
6. State the differences between physiological and biological pollution.
7. What is the basis for identifying different types of pollution?
8. Enumerate the factors that would help you to decide that water from a certain river A is polluted and that from river B is not. Prove this.
9. Define the following terms: BOD, COD, and TOC. How will you determine the COD value of glucose?
10. Which are the two categories of methods used for the biological treatment of organic wastes? State their advantages and disadvantages.
11. Briefly outline the activated sludge process.
12. In the trickling filter method, what is 'filter' and how is it achieved?
13. Define RBC and explain its working methodology in cleaning polluted wastewater.
14. With a neat sketch, show the working principle of the fluidized bed aerobic reactor system for treating liquid effluents.
15. Name at least five popular anaerobic methods used for the treatment of liquid wastes.
16. What is the difference between the fluidized bed reactor and the expanded bed reactor?
17. What are the advantages and disadvantages of using a fluidized bed reactor?
18. With a neat sketch, describe the working principle of UASB.
19. What is MINAS and why is it necessary?

References

Chandler, J.A., W.J. Jewell, J.M. Gossett, P.J. Van Soest, and J.B. Robertson 1980, 'Predicting methane fermentation biodegradability', *Biotechnol. Bioeng. Symp.*, no. 10, pp. 93–107.

Chen, Y.R. and A.G. Hashimoto 1979, 'Kinetics of methane fermentation', *Biotechnol. Bioeng. Symp.*, no. 8, pp. 269–82.

Contois, D.E. 1959, 'Kinetics of bacterial growth: Relationship between population density, and specific growth rate of continuous cultures, *J. Gen. Microbiol.*, vol. 21, pp. 40–50.

Davis, M.L. and D.A. Cornwell 1991, *Introduction to Environmental Engineering*, McGraw-Hill, New York.

Jewell, W.J, S. Dell'Orto, K.J. Fanfoni, T.D. Hayes, A.P. Leuschner, and D.F. Sherman 1980, *Anaerobic Fermentation of Agricultural Residue: Potential for Improvement and Implementation*, Final Report NTIS, vol. 3, US Dept of Commerce, Springfield, VA.

Mahajan, S.P. 1990, *Pollution Control in Process Industries*, Tata McGraw-Hill, New Delhi.

McCarty, P.L. 1964, 'Anaerobic waste treatment fundamentals', Part 1, *Chemistry and Microbiology*, vol. 95, no. 9, pp. 107–12.

Metcalf and Eddy, Inc., 1991, *Wastewater Engineering: Treatment, Disposal and Reuse*, 3rd edn, McGraw-Hill, New York.

Monod, J. 1950, 'La technique de culture continue; theorie et applications', *Ann. Inst. Pasture*, vol. 79, pp. 390–410.

Srma, A. 1990, Biomethanation of Waste Biomass, PhD thesis, Gauhati University.

2

Classification and Characterization of Wastes

Introduction

Chapter 1 explains that bioremediation is the biological treatment and removal of pollutants from the environment. To develop bioremediation processes, the causes of pollution need to be identified. This chapter deals with the classification and characterization of pollutants from different sources.

In general, any biodegradable organic substance is considered a pollutant if its decomposition leads to the production of toxic contaminants such as methane gas. This chapter discusses the different types of waste material responsible for causing environmental pollution, their characteristics, and the biological treatment methods employed for the pretreatment of these substrates or pollutants.

In biological treatment processes, the nature of the pollutants, solid wastes, or liquid effluents, plays an important role. The basic constituents of biodegradable organic waste are usually cellulosic, starchy, lipid, proteinaceous, or mixed in nature. These are readily decomposed by different types of micro-organisms. It is interesting to note that such organic waste can cause environmental pollution if the process of degradation is uncontrolled, and can prove to be economically advantageous if biodegradation is allowed to take place under controlled conditions. For example, biogas is produced as a result of the biodegradation of organic waste under controlled conditions.

The quality of the waste depends on its source. Depending on their origin, wastes have been grouped into seven categories.

1. Agricultural
2. Fruit and vegetable
3. Animal
4. Plant
5. Community or household
6. Industrial
7. Construction material

Waste Material Suitable for Biological Treatment

Table 2.1 lists the different types of wastes suitable for the microbial anaerobic treatment process, which results in the stabilization and biomethanation of these wastes.

Physico-chemical Characteristics of Waste Material

The various biological treatment processes, including biomethanation, involve micro-organisms that derive the requisite amount of nutrients essential for their growth and metabolism from the organic material they act on. Hence an ideal substrate or pollutant, which needs to provide these nutrients, should possess the following characteristics.

1. Proper carbon to nitrogen (C/N) ratio.
2. Appropriate concentration of volatile matter.
3. In the case of solid substrate, the smallest possible particle size.

Table 2.1 Types of wastes suitable for anaerobic digestion

Type of waste	Examples
Crop waste	Sugarcane trash, weeds, corn and related crop stubble, straw, spoiled fodder, spoiled vegetables, rotten grains
Animal waste	Cattle shed waste (dung, urine, litter), poultry litter, sheep and goat droppings, slaughterhouse waste (blood, meat), fishery waste, leather and wool waste, night soil
Plant waste	Aquatic waste (algae, seaweed, water lettuce, water hyacinth), forest litter (twigs, barks, branches, leaves, *Lantana camara*, *Ipomea* sp.)
Industrial waste	Oil cakes, bagasse, rice bran, tobacco waste and seeds, waste from fruit and vegetable processing, press mud from sugar factories, tea waste, cotton dust from textile industries, waste paper, wheat bran, paper mill, and tannery effluents
Municipal waste	Garbage, sewerage sludge, wastewater

C/N ratio

A very high C/N ratio reduces the efficiency of the process due to the limited availability of nitrogen. On the other hand, a very low ratio results in the formation of large quantities of ammonia, which is toxic to the bacterial population. It has been found that in order to achieve an optimum rate of digestion, it is important to maintain the C/N ratio (by weight) close to 30. However, it is pertinent to note that an organic substance which is resistant to microbial attack will not be digested at all, regardless of its C/N ratio. While determining the C/N ratio of a waste material in order to achieve the ideal rate of anaerobic digestion, it is important to consider not only the nature of the waste material but also the relative ease with which it can be digested by microbes. For example, lignin is not very biodegradable. Therefore, its abundance in the plant cell wall protects the cellulose present in the cell wall from bacterial action and thereby reduces the quantity of cellulose available for digestion. Table 2.2 shows the C/N ratios of a variety of waste materials. With the aid of these values, waste materials high in carbon content can be mixed with materials high in nitrogen content to obtain a substrate with the optimal C/N ratio of 30. It has been observed that vegetable matter obtained from young plants is more readily biodegradable than that from older plants; also, dry vegetable matter produces more gas than green vegetable matter. As mentioned, gas production can be increased by supplementing substrates that have high carbon content with substrates rich in nitrogen. For example, a mixture of powdered leaves, straw, and sawdust soaked in animal or human urine followed by dehydration stimulates gas production.

Concentration of volatile matter

In the substrate prepared for feeding to the digester, the concentration of the volatile solid matter is a crucial parameter. The volatile solid content of a waste material is expressed as the difference between the total dry solid content and ash. Usually, 7% to 9% concentration, that is, the presence of 7 to 9 parts of solid in 100 parts of the feed slurry, is considered optimal for efficient biodegradation. A substrate more diluted or concentrated than this optimal range may hamper the rate of biodegradation. For example, to obtain the required solid concentration in a feed slurry containing cattle dung, it is recommended that four parts of dung be mixed with five parts of water. In other words, an approximately 1:1 ratio by volume of raw cow dung and water mixed together gives 10% to 12% total solid concentration, which is the ideal concentration of volatile matter.

It follows from the discussion so far that awareness of the total and volatile solid content in various waste materials is vital for determining the efficiency of the biodegradation process. Table 2.3 lists the approximate percentage of volatile solid content in different types of waste materials.

Table 2.2 Approximate nitrogen content and C/N ratio of various waste materials (dry weight basis)

Material	N (%)	C/N
Animal waste		
Urine	15–18	0.8
Blood	10–14	3
Fish scraps	6.5–10	5.1
Mixed slaughter	7–10	2
Poultry manure	6.3	6
Sheep manure	2.7	16
Pig manure	3.8	14
Horse manure	2.3	25
Cow manure	1.7	18
Farmyard manure (average)	2.15	14
Night soil	5.5–6.5	6–10
Plant waste		
Young grass clippings (hay)	4	12
Grass clippings (average mixed)	2.4	19
Seaweed	1.9	19
Cut straw	1.1	48
Wheat straw	0.3	128
Rotten sawdust	0.25	208
Raw sawdust	0.1	511
Water hyacinth	2.16	20.5
Sorghum	0.5	100.85
Rice straw	1.34	78.58
Household waste		
Raw garbage	2.2	25
Bread	2.1	30
Potato peel	1.5	25
Paper	0.0	-
Municipal solid waste	0.58	26.23

Particle size of substrate

The size of the particles of the substrate, prepared from various waste materials, is another important controlling criterion in the bioconversion process. It has been observed that materials shredded into small pieces ferment better and pose fewer problems than bulky and dense materials. The higher rate of

Table 2.3 Per cent total and volatile solid content in various waste materials

Raw material	Total solids (%)	Volatile solids (%)	N (%)	P (%)	K (%)
Cattle dung (wet weight basis)	20–25	10–15	0.4–0.7	0.1–0.2	-
Night soil and urine	5	3.4	0.57	0.052	0.22
Poultry manure (wet)	17	-	1.2	1.2	-
Agricultural waste (dry basis)					
Linseed stock	-	30	-	-	-
Rice bran	-	60	-	-	-
Cotton	-	90	-	-	-
Groundnut shell	70.50	30	1.21	-	-
Bagasse	-	40	-	-	-
Jute stalk	-	40	-	-	-
Spent coconut shell	71.51	30	1.14	-	-
Rice husk	-	10	-	-	-
Young rye plants	-	50	-	-	-
Mature wheat straw	-	70	0.30	-	-
Soya bean tops	-	50	-	-	-
Young corn stalks	-	60	-	-	-
Mature corn stalks	-	70	-	-	-
Municipal solid waste	68.2	27.57	0.58	0.59	0.67

fermentation increases the rate of gas production, as it increases the surface area exposed to bacterial attack and thereby reduces the retention time in the digester. Moreover, it enables the slurry to flow smoothly to the digester, as less scum is produced. Shredding the waste material can also minimize other problems that arise due to the clogging of the inlet and outlet pipes. Manually operated shredders can also be used, for instance, in rural areas. Though less susceptible to failure, manual shredders are labour intensive.

Availability of Waste Material

Along with the source of the waste material, it is necessary to estimate the quantity of material that would be available from that source. This information is required while designing a waste treatment plant.

The amount of waste material produced depends on various factors. For example, the amount of dung produced per cow per day depends on factors such as age, size, and quality and quantity of fodder. The amount available, in this case, will also depend on whether the cattle are fed in the cattle shed or left to graze outside. However, for the purpose of calculation, usually approximate values (maximum) are taken into consideration. For animals not sheltered in stables, 70% of the values shown in Tables 2.4 and 2.5 may be taken for calculating the approximate quantity of wet manure that will be produced. Table 2.6 gives the availability of crop residues.

Table 2.4 Production of wet manure per cattle per day (Indian conditions)

Animal	Govt dairy (kg)	Private (kg)
Cow	20	10
Bull	30	10
Bullock	30	10
Buffalo cow	30	15
Buffalo bull	40	15
Calves	10	5

Table 2.5 Approximate production of wet manure per animal per day (based on a 500-kg live animal)

Animal	Weight (kg)	Volume (m^3)
Swine	28	0.028
Sheep	20	0.02
Poultry	31	0.028
Horses	28	0.025

Note: The availability of night soil per head per day is 0.4 kg.

Table 2.6 Productivity of crop residues

Crop residue	Residue yield (t/ha)
Rice husk	1–1.3
Wheat straw stubble	3–3.5
Leaves and stalks from corn	6–10
Leaves and stalks from sorghum	6–7
Bagasse	20–25
Pulp from sugar beet	8–15
Groundnut shells (and biomass)	0.6
Coconut shells and waste	0.4
Forest waste	1.2
Sunflower stalks	2.5

Characteristics of Specific Industrial Effluents

This section presents the nature of effluents discharged from different industries such as fertilizer, coke ovens, petroleum and petrochemical, paper and pulp, tanneries, sugar, distilleries, textile, dairy, slaughterhouse, and rubber. The characteristics of the effluent are vital for determining the type of treatment plant to be used.

Let us take the example of a nitrogenous effluent released from a typical large-sized ammonia–urea NPK complex. See Table 2.7 for the characteristics of this effluent. (The MINAS for the discharge of various effluents have already been provided in Chapter 1.)

Table 2.8 lists the concentration of phenol in various liquid effluents from selected industries. For instance, in a petroleum refinery, as much as 2500 t of phenol are discharged along with the effluents in a year; in a chemical manufacturing unit producing bisphenol, the annual phenol loss through the liquid effluent may be of the order of 800 t.

The general characteristics of three types of bioreactors that can be considered for the treatment of effluents containing phenol are listed in Table 2.9. From the data given it is clear that phenolic waste can be treated in all three bioreactors, namely, the continuous stirred tank bioreactor (CSTR), the packed bed bioreactor (PBR), and the fluidized bed bioreactor (FBR). The rate of degradation depends on conditions such as development of biomass, rate of airflow, and concentration of feed. The relative magnitudes of the rate of degradation of phenol for the three bioreactors are in the order FBR > PBR > CSTR. As a

Table 2.7 Characteristics of nitrogenous effluent from an ammonia–urea unit

Characteristic	Value
Temperature	40°C
Flow rate	600–700 m³/day
pH	9.12
Urea	2500 mg/L
Ammoniacal nitrogen	3000 mg/L
Alkalinity	2264 mg/L
Total suspended solids	-
Total dissolved solids	8600 mg/L
Chromate	25 mg/L
COD (chemical oxygen demand)	150 mg/L

Table 2.8 Average phenol concentration in liquid wastes from different processes

Process	Phenol concentration (mg/L)
Metallurgical coke	
(a) Spent liquor after phenol recovery	900–1000
(b) Coke oven effluent	35–250
Coal carbonization	
(a) Low-temperature carbonization	1000–8000
(b) High-temperature carbonization	800–1000
Oil refining	2000
Phenol–formaldehyde resin manufacture	800–2000

Table 2.9 Characteristics of bioreactors used for treatment of effluents containing phenol

Conditions	Continuous stirred tank bioreactor	Packed bed bioreactor	Fluidized bed bioreactor
Maximum phenol degradation rate for inlet phenol concentration of 500 mg/L, 99% conversion	1 g of phenol per day per litre of bioreactor volume	4.7 g of phenol per day per litre of bioreactor volume	8.5 g of phenol per day per litre of bioreactor volume
Maximum phenol degradation rate for 99% of any feed	2.6 g/L/day for inlet conc. of 1400 mg/L and liquid flow of 300 ml/h	4.7 g/L/day for inlet conc. of 500 mg/L and liquid flow of 875 ml/h	11.2 g/L/day for inlet conc. of 260 mg/L and liquid flow of 1800 ml/h
Normal effluent phenol concentration at maximum conversion	0.25–1.0 mg/L	0.25–1.0 mg/L	0.01–0.5 mg/L

typical example, the rates of degradation of phenol at a concentration of 500 mg/L are 1.02, 4.6, and 7.5 g/day/L of reactor volume for CSTR, PBR, and FBR, respectively. Table 2.9 also shows that the lowest concentration of phenol in an effluent is between 0.01 and 0.5 mg/L as seen in the case of the fluidized bed bioreactor.

Fertilizer Industries

Fertilizer plants in India can be classified into the following five categories.

1. Ammonia and urea plants
2. Ammonia, nitric acid, and ammonium nitrate plants
3. Ammonia, urea, phosphoric acid, and complex fertilizer plants
4. Ammonia, urea, ammonium sulphates, and sulphuric acid plants
5. Ammonia, urea, phosphoric acid, nitric acid, sulphuric acid, and complex fertilizer plants

The analysis of typical combined liquid waste from a fertilizer complex, having an average liquid effluent flow of 20,000 t per day, shows that on an average the total nitrogen in the effluent is between 400 and 1000 mg/L, urea-nitrogen between 400 and 800 mg/L, phosphates between 70 and 500 mg/L, fluoride (as F) between 10 and 20 mg/L, and arsenic between 1 and 2.5 mg/L. The acceptable value of nitrogen (in the form of ammonia and urea) in water bodies is 1.2 mg/L, of fluorides (as F) 2 mg/L, and of arsenic (as As) 1 mg/L individually as well as along with other toxic metal compounds.

Some of the major pollutants present in liquid effluents, depending on the category of the fertilizer unit, are listed in Table 2.10. Data on the waste load as well as the effluent volume per tonne of NH_3/day is provided in Table 2.11. Table 2.12 lists the nature, source, concentration (average), effect, and tolerance limit of various pollut ants from fertilizer industries. Table 2.13 lists the effluent limitation parameters—concentration of ammoniacal nitrogen, organic nitrogen, and nitrates and the *p*H range—for these

Table 2.10 Major pollutants in the liquid effluent of a fertilizer unit

	Plant category				
	1	2	3	4	5
*p*H	✓	✓	✓	✓	✓
Ammoniacal nitrogen	✓	✓	✓	✓	✓
Fluorides		✓	✓	✓	✓
Suspended solids			✓	✓	✓
Chromium	✓		✓	✓	✓

Table 2.11 Liquid waste load and effluent volume per tonne of NH_3/day for a typical fertilizer unit (ammonia/urea: approximately 1000 t/day)

Constituent	Loss in kg/tonne of NH_3	Effluent GPD*/tonne of NH_3
Carbon slurry, total carbon content	1.45–1.47	430–450
Scrubber waste		
(a) K_2CO_3	0.065–0.74	5–20
(b) NaOH	0.02–0.03	10–30
(c) As_2O	0.052–0.055	20–160
(d) Methyl ethyl acetone	0.038–0.23	10–160
Ammonia waste (as NH_3)	4.65–10	456–600
Urea plant	15–175	1100–1600
Phosphoric acid (P_2O_5) plant loss in kg/t		
(a) Phosphate	20–24.7	-
(b) Fluoride	6.6–8.0	-

*GPD denotes gallons per day.

Source: NEERI (National Environmental Engineering Research Institute) data

pollutants. As per Indian standards (IS:7968–1976), in the liquid effluent, the *p*H should be between 5.5 and 9, the ammoniacal nitrogen level at 50 mg/L, fluoride at 15 mg/L, suspended solids at 100 mg/L, and chromium at 1 mg/L.

Petroleum Refineries and Petrochemical Units

Petroleum refineries cause pollution during three stages of their operations—crude oil production, transportation, and refining. Tables 2.14 to 2.17 present data related to the processes taking place during the refining operations and the associated effluents released. If the effluent has been treated to the desired standard, then a bioassay test for fish will show 90% survival in 96 hours.

Petrochemical industries are so diverse that it is impossible to generalize the characteristics of the different types of wastes and arrive at any given set of treatment methods. Depending upon the nature of the constituents in the product mix, effluents from different plants would contain fairly high

Table 2.12 Nature, source, concentration, adverse effects, and tolerance limit of various effluents from fertilizer industries

Pollutant	Average concentration in the combined waste (mg/L)	Source	Pollution effects	Tolerance limit
Ammonia	732 (as N)	Ammonia and urea plants	Toxic to fish and aquatic life	1.2 mg/L as N in receiving water, used for fish culture
Urea	614 (as N)	Urea plants	Same toxicity as of ammonia after hydrolysis	Same as for ammonia
Arsenic	1.6 (as As)	Gas purification plants used in manufacture of ammonia and urea	Causes blackfoot disease, accumulative poison; affects plants and crops	1.0 mg/L in industrial effluents individually or along with other metals
Oil	NA	Gas compressor houses used in manufacture of ammonia and urea	Suppresses dissolution of oxygen in water; floats on water and forms ugly and dirty slicks	10 mg/L
Phosphate	77 (as PO$_4$)	Phosphoric acid and complex fertilizer units	Along with ammonia causes algal blooms, which increase cost of water treatment	-
Fluoride	15 (as F)	Scrubber effluents in phosphoric acid and super phosphate manufacture	Causes dental and skeletal fluorosis, affects hatching of fish eggs	2 mg/L as F

Source: NEERI data

concentrations of organic and inorganic pollutants. Hence, each plant effluent requires to be studied separately for determining the quality and quantity of waste involved, in order to select the most suitable technology for its treatment.

Table 2.13 Effluent limitation parameters (concentrations, except *pH* range)

	Effluent limitation parameters* (kg/100 kg of product)			
	Ammoniacal nitrogen	Organic nitrogen	Nitrate nitrogen	*pH* range
Ammonia	0.055	-	-	6.0–9.0
Urea	0.0325 (non-prill)	0.12 (non-prill)	-	6.0–9.0
	0.0325 (prill)	0.35 (prill)	-	-
Ammonium nitrate	0.025 (solution)	-	0.0125 (solution)	6.0–9.0 -
	0.05 (non-solution)	-	0.35 (non-solution)	-
Phosphate	No discharge of process wastewater pollutants			
Nitric acid	No discharge of process wastewater pollutants			
Ammonium sulphate	No discharge of process wastewater pollutants			
Mixed fertilizers	No discharge of process wastewater pollutants			

*Values listed are the maximum allowed average of daily averages for 30 consecutive operating days, i.e., the 30-day maximum averages. Maximum allowed single-day (average) values are equal to twice the 30-day averages.

Source: 'Pollution control in chemical and allied industries', Short-term Course, QIP, IIT Bombay, 1977.

Table 2.14 Refinery processes and associated liquid waste

Refinery process	Liquid waste
Distillation	Effluent with ammonia, hydrogen sulphide, and phenols
Desalting	Sodium chloride, phenols, sulphide, or free hydrogen sulphide
Vacuum distillation	Phenol and oil
Naptha hydrotreating	Sour condensate
Catalytic hydrotreating	Hydrogen sulphide, oil, or phenols
Catalytic or thermal cracking	Phenols, hydrogen sulphide, ammonia, cyanides
Hydrocracking	Ammonia
Catalytic alkylation	Alkalis from washing and acids from drains
Solvent processes	Solvents such as phenol, sulphide, copper acetate
Treating processes	Organic and inorganic pollutants (sulphides, phenols) and emulsified oil

Table 2.15   Constituents of a typical liquid effluent from an Indian refinery

Constituents	Quantity (mg/L)
Free oil	2000–3000
H_2S and sulphides	10–220
Phenols	12–30
Suspended solids	200–400
5-day BOD at 20°C	100–300
Alkalinity	10–20

Table 2.16   Expected standard of a treated effluent from petrochemical industries

Characteristic	Value
*p*H	5.5–9.0
Suspended solids not to exceed	100 mg/L
Dissolved oxygen not less than	5 mg/L
BOD (5-day, 20°C) not to exceed	100 mg/L
COD not to exceed	250 mg/L
Phenolic compound concentration not to exceed	1 mg/L

Table 2.17   Pollution data for treated wastewater from a typical refinery

Characteristics	Fuel refinery		Lube refinery		Limit permitted by Maharashtra State Pollution Board
	Usual range	Maximum	Usual range	Maximum	
*p*H	7.3–7.5	7.5	7.3–7.8	7.5	6.5–8.5
Oil and grease (ppm)	13–14	15	10–14	30	15
Suspended solids (ppm)	300–320	340	280–310	325	100 over influent concentration
5-day BOD at 20°C	50–75	100	50–70	100	100
COD (ppm)	200–230	250	200–220	230	250
Dissolved oxygen (ppm)	3.2–3.5	3.6	3.6–3.8	4.2	Not less than 40% of saturation value at temperature of effluent
Residual Cl_2 (ppm)	-	-	-	-	1
Ammoniacal N_2 (ppm)	0–0.1	0.1	0–0.1	0.1	50
Phenol (ppm)	< 1.0	-	< 1.0	-	2
Sulphides (ppm)	0–0.5	< 1.0	0	< 1.0	2

Paper and Pulp Industries

Different processes, such as the *kraft process*, are used for making pulp in the paper industry. An important aspect of paper production is that the water requirement is very high, as much as 250–450 m^3 per tonne of paper.

In the kraft process, the raw material—bamboo or wood—is cut into small pieces and digested in a solution of sodium hydroxide and sodium sulphate. As the raw material is digested, spent liquor (known as *black liquor*) is produced when the binding material (called *lignin*) present in wood and other compounds dissolves in the cooking liquor. It has been found that the BOD of the discharge from a kraft paper mill is about 25 kg per tonne of pulp product. Table 2.18 lists the characteristics of a liquid effluent from a kraft pulp and paper mill that employs chemical recovery methods in its different units and processes.

In the *sulphate process*, the cooking liquor contains sulphurous acid and calcium or magnesium bisulphate. The spent liquor contains spent and other compounds, which can be separated from the pulp. The BOD level of the spent liquor is around 450 kg per tonne of the pulp produced, which is very high as compared to the BOD level of the discharge in the kraft process. The sulphate process is not popular because the recovery of chemicals from the spent liquor poses many problems.

Another process for pulp making is the *soda process*. In this process the digestion of raw material takes place in a cooking liquor containing mainly NaOH. The raw materials generally used are rice straw, wheat straw, grass, waste paper, and rags, and about 15% of their weight is NaOH. For materials such as rags and waste paper, a low concentration of cooking liquors is used. Table 2.19 gives the characteristics of the spent (black) liquor obtained from the digester of the soda process.

Table 2.18 Chemical characteristics of an effluent from a large kraft (sulphate) pulp and paper mill using chemical recovery methods

Characteristic	Raw material	Effluent from					
		Digester	Pulp washing	Chemical recovery	Pulp bleaching	Paper machine	Combined effluent
Colour	Muddy	Dark brown	Dark brown	Light brown	Brown	Whitish	Brown
pH	6.5–8.0	9–10	8.5–9.5	7–8	6.5–8.0	5.5–8.0	6.5–8.5
Total solids (mg/L)	600–1400	1000–2500	1500–2500	1300–2700	2000–2900	800–1200	1200–2000
Suspended solids (mg/L)	200–500	150–200	400–1000	400–800	150–250	500–900	350–500
COD (mg/L)	200–500	1800–2200	900–1600	300–600	500–800	500–800	500–750
5-day BOD at 20°C (mg/L)	30–50	300–350	250–500	100–200	120–150	80–150	100–250

Table 2.19   Characteristics of black liquor obtained from the soda process

Characteristic	Value
Raw material	Straw
Quantity of black liquor	12 m³/tonne of pulp
pH	10.8–12.0
Total solids	3458 mg/L
Suspended solids	1428 mg/L
COD	2676 mg/L
BOD	780 mg/L
Sodium hydroxide	423 mg/L

Table 2.20   Properties of liquid effluents released from the caustic soda process

Property	Value
Colour	Brown
pH	9.4
Total solids	4800 mg/L
Suspended solids	2000 mg/L
COD	3100 mg/L
5-day BOD	625 mg/L
Chlorides	280 mg/L
Volatile solids	2100 mg/L

The *caustic soda process*, which uses mainly grass, bagasse, and straw as raw materials, discharges liquor effluents having the characteristics listed in Table 2.20.

Table 2.21 gives the characteristics of the liquid waste from a strawboard factory. Table 2.22 gives an overview of the performance of selected methods of treatment of liquid waste from paper and pulp mills, based on NEERI data. These data can be used to select an appropriate method of treatment.

Tanning Industry

This industry manufactures leather and leather products from raw hides or skins, which are usually received in a salted, dry, or wet condition. The raw material is first soaked in water and then washed thoroughly. Next, the skin is subjected to unhairing using lime and sodium sulphide. The unhaired hides

Table 2.21 Characteristics of a strawboard mill effluent

Characteristic	Effluent from			
	Spent liquor	**Washer beater**	**Machine water**	**Combined effluent**
Colour	Turbid orange	Orange	White	Orange
pH	8–11	10–13	7.5–9.0	8–12
Suspended solids (mg/L)	100–2100	3500–6000	50–1000	1000–3000
5-day BOD (mg/L)	2000–7500	4500–8000	150–900	600–3600

Table 2.22 Performance of selected biological treatment methods for paper/pulp mill effluents

Treatment method	Food to micro-organism ratio	BOD load	Duration of treatment
Activated sludge	0.4	3 kg/m^3	8 h
Enhanced/extended aeration	0.2	1 kg/m^3	2 days
Aerated lagoons	0.15	0.18 g/m^3/day	5–8 days
Activated sludge with pure oxygen	0.75	7 kg/m^3	2 h
Rotating biological contactors	0.9	0.06 m^3/day	4 h

are then delimed and washed with selected enzymes and ammonium sulphate. This is followed by vegetable or chrome tanning. Finally, the tanned leather is piled, washed, neutralized, dried, treated with an emulsion of sulphonated oils, and dyed. Table 2.23 details the volume and characteristics of wastewater from a typical tanning industry, while Table 2.24 provides an analysis of composite liquid waste. It is important to note that liquid effluents, if untreated, adversely affect land, streams, and sewers upon discharge into the same.

Sugar Industries

Table 2.25 lists the average quantity of wastewater released in one day per tonne of cane processed for a typical sugar factory. Mill house, filter washing, and boiler house wastewater characteristics are summarized in Table 2.26. The characteristics of the combined effluent, excluding the condenser water, from a standard sugar factory are provided in Table 2.27.

Table 2.23 Volume and characteristics of waste from a tannery

Operation	Vol. of waste (1/100 kg hide)	pH	Total solids (mg/L)	Suspended solids (mg/L)	BOD (mg/L)
Soaking	48	7.7–13.2	21,200	2590	1610
Liming	47	9.9–13.2	48,400	10,332	10,027
Deliming	32	2.4–10.1	5870	1392	1522
Vegetable tanning	36	5.1–6.7	31,800	3510	19,284
Chrome tanning	43	2.6–3.2	11,550	482	1000
Combined washing	343	7.6–12.2	20,000	3170	7000

Source: NEERI data

Table 2.24 Characteristics of composite effluents (all values in mg/L, except for pH)

Characteristic	Value	Tannery waste	
		Vegetable	Chrome
pH	4–6.5	9.0	8.9
Total solids	17,240–63,156	23,700	17,390
Suspended solids	1008–4968	7500	3060
BOD	10,660–24,000	6770	125
COD	23,712–35,000	13,898	4750
Tannin	6800–12,000	2000	-
Alkalinity	-	4385	260
Chlorides	-	7000	8280
Chromium	-	-	180

Source: NEERI data

Table 2.25 Volume of effluents released from a sugar industry

Source of effluent	Average quantity (L/day/tonne of cane)
Mill house	730
Boiler house and floor washings	230
Filter cloth washing	360
Condenser water	1640

Source: NEERI data

Table 2.26   Characteristics of effluents from the sugar industry

	Characteristics			
	*p*H	Total solids (mg/L)	Suspended solids (mg/L)	5-day BOD at 20°C (mg/L)
Mill house	6.7	2760	910	210
Filter cloth washing	9.5	6970	4000	1765
Boiler house	7.2	5130	120	5150

Source: NEERI data

Table 2.27   Characteristics of combined effluent from the sugar industry (excluding condenser water)

Property	Concentration
*p*H	4.6–7.1
Total solids (mg/L)	870–3500
Suspended solids (mg/L)	220–800
Volatile solids (mg/L)	400–2000
5-day BOD at 20°C (mg/L)	300–2000
COD (mg/L)	600–4380
Total nitrogen (mg/L)	10–40

Source: NEERI data

Alcohol Industry

In India, there are a good number of industrial-alcohol manufacturing units, many of them situated near sugar factories. The distilleries are situated in remote areas, where there are no large water bodies into which the effluents can be discharged. Hence, these effluents are discharged onto land, leading to severe land pollution, accompanied by foul smells and dark-coloured effluents. The seepage of such coloured liquids can adversely affect groundwater as well as the quality of the land.

   For the production of industrial alcohol, the diluted molasses are first fermented in the presence of nutrients under controlled temperature and *p*H conditions. Next, the sludge is removed and the clear liquid is distilled to recover alcohol. The remaining spent wash and the sludge are the main causes of pollution. Table 2.28 lists the constituents of spent wash from distilleries using molasses as raw material and their approximate concentrations. As much as 80 L of spent wash is produced per litre of alcohol manufactured. The BOD of the effluent is also very high (15,000–60,000 mg/L depending upon industrial practices).

Table 2.28 Spent wash constituents for molasses distilleries in India (all values, except *p*H and BOD, are expressed as percentages)

Characteristic/constituent	Value/concentration
*p*H	3.0–5.4
Total solids	0.15–10.4
Volatile solids	0.11–7.5
Ash	0.02–2.2
Calcium (as CaO)	-
Potash (as K_2O)	0.03–0.72
Sodium salts (as Na_2O)	-
Iron (hydroxides)	-
Insoluble acids	-
5-day BOD at 20°C (mg/L)	10,000–73,000
Phosphorous (as P)	0.1–1.0
Nitrogen (as N)	0.1–1.5

Source: Data from 15 factories surveyed by NEERI

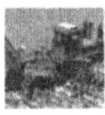

Recommended Effluent Treatment Methods

From the information provided so far, in the form of tables, it can be seen that the characteristics of liquid wastes vary widely, depending upon the type of process industry. Therefore, a good combination of physico–chemical methods (for chemical recovery) and low-cost biological oxidation methods (for the reduction of BOD) is adequate for the treatment of most liquid wastes obtained from process industries. Table 2.29 details the recommended treatment practices followed by some important industries based on the nature of the wastes. It would be helpful to note that, in general, the equipment used for controlling water pollution includes the following.

1. Chemical recovery equipment
2. Coarse screening
3. Fine screening
4. Grit removal equipment
5. Oil removal equipment
6. Flocculators
7. Clarifiers
8. Sludge disposal systems
9. Chemical preparation and dosing systems
10. Mixing/pumping systems
11. Biological oxidation/clarifiers/sludge disposal systems—activated sludge process, trickling filters, rotating biological contactors, fluidized bed contactors.

Table 2.29 Recommended effluent treatment practices for major industries

Industry	Nature of waste	Recommended treatment practices
Cane sugar	Slightly acidic, high solid content, moderate BOD, low nitrogen content	Segregation of waste; reuse of cooling water; marketing of molasses and bagasse; composting of high-strength waste with night soil (to make up for nitrogen deficiency) and refuse (to reduce moisture content); stabilization pond (after achieving C/N of 30); or discharge in municipal sewer (proportioning)
Tanning	Alkaline, high suspended solid content, high BOD, highly persistent colour from vegetable tanners, chromium in chrome tanneries	Segregation of waste; recovery of tannin; lime treatment of spent tan liquor; subsequent aeration (to oxidize catechol and pyrogallol) followed by anaerobic digestion (lagooning) for combined waste; subsequent stabilization pond or oxidation ditch. Alternatively, may be discharged into a municipal sewer after proportioning; process modification to decrease loss of chromium in effluent.
Alcohol	Acidic, high solid content, BOD, COD, chlorides, potash, sulphates, and nitrogen	Segregation of waste; reuse of cooling water; better utilization of raw materials; by-product (e.g., potash and methane) recovery; discharge into municipal sewer; two-stage treatment: anaerobic lagoon and oxidation ditch (stabilization pond)
Fertilizer	Alkaline, high solid content, nitrogen, phosphates, fluorides, and arsenic	Segregation of waste; by-product recovery (oil, chalk and acid sludge, ammonium sulphate, ammonia, fluorides, carbon slurry); chemical precipitation of phosphates and fluorides; concentration of arsenic waste; mixing hay paper or cane sugar waste (to make up for organic carbon deficiency) with ammonia and urea-bearing waste; subsequent stabilization pond treatment for algae harvesting
Paper and pulp	Acidic (sulphate process) or alkaline (other processes), high solid content, moderate BOD, low nitrogen content	Segregation of waste; reuse of white water; recovery of sodium sulphate, caustic soda, and lignin; disposal in municipal sewers; aerated lagoon/ oxidation ditch treatment after nitrogen supplementation
Textile	Alkaline, high solid content, moderate BOD, other chemicals	Segregation of waste; reuse of water; recovery of heat and caustic soda; substitution of chemicals (synthetic dyes in place of acetic acid dyeing, synthetic detergents in place of soap); disposal in municipal sewers; stabilization pond or oxidation ditch

(contd)

Table 2.29   (*contd*)

Industry	Nature of waste	Recommended treatment practices
Dairy	Alkaline, high volatile solid content, moderate BOD, high nitrogen content, grease	Waste volume and strength reduction; segregation of waste; anaerobic lagoon followed by stabilization pond; oxidation ditch; disposal in municipal sewers
Plating	Acidic or alkaline, low concentration of metals in solution, either as ions or cyanide complexes	Waste volume and strength reduction; segregation of waste; alkaline chlorination of cyanide waste; precipitation or ion exchange for removal of metallic ions after neutralization
Canning	High BOD and suspended solid content	Screening for recovery; anaerobic lagoon; floatation; chemical precipitation
Slaughter house	High total and volatile solid content, moderate BOD, grease	Grease recovery; aerobic stabilization ponds
Rubber	Acidic, saline, moderate BOD, coagulated rubber, phenols	Recovery of rubber; adsorption of phenol; oxidation ditch or stabilization pond

Source: 'Pollution control in chemical and allied industries', Short-term Course, QIP, IIT Bombay, 1977. (To be used as a guide only.)

Table 2.30   Water pollution control equipment (partial list of suppliers)

Companies	Trickling filter	Aerator	Clarifier	Flocculator
Hindustan Dorr Oliver, Mumbai	✓	✓	✓	✓
Voltas Ltd, Mumbai	✓	✓	✓	✓
Geo Miller Co., Delhi	✓	✓	✓	✓
KCP Eimco., Chennai	✓	✓	✓	✓
NSE, Mumbai	✓	✓	✓	✓
Ion Exchange, Mumbai	✓		✓	✓
Candy Fillers, Mumbai	✓	✓	✓	✓
Starit, Mumbai			✓	✓
Batliboi/HGE, Mumbai	✓	✓	✓	✓
Cimon Carves, Mumbai		✓		
Parmount Pollution Control, Baroda		✓		

A partial list of water pollution treatment equipment manufacturers in India is provided in Table 2.30. (This list is by no means complete; it is therefore to be used only as a guide.)

Products of Anaerobic Digestion

The main products of anaerobic digestion are biogas and biomanure. We will study each of these in detail in the following sections.

Biogas

The anaerobic digestion of waste material containing starch and cellulose results in the production of biogas. The two main constituents of biogas are methane (CH_4) and carbon dioxide (CO_2). Along with these, hydrogen (H_2) and hydrogen sulphide (H_2S) may also occur in negligible quantities.

The composition and quality of the gas produced depends on the quantity and type of biodegradable organic material added to the digester. Production of gas also depends on the temperature and time period of digestion. The results of the various experiments conducted to investigate the quantity and composition of biogas produced from different waste materials and reported by Khadi and Village Industries Commission, India, are summarized in Tables 2.31 and 2.32.

Methane is the most desirable component of biogas, as it enriches the fuel value of the gas. Carbon dioxide does not give any heat as it is a non-combustible

Household biogas plant

Table 2.31 Yield of biogas from various waste materials
[wet basis, retention time (RT) = 50 days]

Source	Gas per kg (wet basis, m³)
Cattle dung	0.04
Night soil	0.07
Pig manure	0.08
Poultry manure	0.06
Elephant dung	0.02
Duck manure	0.05
Goat droppings	0.05

Table 2.32 Yield of biogas from various waste materials (dry basis)

Raw material	Dry solids (m³/kg)	Temperature (°C)	CH₄ content (%)	RT (days)
Cow dung	0.33	-	-	-
Cattle manure	0.23–0.50	11–31	-	-
Poultry manure	0.26–0.3	32.6	58	10–15
Pig manure	1.02	34.6	68	20
Sugar-beet leaves	0.5	-	-	14
Algae	0.32	45–50	-	11–20
Night soil	0.38	20–26.2	-	21

gas. Therefore, biogas containing more methane is preferred. It has been reported that if the substrate contains mainly cellulosic material, then the gas produced will contain equal percentages of methane and carbon dioxide, each being approximately 50%. The methane content can be improved to up to 70% with the addition of fats and proteins to the substrate. On an average, gas obtained from cattle dung contains 50% to 60% methane (CH_4) and 40% to 45% carbon dioxide (CO_2), with negligible amounts of other components such as hydrogen sulphide (H_2S) and hydrogen (H_2). Whereas the CH_4 content may increase to as much as 65% when cow dung is mixed with night soil, the CO_2 content reduces to 34%, while the H_2S content reduces to 0.6% and the other gases are limited to 0.4% approximately. The amount of gas produced per kilogram of cattle dung varies between 0.03 and 0.06 m³. Table 2.31 lists the gas production rates from various sources as reported by the Khadi and Village Industries Commission, India.

The effect of adding small quantities of plant material to cow dung on biogas production is illustrated in Table 2.33.

Table 2.33 Biogas production at room temperature from the anaerobic fermentation of a mixture of cow dung and agricultural wastes at the end of 24 days

Raw material	Biogas (m³/kg)	CH₄ (%)	H₂ (%)	CO₂ (%)
Cow dung	0.063	60	1.1	34.4
Cow dung + 2.4% fresh leguminous leaves (25% dry matter, 2.31% N)	0.063	61.6	4.0	32.0
Cow dung + 1.2% mustard oil cake (94% dry matter, 4.74% N)	0.063	67.7	-	30.4
Cow dung + 1% cellulose	0.084	52.8	-	44.0
Cow dung + 1% ashes	0.061	60.4	2.9	34.4
Cow dung + 0.4% casein (12.6% N)	0.087	64.0	2.4	32.0
Cow dung + 0.4% cane sugar	0.070	57.6	2.1	38.4
Cow dung + 1% cane sugar + 1% urea (44.5% N)	0.087	68.0	-	30.6
Cow dung + 1% cane sugar + 1% CaCO₃	0.091	70.0	-	28.0
Cow dung + urine at 20 ml/100 g (4% solids)	0.087	67.0	-	32.0
Cow dung + 0.4% charcoal	0.065	65.6	-	32.0
Cow dung + 20% dry, non-leguminous leaves (1.71% N)	0.081	68.0	0.6	28.0

In fermentation tests, it has been found that the gas production at 30°C per kilogram of organic dry material in 10 days is 100 L from cattle droppings, 150 L from rice straw, and 480 L from grass. Table 2.34 presents data on biogas production from various raw materials, which is useful for design calculations.

The composition of raw material before and after digestion may be required sometimes for the purpose of calculations. A general material balance chart is shown in Table 2.35 for mixed raw material (1 kg) on dry basis.

The gas produced by the digestion of organic waste is colourless, flammable, and as already mentioned contains approximately 60% methane and 40% carbon dioxide, with small amounts of other gasses such as hydrogen and hydrogen sulphide. Methane itself is a non-toxic gas and possesses a slight but not unpleasant smell. However, if the conditions of the digestion are such that the gas produced contains a significant quantity of hydrogen sulphide, then the gas will have a distinctly unpleasant odour. As is known, the calorific value of biogas is controlled by the percentage of methane present in the gas. For the aforementioned composition, the calorific value is more than 4500 kcal/m³. Calorific values vary from 4400 to 6200 kcal/m³. The combustion of 30 L of this gas will release an amount of energy equivalent to lighting a 25-W bulb for about 6 hours.

Table 2.34   Biogas production from various raw materials (15-day digestion period)

Biomass	Gas yield (L/kg)	Gas composition (%)	
		CH$_4$	CO$_2$
Sugarcane bagasse	57	45	53
Rice straw	61	55	44
Wheat straw	55	42	46
Rice husk	70	50	49
Corn cob	59	55	43
Banana skin	51	50	50
Algae (mixed species)	77	75	25
Algae (*Spirulina* sp.)	60	77	22
Water hyacinth	115	67	32
Cow dung	102	60	36

Table 2.35   Material balance of mixed raw material (dry basis)

Condition	Energy (kcal)	Energy converted to gaseous form (kcal)	Cellulose (g)	Hemi-cellulose (g)	Crude protein (g)	Lignin (g)	Mineral (g)
Before digestion	2013	–	216	203	189	114	210
After digestion	1108	886.80	105	26	81	114	210

Table 2.36   Biogas consumption pattern for a rural community

Purpose	Consumption
Cooking	0.25–0.30 m^3/day/person
Lighting	0.15 m^3/100 candle power lamp/h
Motive power	0.50 m^3/h.p./h

Among the various potential uses of biogas in the rural community, heating, lighting, cooking, and running irrigation pumps are predominant. The gas consumption pattern for these purposes is given in Table 2.36.

Table 2.37 compares some commonly used fuels with biogas, indicating their replacement values as well.

Biogas from sewage remains in rural India

Table 2.37 Comparison of various fuels

Name of fuel	Calorific value (kcal)	Thermal efficiency (%)	Quantity required to replace 1 m^3 of biogas (based useful energy)
Biogas	4713	60	1 m^3
Kerosene	9122	50	0.620 L
Firewood	4708	17.3	3.474 kg
Cow dung cakes	2092	11	12.296 kg
Charcoal	6930	28	1.458 kg
Soft coke	6292	28	1.605 kg
Indane*	10,882	60	0.433 kg
Furnace oil	9041	75	0.417 L
Coal gas	4004	60	1.177 m^3
Electricity	860	70	4.698 kwh
Vegetable waste	3500	12	6.733 kg
Rice husk	1785	10	15.842 kg
Water hyacinth	2410	12	9.778 kg

*Cooking gas provided by Indian Oil Corporation.

Problems associated with the distribution of biogas

A frequently asked question is: Why is biogas not stored like LPG in cylinders and distributed to the consumers? The main problem encountered in packaging biogas like LPG is that biogas can be liquefied only under conditions of high pressure and low temperature. Table 2.38 gives the pressure and temperature relationship for the liquefaction of biogas. This is one of the main reasons why biogas has not become popular as an industrial commodity. Further, large-scale

Table 2.38 Pressure and temperature required to liquefy biogas

Pressure (kg/cm^2)	Temperature (°C)
80	−70
100	−50
130	−25
160	0
190	25

production of biogas is not undertaken because of the problem of transportation to distant places. The only safe and acceptable means of distribution of biogas is through pipelines, which is a costly proposition.

Biomanure

It is a known fact that 80% of the total population in India is living in villages, mostly involved in farming. The *green revolution* could have substantially improved the living economy but resulted in stagnation in the area of food production. This stagnation was a result of inadequate irrigation facilities and the high cost of chemical fertilizers. Another problem was that the chemical fertilizers needed to be supplemented with a sufficient amount of organic manure to prevent any damage to the land, which may otherwise be caused due to the prolonged and continuous use of chemical fertilizers alone. Accordingly, every possible source of organic matter must be harnessed. In this connection it may be mentioned that the residue obtained from a biogas plant, known as *biomanure*, is an excellent organic manure containing all the desired enriching nutrients required for plant growth as well as humus to improve soil structure. Its use also improves the aeration, moisture-holding capacity, water infiltration capacity, and cation exchange capacity of the soil. Furthermore, the sludge serves as a source of energy and nutrients for the development of the microbial population, which, directly and indirectly, improves the solubility and thus the availability of the essential chemical nutrients contained in soil minerals for higher plants. Biomanure is also known to aid microbial fixation.

Cattle dung has been the farmer's fertilizer in India from time immemorial. It is also being used as a traditional fuel in this country in the form of cow dung cakes. It is estimated that almost one-third of the total cattle dung available in the country is burnt as fuel due to the non-availability of alternative sources of energy. Thus, large quantities of valuable resources of manure are wasted as low-grade fuel. For example, the manure obtained per tonne of wet dung by the traditional method is only 500 kg, compared with the 730 kg produced by a biogas plant. This means that the amount of organic manure obtained by anaerobic digestion in a gas plant is about 43% more than that obtained from a manure pit. Besides, biogas is a much more efficient fuel with a heating value that is 20% more than the useful heat obtainable by burning the entire amount of dung as cake. This is attributed to the fact that in a gas plant the decomposition is selective, while in a traditional manure pit it is more rigorous.

Further, the organic manure obtained from a biogas plant is richer in essential plant nutrients compared to the manure obtained from other sources such as a farmyard and a town compost. Table 2.39 gives the composition of manure obtained from different sources. Another advantage is that the digested residue from the biogas plant can be used along with the irrigation water for crops, as it contains an appreciable amount of total nitrogen in a readily absorbable form. Further, due to the conversion of much of the organic carbon into digester gases, the proportion of inorganic nutrients, which are not affected by anaerobic fermentation, increases considerably in the residue. Accordingly, the increase in phosphate, nitrogen, and potash is reported to be proportional to the reduction in organic carbon. Table 2.40 provides analytical information on a typical dewatered sludge residue obtained in the laboratory and from pilot plant digesters. As only 30% to 50% of the organic components are converted into biogas, the total volume is reduced by only 5%.

The advantage of a biogas plant is that it utilizes all animal waste, including urine, which has high nitrogen content. Further, the anaerobic treatment of

Table 2.39 N, P, and K values of organic manure from different sources

Source	N (%)	P_2O_5 (%)	K_2O (%)
Residue from biogas plant	1.4–2.5	1.0	0.8
Farmyard manure	0.5	0.2	0.5
Town compost	1.5	1.0	1.5

Table 2.40 Concentration of the inorganic components in the digester content

	P_2O_5 (%)	N (%)	K_2O (%)
Total digester content			
Solid material	3.16	2.90	-
Supernatant	0.10	0.50	-
Total suspension	0.24	0.95	0.14
Bottom residues	3.90	4.90	1.80
Heavy metals in solids (%)			
Zn—0.18, Fe—0.60			
Cu—0.04, Pb—0.00			

manure in an almost closed system prevents the loss of vegetable nutritive substances and retains a high proportion of nitrogen, which has a good effect on plant growth. In a conventional manure storage system, 30% to 50% of the total nitrogen content is lost through evaporation. The fermented residue should, therefore, be stored with limited access to air so that the nutritional content is not lost. Hence, the residue is stored in a covered pit, to be used at a later stage. Optimal utilization of the nutritive potential is obtained by using the manure in liquid form. However, it must be noted that the manure value of the fermented residue is insignificant, unless the biogas process itself is nutritive.

A biogas plant can fulfil strict hygienic decontamination regulations as well as aesthetic requirements, e.g., avoidance of irritating smells. Depending on the type of raw material used, the organic fraction of residue sludge from an anaerobic digester (operating on plant and animal wastes) may contain up to 30% to 40% of lignin, undigested cellulose, and lipid materials on a dry weight basis. The remainder consists of substances originally present in the raw material but protected from bacterial decomposition by lignin and cutin, newly synthesized bacterial cellular substances, and relatively small amounts of volatile fatty acids. The bacterial cell mass is small (less than 10% to 20% of the substrate is converted into cells). Therefore, there is less risk of the production of odour and insect breeding problems when anaerobically digested sludge is stored or spread on land than when untreated or partially treated organic waste material is similarly handled or indiscriminately disposed off or stored.

Soil fertility is also maintained by the micro-flora of soil, which transform organic manure (farmyard, compost, or green) into humus in the soil. Some of the advantages of biomanure are the following.

1. Due to dark colouration, it helps the soil to absorb sunlight and become warm.
2. Helps in soil aggregations.
3. Makes soil porous and hence facilitates aeration and crop root penetration.
4. Increases water-holding capacity.
5. Prevents leaching of nutrients.
6. Acts as a buffer to pH change.
7. Forms complexes with toxic elements such as copper and aluminium, and minimizes plant toxicity.
8. Supplies nutrients to beneficial soil microbes.
9. Changes the membrane permeability of root hair and enhances nutrient uptake.

It is very important to note that no synthetic fertilizer plays such a multifarious role in enhancing soil fertility. The N, P, and K values for some biomanures are listed in Table 2.41.

Some of the comparative merits and demerits of the two processes used to produce biomanure, namely, composting and anaerobic digestion, are enumerated in

Table 2.41 N, P, K content in biomanures

Manure	N (%)	P_2O_5 (%)	K_2O (%)
Fresh cattle dung	0.3–0.4	0.1–0.2	0.1–0.3
Farmyard manure	0.4–1.5	0.3–0.9	0.3–1.9
Compost	0.5–1.5	0.3–0.9	0.8–1.2
Poultry manure	0.1–1.8	1.4–1.8	0.8–0.9
Cattle urine	0.9–1.2	Trace	0.5–1.0
Paddy straw	0.3–0.4	0.08–0.1	0.7–0.9
Wheat straw	0.5–0.6	0.1–0.2	1.1–1.3

Table 2.42 Comparison between composting and anaerobic digestion

Composting	Anaerobic digestion
Unselective degradation by both aerobic and anaerobic microbes causing more weight loss without proportional enrichment of fertilizer value. Both carbon and nitrogen are consumed during degradation. One tonne of wet dung yields 500 kg wet compost manure.	In selective anaerobic digestion, the weight loss is less. Carbon is consumed mainly, producing methane. One tonne of wet dung yields 730 kg of residue enriched in N, P, and K.
Proper degradation time is 3–4 months.	Proper digestion time is 30–40 days.
About 30%–50% nitrogen is lost by evaporation as ammonia.	Loss by evaporation is negligible.
Compost has a bad odour and provides a breeding ground for insects and flies.	Odour is minimal. No insect and fly breeding.
N, P, K enrichment is 10%–15%.	N, P, K enrichment is 30%–40%.

Table 2.42. It can be concluded from this table that anaerobic digestion is far more profitable than composting and is a useful technique that can be utilized to solve the energy and fertilizer problems faced by farmers in rural areas. For this reason, biogas technology or anaerobic digestion of waste materials is said to provide an economical means of pollution abatement.

Many methods can be adopted to ensure better utilization of biogas plant residues. Two such measures are listed here.

1. Immobilization of volatile nitrogen by growing fungus, especially mushrooms.
2. Supplementary addition of synthetic micro- or macro-elements, after testing fertility.

These residues can also be used as plankton or fish feed.

Summary

This chaper has primarily focused upon the nature and characteristics of solid and liquid wastes released from domestic, agricultural, and various chemical industries. This information is vital when deciding the design and development of appropriate waste treatment plants, which will be discussed further in the subsequent chapters.

However, before design and development, the logical next step in our study would be awareness of the basic scientific aspects of waste treatment, such as microbiological, biochemical, strain improvement, and the environmental parameters influencing the degradation of wastes, which will be dealt with in the next chapter.

Review Questions

1. List the factors considered for classifying waste materials that otherwise cause pollution due to uncontrolled microbial attack.
2. For the design of a biological waste treatment plant, quantitative information regarding the physico-chemical characteristics of the waste material under consideration is essential. What are these characteristic parameters and how does their magnitude influence the biological treatment process?
3. It is known that for a given waste material, a C/N ratio (by weight) close to 30 is considered ideal for microbial degradation. Also the organic carbon and nitrogen contents of wastes from different sources are different. The C/N ratios of three different waste materials are given below:

Material	N (%)	C/N
Rice straw	1.34	78.58
Cow manure	1.70	18.00
Seaweed	1.90	19.00

Considering the fact that the supply of seaweed is limited, determine the proportion of each of the three wastes in the blended composition such that the C/N ratio is close to the desired value of 30 (with minimum seaweed content).
4. Discuss the effect of particle size on the rate of digestion of a waste material.
5. What is the volatile solid content in a waste material and what should its concentration in the feed of the digester be?
6. As an environmental biotechnologist, how will you utilize the data given in Tables 2.8 and 2.9?
7. (a) In a 1000-t-per-day ammonia production unit, calculate the ammonia loss as well as the concentration of ammonia in the liquid effluent.
   (b) What is the concentration of urea in the combined waste from a fertilizer plant?
   (c) In a petroleum refinery, what kind of wastes are generated in the vacuum distillation unit, in the catalytic or thermal cracking unit, during hydrocracking, and in the solvent processes?
   (d) List the contents (average) of the liquid effluent from a typical Indian refinery.
   (e) What is the water requirement of the paper and pulp industry?
   (f) What is the BOD of the discharge from a kraft paper mill?

(g) What are the characteristics of the spent (black) liquor obtained from the digester of the soda process used for pulp making?

(h) What is the BOD value of and tannin content in the spent vegetable tanning liquor?

 (i) What would be the volume of the effluent from the mill house of a sugar industry? Enumerate its characteristics.

 (j) What is the volume of spent wash per litre of alcohol produced? What will the corresponding range of BOD values be?

8. Discuss the importance of knowledge about the characteristics of different industrial effluents and their practicality.

9. Write notes on the nature of wastes from different industries such as sugarcane, tanning, alcohol, fertilizer, paper and pulp, textile, dairy, plating, canning, slaughterhouse, and rubber.

References

Mahajan, S.P. 1990, *Pollution Control in Process Industries*, Tata McGraw-Hill, New Delhi.

Bhattacharyya, B.C. 1979, 'Design and development of gobar (bio) gas plant for rural population', Monograph, Chemical Age of India.

ICAR 1969, 'Manures and Fertilizer', *Handbook of Agriculture*, ICAR, New Delhi, Chapter 3, pp. 98–110.

Scientific Aspects of Biological Waste Treatment for Biofuel Production

Introduction

Chapter 2 dealt with the classification and characterization of wastes obtained from different sources such as agricultural, household, and industrial enterprises. The study also included the BOD (biological oxygen demand) values as an indicator of the level of pollution caused by the different wastes. As mentioned previously, the BOD level of industrial and other wastes needs to be reduced in order to stabilize the wastes and thereby convert them into non-polluting and harmless organic materials.

It is pertinent to recapitulate that the biological treatment methods used for stabilization of polluting organic wastes can be broadly classified as aerobic and anaerobic treatment processes. Aerobes and anaerobes control these two waste treatment methods, respectively. Micro-organisms are, therefore, key elements for bioremediation. The scientific aspect of waste treatment processes includes various facets as listed here.

1. Selection of suitable micro-organisms.
2. Study of microbial characteristics for improving their biodegradation efficiency.
3. Improving the environmental conditions that influence the microbial effectiveness in product formation during the biodegradation of organic wastes.

This chapter will focus on these scientific aspects of waste treatment processes, with special reference to the anaerobic method of treating wastes. This is because the anaerobic treatment process is economically more advantageous than its counterpart, the aerobic treatment method, as the former method produces commercially viable biofuels such as bioethanol, biohydrogen, and biogas.

Micro-organisms

The degradation of organic waste occurs under the influence of micro-organisms. Therefore, in order to understand the mechanism of waste degradation, it is a prerequisite to have an understanding of the functioning of a micro-organism at the cellular level. We all know that micro-organisms are too small to be perceived by the unaided human eye. The credit for the discovery of the microbial world goes to the Dutch merchant, Anton van Leeuwenhoek, who invented the microscope in 1665. The existence of the microbial world came to fore with the invention of the microscope.

Depending upon the structure of the organism, especially on the presence or absence of membrane-bound organelles, the entire microbial kingdom is subdivided into prokaryotes and eukaryotes.

The *prokaryotes* are characterized by the presence of a peptidoglycan cell wall, 70s ribosomes, and a nucleoid. They are also characterized by the absence of cytoskeleton, microtubules, 80s ribosomes, membrane-bound organelles such as mitochondria, chloroplast, Golgi apparatus, a well-defined nucleus, the histone–DNA complex, and introns in genes. Further, these microbes show only mitotic form of reproduction. On the other hand, the *eukaryotes* are characterized by the presence of a cell wall, lipo-proteinaceous cell membrane, microtubules, cytoskeleton, 80s ribosomes, and membrane-bound organelles such as mitochondria, chloroplast, Golgi bodies, well-defined nucleus, histone–DNA complex, and more than one gene with introns. The general structure of a prokaryote such as bacteria is given in Fig. 3.1. Among these micro-organisms, methanogenic bacteria (prokaryotic by nature) are primarily responsible for anaerobic digestion. Therefore, we will now study in detail the characteristics of methanogenic bacteria, as our main focus is on anaerobic digestion.

Bacteria

Bacteria are very minute organisms, often referred to as animalcules, which are barely visible under the light microscope. The smallest bacteria are about 0.1 micron in diameter, while the diameter of the largest ones may be $60 \times 6 \ \mu m^2$. Bacteria can be both beneficial as well as harmful to humans. Beneficial bacteria help in various processes such as compost-making and production of biogas, vinegar, curds, ethanol, and hydrogen, whereas harmful or pathogenic bacteria cause diseases such as cholera, typhoid, diphtheria, tetanus, and tuberculosis.

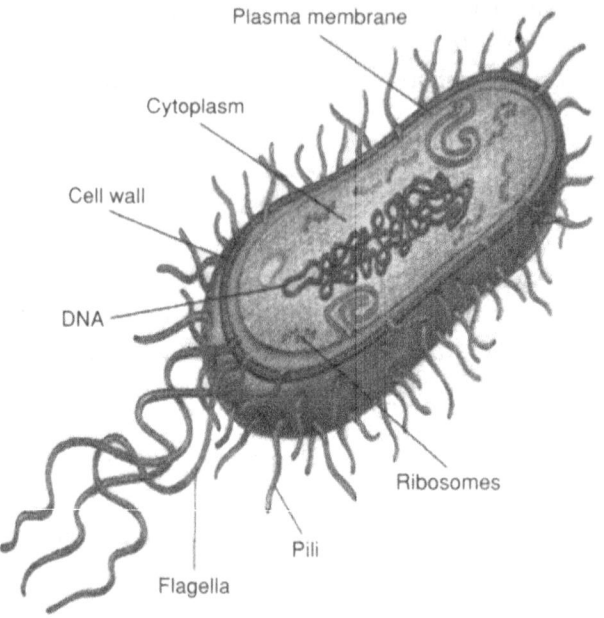

Fig. 3.1 General structure of an eukaryotic bacterium

Bacteria can be classified on the basis of different characteristics such as shape, source of energy, tolerance to temperature, mode of nutrition, and so on. All these aspects have been touched upon briefly in the next few sections.

Shape

Bacteria occur mainly in three shapes, namely, spherical, rod-shaped, spiral or comma-shaped. Spherical bacteria are called *cocci* (singular—coccus). The cells may occur in pairs, in bunches in bead-like chains, or as cubical structures of eight or more cells. Rod-like bacteria are called *bacilli* (singular—bacillus). They generally occur singly, but may occasionally be found in pairs or chains. Spiral-shaped bacteria are called *spirilla* (singular—spirillum). The comma-shaped bacteria are also known as *vibrios*.

Source of energy

Bacteria are also classified depending upon their source and requirement of energy. Accordingly, we have three different classes of bacteria as listed here.

1. *Phototrophic* Photosynthetic bacteria need radiant or light energy for their growth.
2. *Chemotrophic* Chemosynthetic organisms use oxidation–reduction reactions as a source of energy. They do not require light energy.
3. *Lithotrophic* These organisms utilize inorganic electron donors such as hydrogen sulphide, ammonia, or sulphur to derive energy.

4. *Organotrophic* This class of bacteria requires organic compounds as electron donors to gain energy.

Temperature tolerance

Bacteria can also be arranged into three general groups on the basis of their tolerance for different ranges of temperature. However, as the division between these groups is not sharp and somewhat arbitrary, there is a degree of overlapping.

1. *Psychrophilic* (cryophilic) or cold-loving bacteria are microbes that play an important role in the spoilage of food in refrigerators and cold storage plants.
2. *Mesophilic* or mid-range temperature loving organisms are the most commonly found group and thrive in medium-range temperatures. This group includes organisms from soil and water sources as well as pathogens found in human beings and lower animals.
3. *Thermophilic* (heat-loving) micro-organisms grow well at high temperatures, which otherwise are lethal to most other bacteria.

Table 3.1 shows the temperature tolerance (Atlas 1984) of different types of bacteria.

Oxygen requirement

Depending on their oxygen requirement, bacteria can further be divided into three major groups as listed here.

1. *Aerobic* group of bacteria, which function in the presence of oxygen.
2. *Anaerobic* group of bacteria, which function in the absence of oxygen.
3. *Facultative aerobes*, which can function in both the presence and the absence of oxygen.

Composition of cell wall

The entire bacterial world can also be classified into two groups depending upon cell wall composition. One group which takes gram staining is called *gram-positive* bacteria, while the other which does not take gram staining is termed *gram-negative* bacteria.

Table 3.1 Classification of bacteria based on their temperature tolerance

	Optimum temp. (°C)	Minimum temp. (°C)	Maximum temp. (°C)
Psychrophilic (cryophilic)	15–20	0	30
Mesophilic	25–40	15–25	50
Thermophilic	45–55	25–45	55–85

Prokaryotes—archaebacteria and eubacteria

Besides this broad classification, bacteria have been further divided into different groups and subgroups based on their biochemical properties, habitat, pigmentation, capsular materials, and cellulosic properties, i.e., surface antigens, GC content, DNA–DNA hybridization, sequence of 23s, 16s, and 5s rRNAs, or antibiotic sensitivities.

Bergey's Manual of Systematic Bacteriology is the most widely accepted and reliable literature for bacterial classification. It contains the morphological and physiological properties with appropriate literature citations for classification of new isolates based on the 16s rRNA classification system (which is a part of the phylogenetic tree). For instance, different and distinctive characteristics have been noted between two morphologically and physiologically distinct bacteria, *E. coli* and *methanobacteria*, which are both prokaryotes. Prokaryotes can be once again divided into two groups, namely,

1. archaebacteria—the most primitive ones (early cell) and
2. eubacteria—progenotes (developed cell).

It is worth mentioning here that methanobacteria fall under the group archaebacteria and are easily distinguishable from eubacteria.

Archaebacteria can once again be divided into different groups depending upon their oxygen uptake characteristics.

1. Some groups are aerobes, e.g., *halobacteria, thermoplasma, thermes, sulphococcus, halococcus.*
2. Some groups are facultative anaerobes, which grow in the absence of oxygen, e.g., *thermoproteus, pyrodictium, desulfurococcus, pyrococcus, thermococcus, thermodiscus.*
3. Some groups are strictly anaerobic in nature, e.g., *methanothermus, methanothrix, methanobacterium, methanococcus.*

Archaebacteria

Archaebacteria can be distinguished from eubacteria by their extreme and special habitats. The group includes lithoautotrophs and heterotrophic aerobes and anaerobes. It is a well-known and established fact that these groups of bacteria have many properties in common which distinguish them from eubacteria. Moreover, as archaebacteria are a large group, they have different shapes of cells, cell components, and varied metabolic pathways, which are sometimes common with and at times distinct from the eubacterial group.

The present day system of classification divides archaebacteria into the following three categories.

1. Methanobacteria
2. Halophilic bacteria
3. Thermoacidophilic bacteria

Archaebacteria can stain either gram-positive or gram-negative and may be spherical, rod-shaped or spiral, irregular, plate-shaped, or pleomorphic. They are usually found as a single entity, in aggregates, or in filamentous form. In general, their diameter varies from 0.1 to 15 μm but some filaments can grow up to 200 μm in length. Archaebacteria are mostly found in extreme aquatic and terrestrial habitats.

The components in the cell wall of archaebacteria are distinctly different from the cell wall constituents found in eubacteria. Further, there is a considerable variety in the structure of the cell wall of archaebacteria. This is evident from the fact that they can stain both gram-positive and gram-negative. Unlike in eubacteria, the cell wall of archaebacteria contains chemicals such as muramic acid and D-amino acid. Also the cell wall lipid transcription and translation apparatus, prosthetic group, and coenzyme are different in both the groups. The archaebacterial cell wall does not contain any peptidoglycan chains and in the case of methanobacterium and other methanogens, the cell wall contains pseudomurin, a peptidoglycan polymer having L-amino acid cross-linking. It has been found that gram-positive archaebacteria have a single, thick homogeneous layer like that in gram-positive eubacteria, whereas gram-negative archaebacteria do not have any outer membrane and the complex peptidoglycan network with protein and glycoprotein subunits is also absent. Therefore, it can be concluded that archaebacteria are not sensitive to antibiotics such as penicillin and cephalosporin.

In archaebacteria, the cytoplasmic membrane contains glycerol ethers and C-20 (phytanil) and C-40 (biphytanil) alkyl isoprenoids, instead of fatty acid glycerol esters. The archaebacterial cell membrane may also contain a mixture of diethers, tetraethers, and lipids. Neutral lipids are also found in the form of C-15 and C-30 isoprenoid hydrocarbon molecules. Based on the type of sequences, the latest system of classification divides the microbiological world into three major groups, i.e., bacteria, archaebacteria, and eukaryotes.

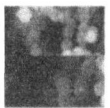

Metabolism of Anaerobic Organisms

In this section the different metabolic pathways (Stanier et al. 1987) followed by the biochemical reactions occuring in micro-organisms are discussed. These chemical pathways play an important role in the degradation of organic wastes and other intermediates.

The term metabolism refers to the complete set of biochemical reactions occurring within a cell. The metabolic reactions which occur sequentially in the cell constitute the metabolic pathway; different pathways are often interrelated. At each step, a pathway involves one or more enzyme-catalysed reactions resulting in the formation of useful products. In other words, it can be said that with the help of metabolic pathways a micro-organism produces order from its disorderly surroundings.

Another important aspect is that these metabolic pathways show how energy from the environment is used to drive the different metabolic processes. Studies conducted on various metabolic pathways also help to explain the major differences in the functioning of a cell in the presence and absence of oxygen. As we know, these two conditions are called aerobic and anaerobic, respectively. That is, while some cells (obligate or strict anaerobes) do not use free O_2, for some other cells (obligate aerobes) free O_2 is an essential requirement. However, there is a third group of cells (facultative) which can grow in either environment. It is interesting to note that an important catabolic function, i.e., respiration in anaerobic organisms, involves the same biochemical pathway as that for aerobic metabolism in heterotrophs. The major difference between them lies only in the compounds which serve as the terminal electron acceptors in the electron transport chain. It means that in the case of anaerobic respiration, molecular O_2 does not play the role of the terminal electron acceptor. The corresponding role is actually played by compounds such as nitrates (NO_3), sulphates (SO_4^{2-}), and fumarates. In the case of (SO_4^{2-}) or NO_3, the products of their reduction are also capable of acting as the terminal electron acceptors, and thus can play a role in anaerobic respiration.

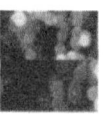

Source of Energy for Anaerobic Digestion

Biodegradation of waste means the breaking down of the different types of polymeric substances present in that waste. The hydrolysis of these polymers requires energy in order to break the chemical bonds between the molecules of the polymeric compounds (Beiley & Ollis 1987; Samson et al. 2004). The question is from where do the microbes derive this energy? As mentioned earlier, there are different types of oxidation–reduction reactions taking place inside a cell. The oxidation reactions generate energy, while energy is consumed during the reduction reactions. Thus, through the oxidation process, energy gets stored in the form of ATP. It is interesting to note that this energy is utilized more economically and efficiently by the cells as compared to the energy utilization by many man-made machines and therefore this process maintains a thermodynamically favourable situation in the biological system.

The production of ATP is generally associated with the cytoplasmic membrane of bacteria. Studies reveal that upon hydrolysis, ATP is first converted to ADP and then to AMP, accompanied by the release of energy at each step. Of the three variants, ATP has the highest energy phosphate bond with a free energy, ΔF, of about -7.3 kcal/mol. ADP also has moderately high energy bonds, having a capability of releasing ΔF of about 7 kcal/mol upon hydrolysis, while AMP is the lowest energy compound, which gives ΔF of about 3 to 4 kcal/mol. Other oxidizing agents present inside the cell such as NAD^+, $NADP^+$, and FAD also play a vital role by participating in the different hydrogen transfer reactions.

Let us take the example of the TCA cycle. One of the important products in this cycle is isocitrate, which is converted into oxalosuccinate with the transfer of an electron from NAD to NADH. As already mentioned, the process of respiration is associated with the transport of electrons. During this process lot of energy is released and conserved in the form of ATP molecules. This entire process is called *oxidative phosphorylation*, where the coenzymes NADH and $FADH_2$ become NAD^+ and FAD, resulting in the production of 3 mols and 2 mols of ATP, respectively. In the case of aerobes, O_2 is supplied to the medium, which helps in phosphorylation. However, in the case of strict anaerobes, as free oxygen cannot be made available, the process of oxidative phosphorylation becomes less efficient, resulting in a low yield of ATP molecules per mole of NADH reoxidation.

As the oxidation of NADH to NAD^+ becomes difficult in the absence of oxygen, the anaerobes have to adopt some indirect means to achieve the desired reoxidation of the reduced cofactor. This comparatively less efficient process of oxidative phosphorylation results in the accumulation of reduced metabolites during the process of degradation of any substrate. Since, in anaerobes there is a relatively low ATP yield from substrate degradation, the resultant accumulation of reduced metabolites is bound to be large compared to the cell material synthesized. This has been cited as one of the probable reasons for the slower growth rate of anaerobes as compared to aerobes.

The hydrolytic product of carbohydrates, including cellulose, is mainly glucose. It is the major source of carbon for the process of metabolism in organisms. Various pathways have been identified in the processes of both aerobic and anaerobic metabolism, which describe the intermediate and end products generated by individual organisms. The next few sections deal with some commonly known pathways of anaerobic metabolism.

The Embden–Meyerhof–Parnas (EMP) pathway, shown in Fig. 3.2, is the most commonly found pathway for the conversion of glucose to

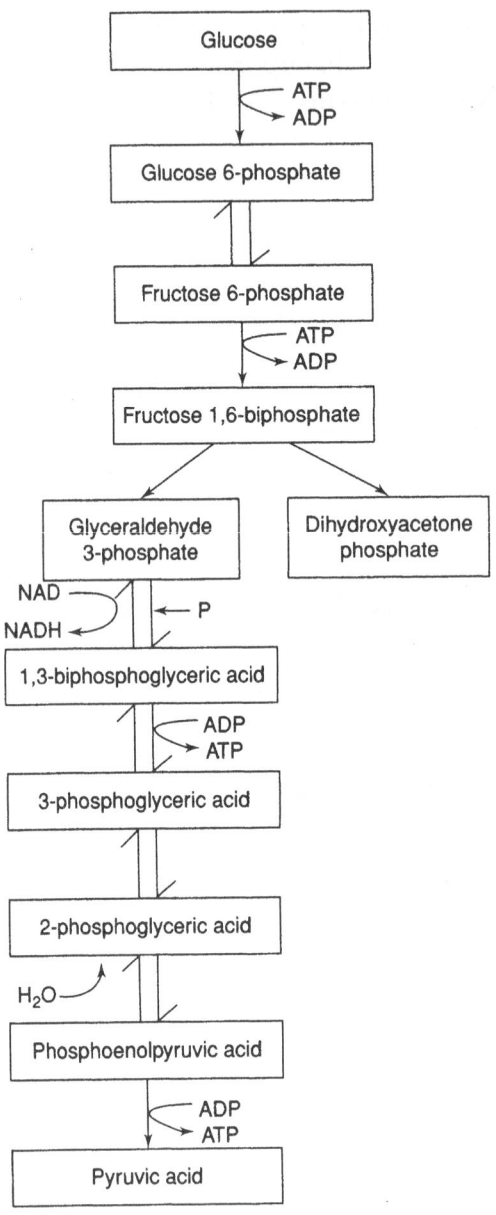

Fig 3.2   Embden–Meyerhof–Parnas pathway of conversion of glucose to pyruvic acid

pyruvic acid. It has been seen in many of the bacteria that produce fermentative end products other than lactic acid and alcohol. Figure 3.3 shows the various pathways of pyruvic acid metabolism. The end products of this bioprocess may be propionic acid, isopropanol, butyric acid, butanol, butanediol, etc.

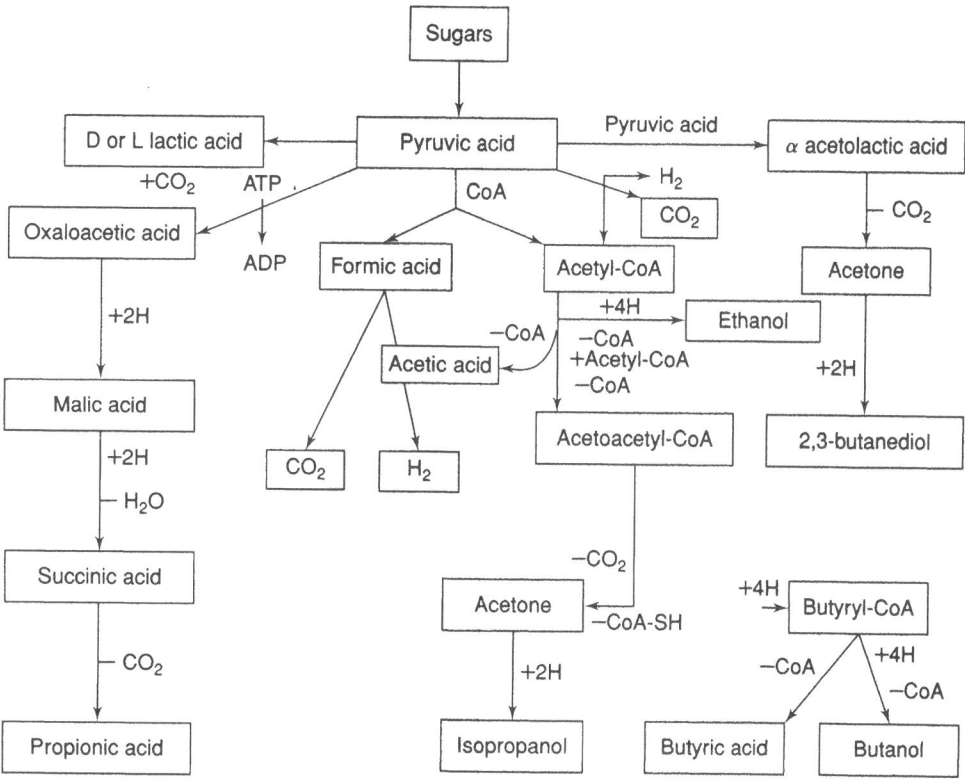

Pathways showing some major end products of bacterial fermentation under anaerobic conditions

Microbiological and Biochemical Aspects of Anaerobic Digestion

The major constituent of biogas, i.e., methane, is generated in anaerobic environments such as bottom moods, paddy fields, and the ruminant stomach as a result of anaerobic breakdown of organic matter. The microbiological and biochemical processes involved in the production of methane gas by methanogenic bacteria, which are discussed in this chapter, play a significant role in the carbon cycle, with over 50% of the annual carbon input into the aquatic environment being regenerated as methane (Hanson 1980). It has been calculated that methanogenesis results in the generation of some 800 million t of methane per annum (Higgins et al. 1981). Aerobes play a key role in the degradation of organic substances, leading to the eventual regeneration of CO_2.

We have already seen that the production of methane through anaerobic digestion is a mixed culture process. However, the biochemistry of the mixed culture fermentation system is not yet fully known. However, significant

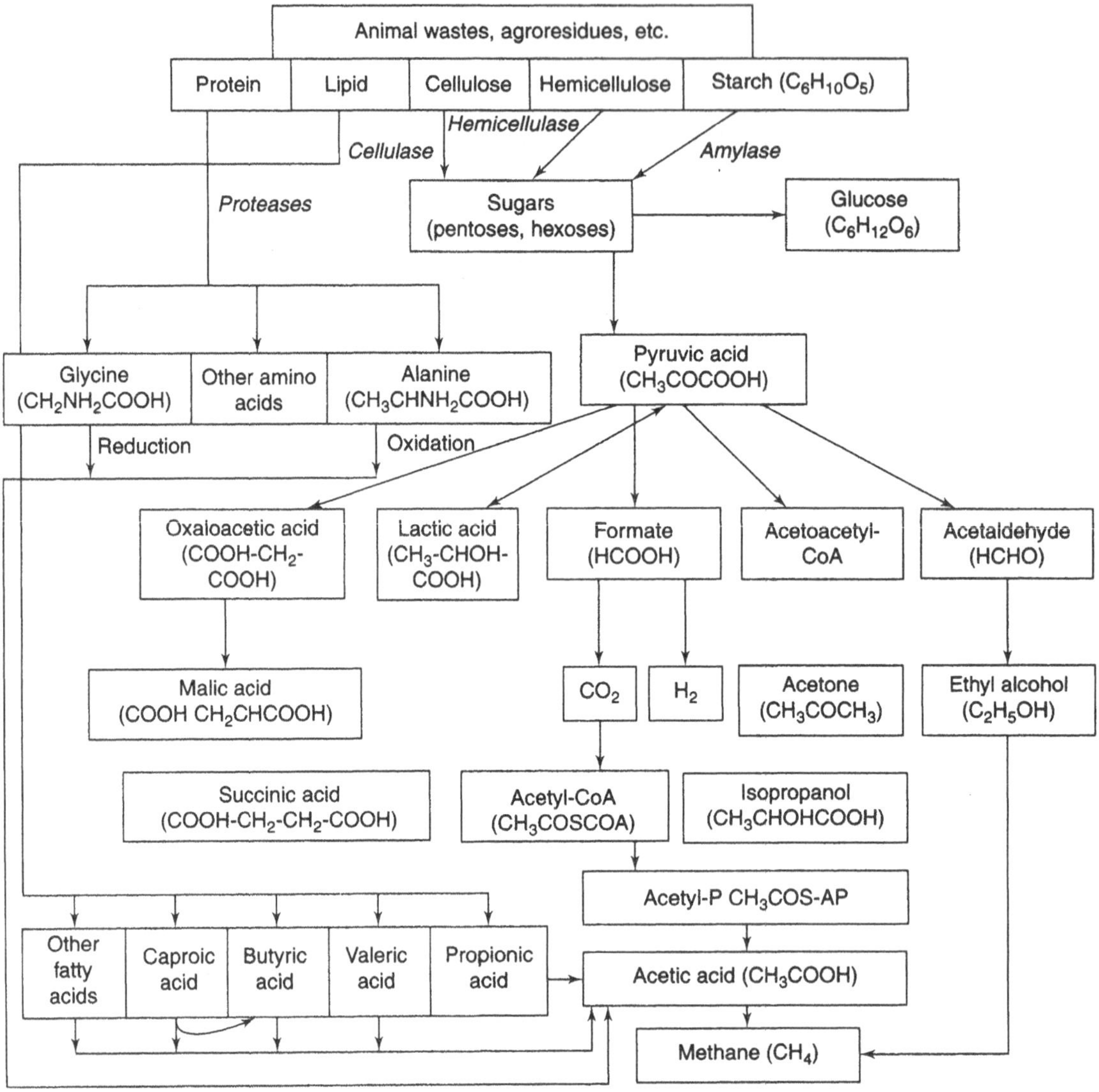

Fig. 3.4   Schematic presentation of the biochemical process involved in the anaerobic digestion of biopolymers into methane

understanding of the complex interrelated mechanism involved in the process has led to greater confidence in designing anaerobic digestion plants for treatment of waste biopolymers. Figure 3.4 shows the different biochemical processes taking place in bacteria-dominated biomass. These processes are responsible for the conversion of complex organic material such as polysaccharides, lipids, and proteins into methane and carbon dioxide. The symbiotic microbial communities, by virtue of their ability to alter fermentation pathways,

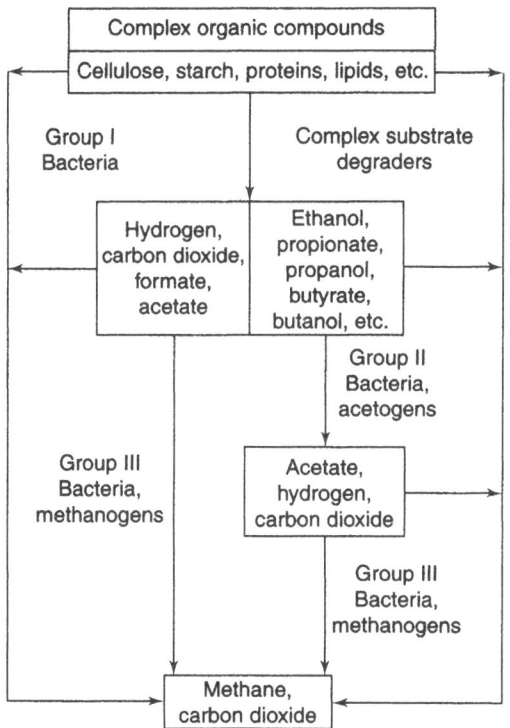

Fig. 3.5 Schematic diagram showing bacterial reactions in stages of anaerobic digestion

function as unified self-regulating systems and thus maintain the conditions of *p*H, oxidation–reduction potential (redox), and thermodynamic equilibria to optimize growth and maintain the stability of the digester.

According to their tropic requirements, the bacteria responsible for anaerobic digestion may conveniently be divided into three broad groups. Thus the process of anaerobic digestion is now known to include three stages and three different groups of anaerobes as shown in Fig. 3.5.

1. Group I organisms are known as *degraders* or *hydrolytic* bacteria. This group facilitates the hydrolytic breakdown of polymers such as proteins, lipids, nucleic acids, and polysaccharides to produce H_2, CO_2, monosaccharide, formate, acetate, ethanol, propionate, butyrate, etc.
2. Group II organisms are known as *acitogens*. They further ferment ethanol, propionate, butyrate, etc. to yield acetate, hydrogen, and carbon dioxide as major end products.
3. Group III organisms are the *methonogens* and degrade acetate to yield CH_4 and CO_2.

Some of the anaerobes may further be subdivided into hydrogen utilizers (*lithotrophs*) and users of acetic acid, i.e., *acitotrophs*. Feedstocks containing oxidized sulphur and nitrogen may give rise to other additional groups of bacteria such as sulphate reducers and denitrifiers.

Degraders or Hydrolytic Bacteria (Group I)

In the first stage this group of bacteria is responsible for the hydrolytic degradation process of macromolecules into soluble products such as sugars, amino acids, and fatty acids. They may be either strict anaerobes or facultative (MACS 1986) in nature (under group I). Hydrolytic bacteria belong to the genera *Actinomyces, Acetivibrio, Aerobacter, Aeromonas, Alcaligenes, Butyrivibrio, Cellulomonas, Citrobacter, Clostridium, Corynebacterium, Desulphovibrio, Enterobacter, Escherichia, Flavobacterium, Klebsiella, Lactobacillus, Laptospira, Listeria, Micrococcus, Neisseria, Nocardia, Parapolobacterium, Peptococcus, Pediococcus, Proteus, Pseudomonas, Ramibacterium, Rhodopseudomonas, Ruminococcus, Sarcina, Staphylococcus, Streptococcus, Streptomyces, Vibrio.*

Hydrolytic bacteria utilize a range of exo-enzymes (Bhattacharyya 1979) such as proteases, amylases, cellulases, lipases, and pectinases for effecting the dissolution of polymeric substrates. Enzymes are frequently responsible for the process of hydrolysis and hence degrade neutral substrates such as proteins, lipids, and homo- and hetero-polysaccharides such as cellulose and starch. It has been found that all substances do not respond to anaerobic digestion. For example, lignin is not digested under anaerobic conditions as this process of degradation requires the presence of high oxidative conditions. In addition to the natural substrates, anaerobic microbes degrade phenol and sulphur compounds derived from sulphur pulping processes, coal gasifier effluents, and the wastewater of petrochemical refineries. The products of degradation may vary with the species and strain of bacteria, constitution and amount of feedstock, and other culture conditions such as pH, temperature, and redox potential.

Acetogenic Bacteria (Group II)

In the second stage, the acetogens ferment the end products of the first stage—butyrate, propionate, caproate, glucose, amino acids, and ethyl alcohol—to acetate, H_2, and CO_2.

An important factor affecting the production of hetero acid is the concentration of hydrogen in the reactor. This concentration affects both the pH and the oxidation–reduction potential. In the event of hydrogen stress, the organisms adopt alternative fermentation pathways in order to utilize more number of reduced compounds as oxygen sinks and thereby try to control the decreasing concentration of hydrogen. For example, let us consider a typical chemical reaction. We know that hydrolysis of glucose results in the formation of 4 mols of hydrogen gas and 2 mols of acetic acid per mole of substrate as shown here.

$$C_6H_{12}O_6 + 2H_2O \longrightarrow 2CH_2COOH + 4H_2$$

Soluble carbohydrates, starch, and pectin are fermented by many species of *Clostridium* (Stanier et al. 1987) resulting in the formation of acetic acid, butyric acid, CO_2, and H_2.

In the case of butyric acid fermentation, the conversion of sugar to pyruvic acid follows the Embden–Meyerhof–Parnas pathway. Pyruvate is oxidized to 2 acetyl-CoA (with the loss of CO_2) and isopropanol. Initially, 2 acetyl-CoA is converted to acetoacetyl-CoA. The reactions are then bifurcated into two pathways as shown in Fig. 3.6. In one pathway, the sequential biochemical reactions are accompanied by the oxidation of NADH to NAD, which in turn is again reduced to NADH by the utilization of the H_2 produced initially, and ultimately butanol is produced. In the other pathway, acetoacetyl-CoA is converted to acetoacetic acid (the presence of acetate in the surroundings is essential for this step), and further conversion to acetone is accomplished through the process of decarboxylation. In the last step, conversion of acetone to isopropanol is associated with the oxidation of NADH to NAD^+. The accumulation of neutral end products is favoured by the maintenance of high partial pressure of H_2 in the culture, and it can be largely prevented if hydrogen is removed as soon as it is produced in the system.

Dr L.H. Strickland et al. worked on a particular species of *Clostridium* (Stanier et al. 1987) and formulated the *Strickland reaction*. Their work illustrated that after fermentation, glycine and alanine produced ammonia, CO_2, and 3 mols of acetic acid. In this reaction (Fig. 3.7) alanine, the electron donor, is oxidatively deaminated to form pyruvic acid, which later on undergoes thiolytic cleavage to produce acetyl-CoA accompanied by the evolution of CO_2. In the subsequent reactions it forms acetic acid and one molecule of ATP. On the other hand, glycine is deaminated and NAD^+ is reduced to NADH. This reaction does not occur on its own and hence allows all the constituents of amino acids present in the proteins to be utilized as sources of energy.

Glucose is also utilized by other *Clostridium* species such as *C. thermoaceticum*, which produces acetate from glucose, quantitatively (see Fig. 3.8). In this, three molecules of acetate are produced per molecule of glucose oxidized. Glucose undergoes fermentation through the glycolytic pathway to produce 2 pyruvate (as already seen in Fig. 3.2). In fact, two-thirds of the total quantity of acetate is produced by synthesis from CO_2, which undergoes reduction to form

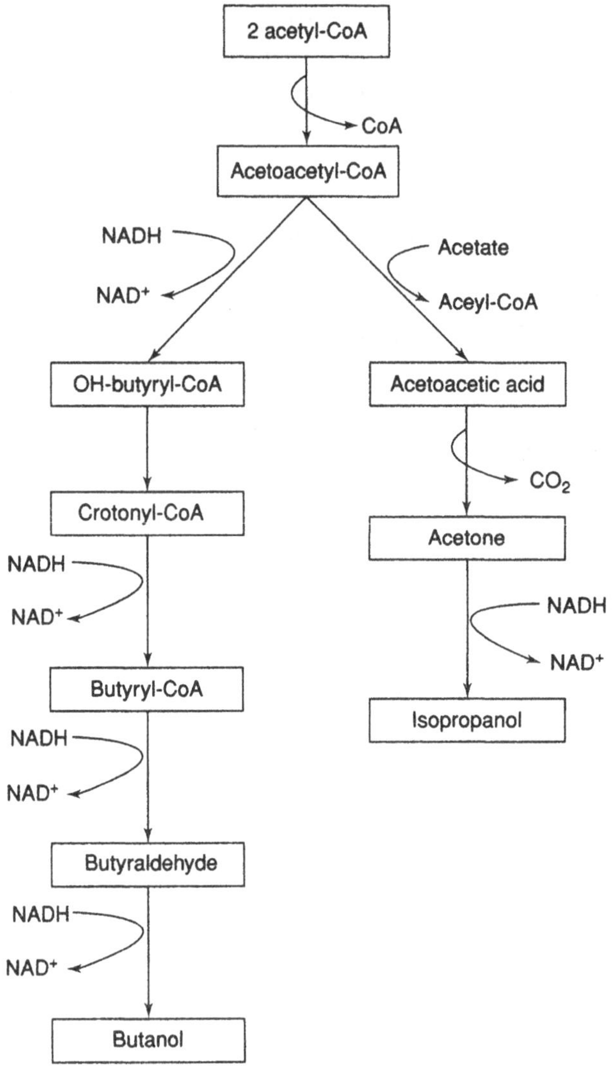

Fig. 3.8   Pathways for the formation of butanol and isopropanol from acetyl-CoA

acetyl-CoA. In this step energy is received due to the hydrolysis of ATP to ADP, while reduction of CO_2 occurs by the oxidation of NADH to NAD^+. The next step requires the presence of inorganic phosphate. In the final step acetate is formed with the release of energy through the phosphorylation of ADP to ATP.

Studies conducted on anaerobic homoacetogenic bacteria such as *Clostridium thermoaceticum* and *Clostridium thermoautotrophicum* have shown that these bacteria have the potential to be used for large-scale industrial production of acetate. Homoacetogenic bacteria are unique in that they can convert 1 mol of

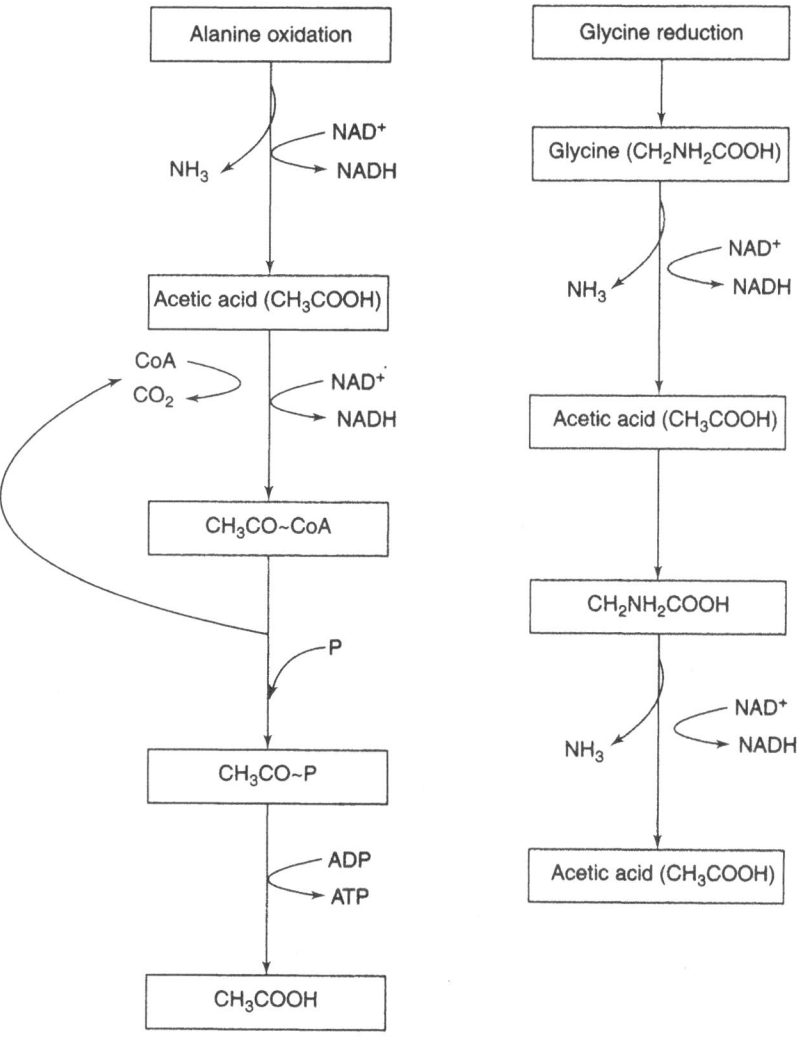

Mechanism of the Strickland reaction

glucose to 3 mols of acetate. Of these 3 mols, 2 mols are formed by fermentation of glucose, whereas 1 mol is formed by the fixation of CO_2.

The synthesis of acetate from CO_2 occurs via the CO_2 fixation (Fuchs 1986; Ljungdahl 1986; Wood 1986) pathway, which leads to the fermentation of acetyl-CoA as shown in Fig. 3.8. Many organisms belonging to the homoacetogenic group grow on a gas mixture of H_2 and CO_2 or CO with the reduction of CO_2 (Fuchs 1986).

$$2CO_2 + 4H_2 \longrightarrow CH_3COOH + 2H_2O$$
$$4CO + 2H_2O \longrightarrow CH_3COOH + 2CO_2$$

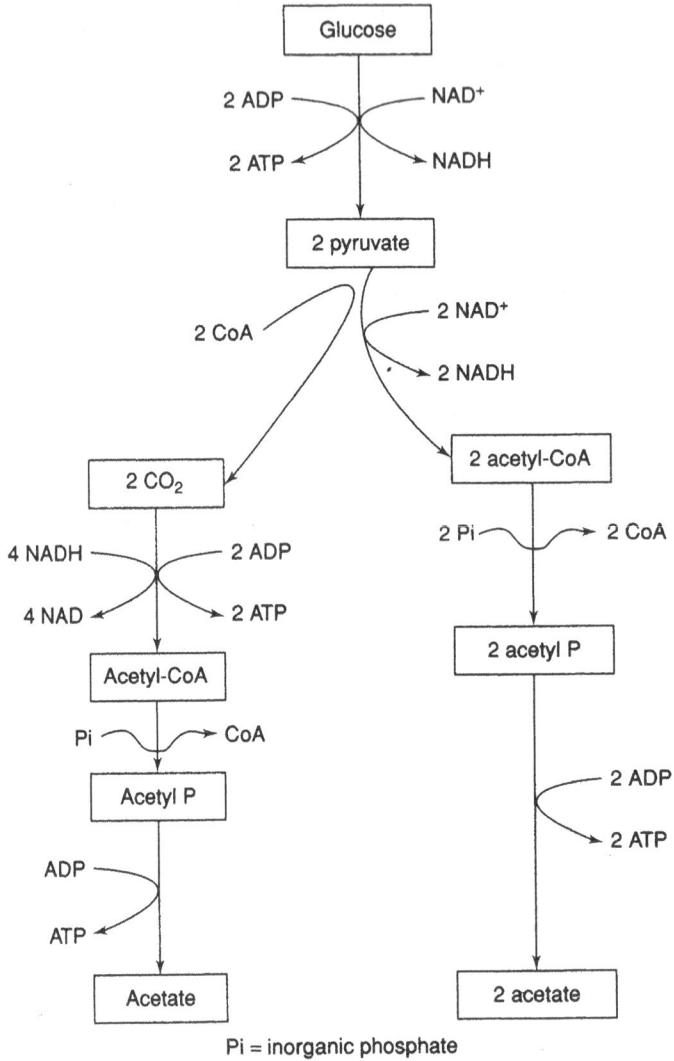

Fig. 3.8   Homoacetate fermentation of glucose by *Clostridium thermoaceticum*

These sources also include formate (Wiegel et al. 1981), methanol (Gottwald 1973; Wiegel & Garrison 1985), and methyl phenol esters (Bache & Pfennig 1981; Winters-Ivey et al. 1985). Methanol fermentation may be summarized as follows:

$$CH_3OH + H_2O \longrightarrow CO_2 + 6H^+ + 6e^-$$
$$3CO_2 + 6H^+ + 6e^- \longrightarrow 3CO + 3H_2O$$
$$3CH_3OH + 3CO \longrightarrow 3CH_3COOH$$

Net reaction: $4CH_3OH + 2CO_2 \longrightarrow 3CH_3COOH + 2H_2O$

Methanogenic Bacteria (Group III)

This group of bacteria is involved in the third stage of bioconversion of organic substrates into methane. Methanogens, belonging to the group archaebacteria, lack the peptidoglycan component in their cell wall and have certain other characteristics with respect to the mechanism of protein synthesis, which distinguishes them from the majority of eubacteria. They are generally found as a very restricted and specialized group living in waterlogged soils, guts of animals, sewage sludge, rotting vegetation, and aquatic sediments. Three orders of strict (obligate) methanogens have been recognized so far. These are the *methane bacteria* including some species of *Methanobacterium* and *Methanobravibacter*; the *Methanococcales* including species of *Methanococcus*; and *Methanomicrobiales* including species of *Methanomicrobium, Methanoganium, Methanospirillum,* and *Methanosarcina.*

Methanogens are mostly non-motile and are amongst the strictest anaerobes. Even 0.01 mg/L of O_2 inhibits their activity. Nitrate and sulphate also have an inhibitory effect, while ammonia and CO_2 are essential for their growth. H_2S generally has a stimulatory effect. Until recently, all methanogens were believed to have a very slow rate of growth, the reproduction time being in excess of 10 hours, even under optimal conditions. However, some isolates (such as *Methanococcus jannaschii*) have demonstrated a much more rapid rate of growth, with doubling time of little more than 30 minutes.

Substrates such as CO_2 and H_2 have proved to be most ideal for the growth of the majority of methanogens. The energy necessary for their growth is derived from the reduction of CO_2 to methane. Some methanogens utilize formate instead of H_2 and CO_2. A small number of species have been found to grow well by using other substrates such as formate, methanol, or methylamine. The most versatile species, in terms of utilization of different types of substrates, are the different strains of the group *Methanosarcina*, as they utilize H_2, CO_2, methanol, methylamine, and acetate. In fact acetate is a substrate which no other species has been found to utilize till date.

In general, any organism grows using CO_2 as its apparent substrate of growth is strictly termed an autotroph. However, methanogens are not placed in this category since their utilization of CO_2 is quite different from other autotrophs. No evidence has been found to show that the assimilation of CO_2 in methanogens occurs through the RBP (ribulose-biphosphate pathway), as the crucial enzyme ribulose-biphosphate carboxylase, a prerequisite for autotrophic CO_2 fixation via the RBP, is totally absent.

Some commonly found species of methanogenic bacteria are *Methanobacterium bryantii, M. formicicum, M. soehngenii, M. thermoautotrophicum, Methanobrevibacter arboriphilus, Methanococcus vannielii, Methanococcus voltae, Methanogenium aggregans, Methanogenium cariaci, Methanogenium marisnigri, Methanomicrobium*

Table 3.2 Characteristics of methanogenic species in pure culture

Species*	Morphology	Gram strain	Growth substrate	Motility	Optimum temp. (°C)	Optimum pH	Special growth requirement
Methanobacterium formicicum	Long rod to filament shaped	Variable	$H_2 + CO_2$, formate	−	37–45	6.6–7.8	
M. bryantii	Long rod shaped	Variable	$H_2 + CO_2$	−	37–39	6.9–7.2	
M. thermoautotrophicum	Long rod to filament	+	$H_2 + CO_2$	−	65–70	7.2–7.6	
Methanobrevibacter ruminantium	Lancet-shaped cocci	+	$H_2 + CO_2$, formate	±	37–39	6.3–6.8	Acetate, coenzyme M, digested fluid
M. smithi	Lancet-shaped cocci	+	$H_2 + CO_2$, formate	−	37–39	6.9–7.4	Acetate
M. arboriphilus	Short rod shaped	+	$H_2 + CO_2$	−	37–39	7.5–8.0	Digested fluid
Methanomicrobium mobile	Short rod shaped	−	$H_2 + CO_2$, formate	+	40	6.1–6.9	Digested fluid
Methanogenium cariaci	Irregular, small cocci	−	$H_2 + CO_2$, formate	+	20–25	6.8–7.3	Acetate, yeast, NaCl
M. marisnigri	Irregular, small cocci	−	$H_2 + CO_2$, formate	+	20–25	6.0–6.6	NaCl
Methanospirillum hungatei	Short to long, wavy spirillum	−	$H_2 + CO_2$, formate	+	30–40	6.8–7.5	
Methanosarcina barkeri	Pseudosarcina	+	$H_2 + CO_2$, formate	−	35–40	6.7–7.2	
Methanococcus vannielii	Irregular, small cocci	−	$H_2 + CO_2$, formate	+	36–40	7.0–9.0	
M. voltae	Irregular, small cocci	−	$H_2 + CO_2$, formate	+	36–40	6.7–7.4	NaCl

* M in this column represents the genus *Methanobacterium*.

mobile, *Methanoplanus limicola, Methanosarcina barkeri, Methanospirillum hungatei,* and *Methanothrix concillii.*

Table 3.2 presents the characteristics of some methanogenic species in pure culture.

Utilization of volatile acids by methanogenic bacteria

Studies have shown that methanogenic bacteria ferment only a few compounds, most of them being products of other bacterial fermentation, such as alcohol, volatile acids, and few gases (Fig. 3.9).

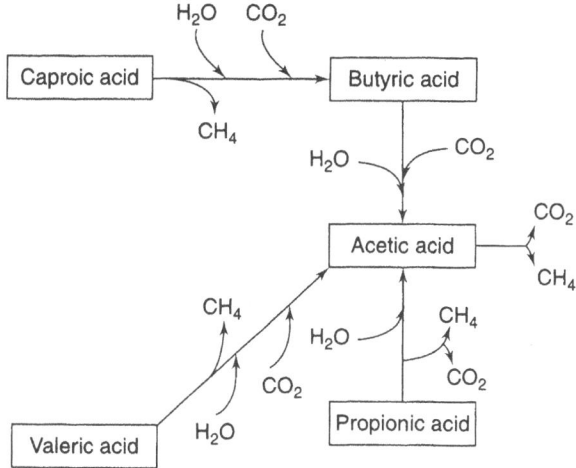

Fig. 3.   Volatile acid conversion by some methanobacterium

Methane can also be produced from long chain fatty acids (stearic, palmitic, etc.) by beta oxidation. In β-oxidation, the β-carbon is attacked resulting in the cleavage of two fragments of carbon and acetic acid, until the molecule is degraded. Here CO_2 serves as the hydrogen acceptor. Some species of methanobacteria can produce not only CH_4 and CO_2 but also other volatile acids (Daniel & Drake 1998).

As mentioned earlier, chemolithotrophic bacteria convert H_2 and CO_2 into methane, using gaseous hydrogen as an electron donor and incorporating the hydrogen obtained from water.

$$CO_2 + 4H_2 \longrightarrow CH_4 + 2H_2O$$

The reaction is thermodynamically favourable as it converts one ADP to ATP. Currently, another group of methanogenic bacteria is known to metabolize acetate, formate, and polymethylamine in methanogenesis. The acetate conversion may be written as

$$CH_3COOH \longrightarrow CH_4 + CO_2$$

This reaction generates only 0.25 mol of ATP, and thermodynamically is relatively less favourable.

It has been reported that approximately 70%–75% mole per mole of methane produced in anaerobic fermentation is derived from the acetate pathway. This means that approximately 25%–30% mole per mole of methane is derived from autotrophism and catabolism of compounds. Methane derived from methanogenic pathways in a properly designed digester should therefore contain

$$70\% \ (CH_3COOH) \longrightarrow CH_4 + CO_2 \qquad \text{(a)}$$
$$30\% \ (CO_2 + 4H_2) \longrightarrow CH_4 + 2H_2O \qquad \text{(b)}$$

If all the carbon dioxide used in Eqn (b) is derived from Eqn (a), then the mole ratio of methane to carbon dioxide will be 50:20 or 2.5:1. This ratio is in tune with the conventional values, which range between 1:1 and 3:1. This biogas also includes water vapour, the concentration of which is dependent upon the partial pressure of water under the digester conditions of pressure and temperature. When CO_2 exists in a ratio greater than predicted, it may be attributed either to the lack of chemolithic methanogens in the biomass or more probably to contribution from other groups. Fermentation of substrates such as hexoses, pentoses, and cellobioses by acidogens such as Lactobacillus brevis produces CO_2 of the order of 80–131 mols per 100 mols of substrate.

The ratio of CH_4 to CO_2 in biogas is further influenced by the physico-chemical balance of hydroxide, carbonate, and bicarbonate in the digester. If a hydroxide, such as sodium hydroxide, is used for *p*H correction, it will absorb more CO_2 than if carbonates and bicarbonates are used. The concentration of methane in biogas will, therefore, depend not only on the microbial population, feedstock, and culture conditions such as buffering *p*H and redox potential, but also on influent composition.

Equations relating to biogas production for a given set of substrates can be based on the elemental composition of the feedstock and its potential bio-degradability or recalcitrance. However, it is difficult to account for additional variations as discussed previously. Modern process design criteria are mostly based on bench or pilot-scale digesters. Methane production commonly varies between 0.3 and 0.5 m^3 of CH_4 per kg of COD (chemical oxygen demand) degraded.

Methanogens, culturally, are the most fastidious group in the symbiosis of anaerobic digestion. In order to grow they require a wide variety of nutrients including carbon, sodium, organic nutrients such as amino acids and vitamins, and trace metals. Recent understanding of the trace metal requirement of methanogens has helped to improve the conditions required for growth, both in pure culture and mixed culture fermentations. In addition to iron, zinc, and manganese, it has been established that methanogens also require trace amounts of cobalt, molybdenum, and nickel. As already mentioned, hydrogen and CO_2 are clearly essential for chemolithotrophic growth.

The empirical equation for anaerobic biomass is $C_5H_0O_3N$, i.e., the molar ratio of C:N is 5:1, which is also approximately equal to the weight ratio. A ratio of N:P of 5:1 is widely applicable to both anaerobic and aerobic biomasses. A ratio of C:N:P for feedstock can therefore be calculated and a mean yield coefficient for the digester estimated. This will be useful in order to differentiate between carbon incorporated into biomass and that evolved as CO_2 and CH_4.

It has been found that a majority of mesophilic methanogens do not grow at *p*H values below 5.5. Empirical results have additionally shown that an upper *p*H limit of 8 is desirable. To maintain this optimum pH range, it can be obviously concluded that digester stability will prove to be a good buffering system. However, when the feed has a high concentration of nitrogen, this buffering may be maintained naturally by the ammonia/ammonium couple ($NH_4^+ \longleftrightarrow NH_3^+ + H^+$). This buffering system is useful, provided that the concentration of free ammonia does not reach toxic proportions.

The different reactions involved in an anaerobic digestion process are as follows:

$$2C_6H_{12}O_6 + 2H_2O \longrightarrow 5CH_3COOH + 4H_2 + 2CO_2$$

$$5C_6H_{12}O_6 + 17H_2 + 3CO_2 \longrightarrow 11CH_3CH_2COOH + 14H_2O$$

$$C_6H_{12}O_6 \longrightarrow CH_3CH_2CH_2COOH + 2CO_2 + 2H_2$$

$$CH_3CH_2COOH + 4H_2O \longrightarrow 2CH_3COOH + 2CO_2 + 6H_2$$

$$CH_3CH_2CH_2COOH + 2H_2O \longrightarrow 2CH_3COOH + 2H_2$$

$$2CO_2 + 4H_2 \longrightarrow CH_3COOH + 2H_2O$$

$$4CO + 2H_2O \longrightarrow CH_3COOH + 2CO_2$$

$$4CH_3OH + 2CO_2 \longrightarrow 3CH_3COOH + 2H_2O$$

$$CO_2 + 4H_2 \longrightarrow CH_4 + 2H_2O$$

$$CH_3COOH \longrightarrow CH_4 + CO_2$$

Molecular biology of methanogens

Methanogenesis (Reeve 1992; Ferry 1999; Shima et al. 2002) is the process by which methane gas is produced. Methanogens are strict anaerobes and belong to the phylum Euryarchaeota of the superkingdom Archea. These microbes release methane as a waste product of their cellular metabolism. Hence, they can be used to produce methane (biogas) from biomass. They can degrade and detoxify agricultural, municipal, and industrial wastes. The study of methanogens, which are mainly considered to be extremophiles, at the molecular level has two facets to its understanding. The first step involves evaluation of the results for their intrinsic novelty and importance. In the second stage, these data is compared to the established norms for bacteria and higher organisms.

There are many different species of methanogens having different properties. Several groups of scientists have attempted to isolate mutants to evaluate the process of transformation and thereby establish a protocol for strain improvement. Several biochemical pathways have been identified that are enzyme-mediated and such enzymes can be considered a product of *methane genes*. These methane genes encode enzymes, which, in the presence of various cofactors, facilitate the supply and subsequent metabolism of the precursors to produce methane. It has been found that methyl groups and electrons are the main precursors for methanogens. Figure 3.10 gives an overview of substrate metabolism catalysed

C1 substrate metabolism

Fig. 3.8

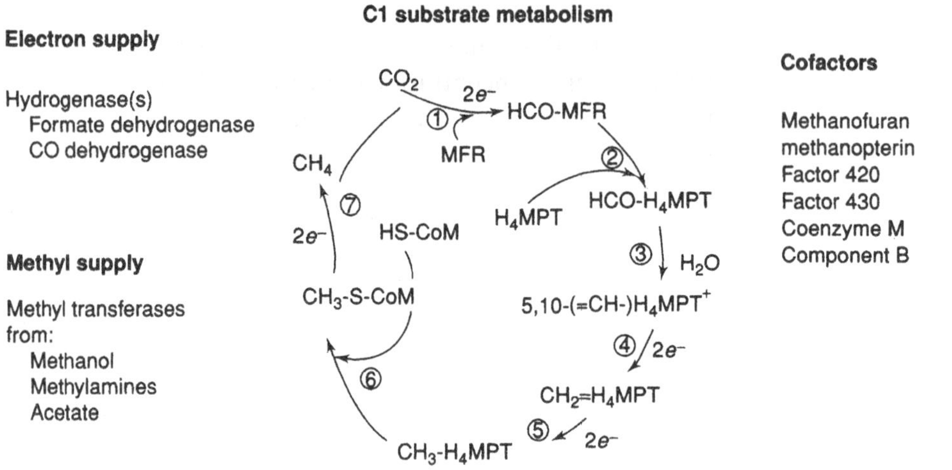

Fig. 3.   An overview of substrate metabolism catalysed by enzymes encoded by methane genes

by enzymes coded by methane genes. The enzymes involved by listed in Table 3.3. Figure 3.11 illustrates the methanogenic pathway utilizing CO_2 and molecular hydrogen.

Studies have shown that the genome of methanogenic bacteria is a circular double-stranded DNA molecule of 1.9 Mbp in length or approximately 45% of the size of the *E. coli* genome (Atlas 1984; Beiley & Ollis 1987; Stanier et al. 1987). With the Southern hybridization experiments, it was found that the major portion of this archaeal genome has no repetitive DNA sequences. Therefore it appears to be composed of some unique sequences, presumed to be coding regions and

Table 3.3   Enzymes involved in the reduction of CO_2 to CH_4

Enzymes	Cofactors
Formylmethanofuran synthetase	Methanofuran
Formylmethanofuran tetrahydromethanopterin formyl transferase	Methanopterin
Methyl tetrahydromethanopterin cyclohydrolase	Methanopterin
Methyl tetrahydromethanopterin coenzyme M-methyl transferase	Methanopterin, coenzyme M, corrinoid
Methylene tetrahydromethanopterin reductase	Methanopterin
Methyl coenzyme M-reductase	Coenzyme M, F430, B12, FAD^+, ATP, Mg^{2+}
Methylene tetrahydromethanopterin F420 oxidoreductase	Methanopterin, F420

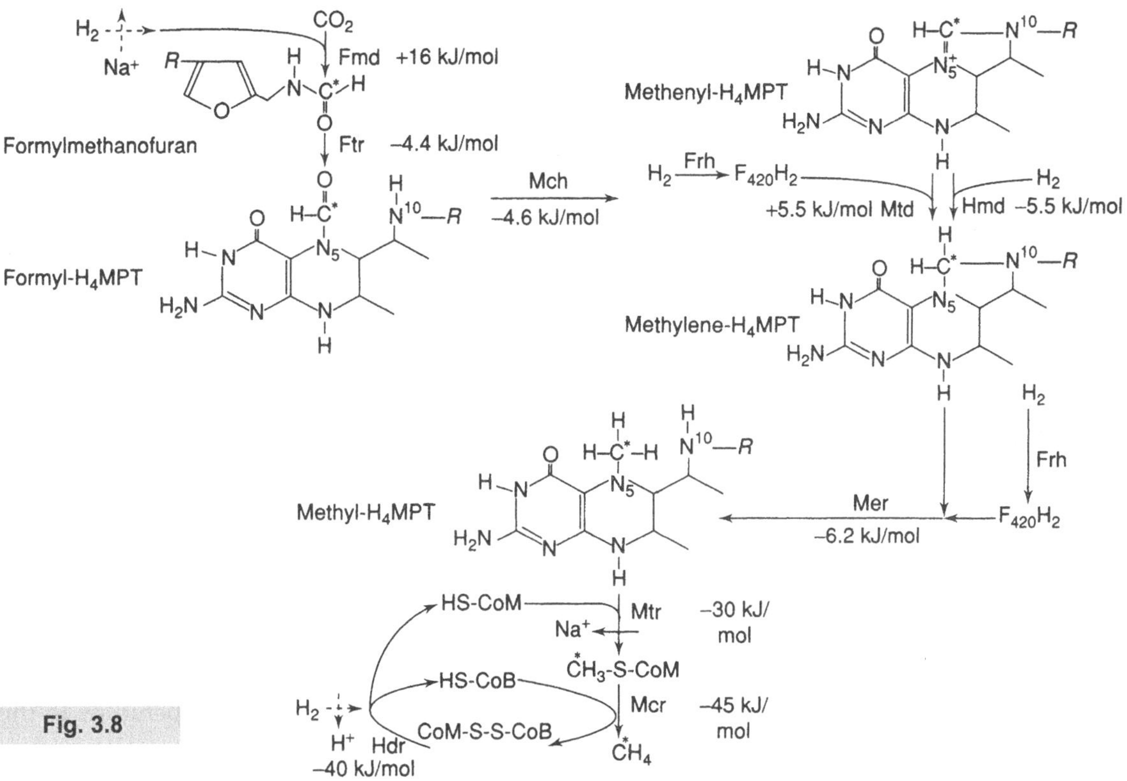

Fig. 3.8

Fig. 3. Methanogenic pathway utilizing carbon dioxide and molecular hydrogen. H₄MPT, tetrahydromethanopterin; HS-CoM, coenzyme M; HS-CoB, coenzyme B; CoM-S-S-CoB, heterodisulphide of coenzyme M and coenzyme B; Fmd, formylmethanofuran dehydrogenase; Ftr, formylmethanofuran: H₄MPT formyltransferase; Mch, methenyl-H₄MPT cyclohydrolase; Mtd, F_{420}-dependent methylene-H₄MPT dehydrogenase; Hmd, H₂-forming methylene-H₄MPT dehydrogenase; Mer, methylene-H₄MPT reductase; Mtr, methyl-H₄MPT: coenzyme M methyltransferase; Mcr, methyl-coenzyme M reductase; Hdr, heterodisulphide reductase; Frh, F_{420}-reducing hydrogenase. The $\Delta G^{o\prime}$ values are from Shima and Thauer. Sodium ion or proton translocations coupled to the reactions are indicated by arrows.

intergenic regulatory sequences, as found in bacterial genomes. *BamHI, BglII, BclI,* and *PvuI* can be used for the construction of physical maps, due to the under-representation of the sequence 5′GATC in the genome of M. voltae. Methanogen genomic DNAs have about 26 to 68 mol % of G + C, although intergenic regions are frequently more AT rich than the average value for a genome. Several restriction enzymes have been isolated from methanogens, and some have been synthesized in *E. coli* by expression of the cloned encoding genes for development as commercial products. Codons such as AUA, AGA, and AGG are frequently found in methanogens but rarely in *E. coli*. The high levels of A's and T's in the third position, often found in methanogens, may pose a problem for proper expression in other organisms. The *Methanococcus* genome is an example of this. The circular chromosome has a G + C content of only 31.4%. Natural extra-chromosomal elements and other mobile genetic elements found in methanogens may be useful in the construction of functional genetic vector systems, and have been identified in several methanogens.

Many of the small, basic DNA-binding proteins found in the genome of methonogens have been isolated and sequenced. This information contributes to the architecture of the methanogen genome in vivo, although this hypothesis is yet to be investigated directly. The genes HMf and HMt, which encode the subunits of the DNA-binding proteins in *Methanothermusfervidus* and the *M. thermoautotrophicum* strain AH, respectively, have been cloned and sequenced. The amino acid sequences deduced for HMf and HMt are more than 80% identical to each other and nearly 30% identical to a consensus sequence for eukaryal core histones. HMf and HMt bind to DNA in vitro, forming nucleosome-like structures in which the DNA molecule is found wrapped in a positive supercoil (it is not negatively supercoiled as in eukaryal nucleosomes). The overall superhelicity of a methanogen genome in vivo must, however, reflect not only the effects of DNA-binding proteins, but also the activities of enzymes such as DNA polymerase, topoisomerases, and RNA polymerase (RNAp) along with the contribution of physical parameters such as internal salt concentration and temperature. In this regard, hyperthermophilic methanogens contain an unusual topoisomerase, designated reverse gyrase, and a very high intracellular concentration of an unusual salt, potassium 2', 3'(cyclic)-diphosphoglycerate. It is believed that these novel components and the DNA-binding proteins, such as HMf, may have evolved to protect the genomes of these hyperthermophiles from heat denaturation.

The plasmid pME2001 has been used for the construction of cloning vehicles and shuttle vectors, which carry selectable markers plus origins of replication from bacteria and *Eukarya*. Restriction maps have been constructed for pME2001, a cryptic plasmid having 4439 bp of DNA sequence isolated from the *M. thermoautotrophicum* strain Marburg. No recognizable phenotype is associated with pME2001 although it contains several open reading frames (ORFs) and is transcribed in vivo, which is evidenced by the viability of the Marburg isolate, without pME2001.

After this brief discussion on the mechanism of methanogenesis, let us now focus on the microbiological and biochemical aspects. In this section, the genes and the enzymes involved in the mechanism of methane formation by the utilization of CO_2 and H_2 from the environment will be given the main priority. There are seven different enzymes involved with important roles in this pathway. These are Fmd, Ftr, Mch, Mtd/Hmd, Mer, Mtr, and Mcr. In the next few sections, the role of these enzymes and the genes responsible for such bioconversions will be discussed in detail.

Formylmethanofuran dehydrogenase (Fmd)   The enzyme formylmethanofuran dehydrogenase is the first enzyme of the pathway and plays an important role in the fixation of CO_2 to formylmethanofuran. During this process, it utilizes CO_2, instead of HCO_3^-, as the active species and catalyses a methanofuran (MF)-dependent exchange between CO_2 and the formyl group of formyl-MF, consistent with N-carboxymethanofuran as an intermediate. This is a regulatory enzyme having five subunits and contains molybdoterin guanine dinucleotide, 30 non-haeme iron, and 30 acid labile sulphide, where iron and molybdoterin cluster act as a prosthetic group. The genes encoding the five subunits form a single transcription unit (*fmdEFACDB*) and are co-transcribed with a gene, *fmdF*, predicted to encode a polyferredoxin possibly containing eight [4Fe-4S] clusters. Whereas, FmdB is the catalytic subunit having the molybdopterin cofactor and has a deduced sequence similar to that of formate dehydrogenase of *Methanobacterium formicicum*. In addition, FmdB has the potential to bind one [4Fe-4S] cluster. For proper functioning of the enzyme, either tungsten or molybdenum is essential but in few species of methanogens both the cofactors are present and play an important role. Generally, Fmd has five subunits but in the case of *M. thermoautotrophicum* tungstoenzyme, it is composed of four subunits (FwdABCD), the genes for which are co-transcribed with three other genes (*fwdHFGDACB*), and *M. thermoautotrophicum* molybdoenzyme contains three subunits (FmdABC) encoded on a transcriptional unit (*fmdECB*). Southern blotting indicates only one DNA sequence encoding the N-terminal sequence leading to the proposal that the enzymes having tungsten and molybdenum as the cofactor share this subunit, which is also consistent with the genomic sequence. Using heterologous oligonucleotide probes for *fmdB* from *M. thermoautotrophicum* and *fmdB* from *M. barkeri*, two genes were isolated from *Methanopyrus kandleri* with deduced sequences, which suggested that this hyperthermophile contains two tungsten isozymes of formylmethanofuran dehydrogenase, one of which is a novel selenoenzyme. This reaction is endergonic ($DG^{o'} = +16$ kJ/mol), which indicates that the reaction requires external energy, which may be fulfilled by the Na^+ ion membrane potential.

Formylmethanofuran tetrahydromethanopterin formyltransferase (Ftr)   This enzyme catalyses the transfer of the formyl group on formylmethanofuran to the second C1 carrier tetrahydromethanopterin (H_4MPT). The purified enzyme contains one subunit with a molecular mass of approximately 32 kDa and exhibits a sequential kinetic mechanism consistent with the formation of a ternary complex. A study of the crystal structure of this enzyme has revealed that the surface of the protein contains an unusually high number of negatively charged residues. This is proposed to account for the tolerance to high salt conditions. However, the structures of the enzyme–substrate complexes have not been studied so far, which may otherwise give the accurate identification of an active site or the proposal of the mechanism.

The *ftr* mRNA from *M. barkeri* and *M. fervidus* is monocistronic in nature and does not exist as an operon, as has been suggested for the *ftr* from *M. thermoautotrophicum*. The genomic sequence of *M. thermoautotrophicum* predicts two *ftr* genes. However, the genomic sequence of *M. jannaschii* predicts only one, suggesting that only one formyltransferase is essential. *M. extorquens* also contains a methenyl-H_4MPT cyclohydrolase encoded by an open reading frame with a deduced sequence strongly identical to cyclohydrolases from methanoarchaea. The gene (*ftr*) that encodes formylmethanofuran tetrahydromethanopterin formyl transferase (the enzyme that catalyses this reaction) has been cloned and sequenced from the *M. thermoautotrophicum* strain AH and functionally expressed in *E. coil*. It appears to be the promoter distal gene in an operon, but the other genes within this unit have not been identified. The reaction carried out by this enzyme is an exothermic reaction ($\Delta G^{o'} = -4.4$ kJ/mol).

N^5, N^{10}-methenyltetrahydromethanopterin cyclohydrolase (Mch)   Mch catalyses the conversion of formyl-H_4MPT into N^5, N^{10}-methenyl-H_4MPT. The enzyme from *M. thermoautotrophicum*, *M. kandleri*, and *M. barkeri* contains two identical subunits of 37–41 kDa and has no identifiable prosthetic groups. The gene encoding the enzyme from *M. thermoautotrophicum* is apparently transcribed monocistronically.

N^5, N^{10}-methylenetetrahydromethanopterin dehydrogenase (Mtd/Hmd)   This enzyme catalyses the reversible reduction of N^5, N^{10}-methenyl-H_4MPT and reduces F_{420} ($F_{420}H_2$) to N^5, N^{10}-methylene-H_4MPT. The reduced $F_{420}H_2$ is regenerated by the reaction catalysed by F_{420}-reducing hydrogenase (Frh). Two genetically distinct dehydrogenases have been described. One of them is coenzyme F_{420}-dependent (*mtd*) and the other is H_2-dependent (Hmd). Coenzyme F_{420} is an obligate two-electron carrier (redox midpoint potential near −350 mV) that donates or accepts a hydride ion. The F_{420}-dependent enzyme has been purified from *M. barkeri*, *M. thermoautotrophicum*, and *M. kandleri*. Both dehydrogenases are composed of one type of subunit (30–36 kDa), either as a hexamer or as an octamer, with no detectable prosthetic group. The catalytic mechanism is ternary, a result consistent with a direct hydride transfer either to or from F_{420}. The *mtd* genes from these two methanoarchaea are transcribed monocistronically. The sequence analysis of *mtd* from *M. thermoautotrophicum* suggests a potential F_{420} binding site in the N-terminal.

The reaction that catalyses the conversion of methenyl-H_4MPT to methylene-H_4MPT is a reversible reaction in which two different enzymes participate actively. F_{420}-dependent methylene-H_4MPT dehydrogenase (*mtd*) catalyses the reversible reduction of N^5, N^{10}-methenyl-H_4MPT and reduces F_{420} ($F_{420}H_2$) to N^5, N^{10}-methylene-H_4MPT. F_{420} is a coenzyme for hydride transfer, which is endergonic in nature. The change in free energy is +5.5 kJ/mol. F_{420}-independent methylene-H_4MPT dehydrogenase, which is designated H_2-forming methylene-H_4MPT dehydrogenase (Hmd), catalyses the reversible reduction of N^5, N^{10}-methenyl-H_4MPT and molecular hydrogen to N^5, N^{10}-methylene-H_4MPT. This reaction is exergonic in nature; the free energy released is −5.5 kJ/mol.

In fed-batch culture, where H_2 is abundant, transcription of *hmd* (encoding the H_2-dependent dehydrogenase) is favoured over transcription of the *mtd* (encoding the F_{420}-dependent dehydrogenase), whereas the opposite phenomenon is observed under culture conditions with limited supply of H_2. Under culture conditions that favour the expression of F_{420}-dependent dehydrogenase, the oxidation of H_2 is accomplished by the nickel-containing F_{420}-dependent hydrogenase, which has a 10-fold higher affinity for H_2 (K_m = 0.02 mM) as compared to the H_2-dependent dehydrogenase (K_m = 0.2 mM). These results have led to the conclusion that under H_2-non-limiting conditions, the H_2-dependent dehydrogenase substitutes for the F_{420}-dependent dehydrogenase, catalysing the reduction of CH H_4MPT$^+$ to CH$_2$ H_4MPT.

N^5, N^{10}-methylenetetrahydromethanopterin reductase (Mer) This reductase catalyses the reduction of N^5, N^{10}-methylene-H_4MPT to N^5-methyl-H_4MPT at the expense of $F_{420}H_2$. The enzymes purified from *M. thermoautotrophicum*, *M. kandleri*, and *M. barkeri* are F_{420}-dependent and contain one subunit (35–38 kDa) with no discernable prosthetic groups. They also exhibit a ternary complex kinetic mechanism, suggesting direct hydride transfer. The genes (*mer*) encoding the reductases from *M. thermoautotrophicum* strains Marburg and AH are transcribed monocistronically. Significant similarity has not been found with any of the published sequences for F_{420}-dependent enzymes from microbes in the archaea and bacterial domain.

N^5-methyltetrahydromethanopterin coenzyme M-methyltransferase (mtr) The methyltransferase catalyses the transfer of the methyl group of the N^5-methyl-H_4MPT to coenzyme M, which is the third C1 carrier of this pathway. 5-hydroxy benzimidazolyl cobamide is the prosthetic group for this enzyme. The molecular weight of the enzyme, which is an integral membrane protein complex composed of eight non-identical subunits (mtrA–H), is 670 kDa. The free energy change of this reaction is estimated to be −30 kJ/mol. The Na$^+$ ion membrane potential conserves the negative free energy change of this reaction. The genes encoding the eight non-identical subunits form an operon (*mtrEDCBAFGH*). This operon is located between the methyl-coenzyme M reductase I operon (*mcr*) and a downstream open reading frame. The downstream open reading frame is predicted to encode the Na$^+$/Ca^{++}, K$^+$ exchanger, which creates the Na$^+$ ion membrane potential.

All genes encoding methanogenic methylamine methyltransferases contain an in-frame amber (UAG) codon that is read through during translation and does not appear to stop translation during protein synthesis. The UAG-encoded residue in a 1.55 Å resolution structure of the *Methanosarcina barkeri* monomethylamine methyltransferase (MtmB) was identified to be structurally like a homohexamer protein comprised of individual subunits with a triosephphateisomerase (TIM) barrel fold. The electron density for the UAG-encoded residue is distinct from any of the 21 natural amino acids; this residue was named L-pyrrolysine (Hao et al. 2002).

Methyl-coenzyme M reductase (Mcr)   Mcr, catalysing the reduction of methyl-coenzyme M to methane, is common to all methanogenic pathways. The enzyme has nickel porphynoid F_{430} as the prosthetic group. The electron donor for all reductases is coenzyme B, which is mercapto-heptanoylthreonine phosphate (CoB), and heterodisulphide CoM-S-S-CoB is the other product, in addition to CH_4. The genes encoding several reductases from phylogenetically diverse methanoarchaea (*mcrBGA*) are co-transcribed in operons (*mcrBDCGA*) with two additional genes. Factor F_{390}, a degradation product of F_{420}, produced in cells under oxidative stress is proposed to play a role in the regulation of both methylreductase and F_{420}- and H_2-dependent dehydrogenases. The recent discovery of the crystal structure of the MCRI isozyme from *M. thermoautotrophicum* has shed light on the active site and mechanism. The two F_{430} cofactors are positioned at the bottom of identical narrow channels that are formed by residues from the subunits.

Several other enzymes involved in the reductive pathway of CO_2 to CH_4 have also been purified. There is a much greater possibility of cloning of their encoding genes now, especially as N-terminal amino acid sequences are already available. Standard molecular biology methods for genetic modification have been successfully applied for deciphering the genomes of methanogens such as *Methanococcus jannaschii* and *M. thermoautotrophicum*. Currently, many studies are being undertaken to understand gene regulations and expressions in order to effect genetic modifications in methanogens for achieving improved productivity. It is pertinent to note that though archea are anaerobic in nature, the standard protocols used for simple DNA and RNA work can normally be employed in their case also. At the same time, special procedures may be necessary to modify the permeability of the cell wall, as the composition of the cell wall is quite different from that of eubacteria. The use of shuttle and integration vector systems has been found efficient for genetic manipulation. These systems have yielded positive results for mesophilic methanogens, but not for the thermophilic species.

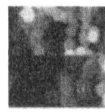

Microbial Strain Improvement for Anaerobic Processes

Genomic structure is responsible for the metabolic function of an organism. Studies have shown that the metabolic efficiency of wild strains is normally low. Therefore, to improve their efficiency, the genomic structure needs to be modified. The strain improvement methodologies are essentially the same for aerobic or anaerobic micro-organisms. However, there may be some differences in the environmental conditions under which they are isolated. The main objective of undertaking the process of improving strains of industrial and commercial importance is to improve their efficiency during the bioconversion process. We know that the different microbial activities are controlled by the microbe's genomic characteristics. Therefore, microbial functions can be suitably modified by effecting certain desired changes in the nature of the genes. To some extent the culture conditions can also be optimized to improve the product yield. However, as is well-known, it is again the genetic make-up of the microbe that controls the productivity. Therefore, to improve the potential activity, the genome of the organism must be modified. This can be accomplished mainly in two ways, namely, *mutation* and *recombination*. At this stage, it will be pertinent to discuss briefly the gene and its characteristics, which will help us to better understand these two techniques of strain improvement.

A gene is a region of a chromosome and is responsible for the behaviour and function of an organism. The chemical structure of a gene is now well established. Genes are made up of nucleic acids and are segments of DNA (deoxyribose nucleic acid). Usually, more than one DNA molecule exists in an organism and each such molecule comprises of a number of genes which are responsible for individual or specific functions of organisms. For example, the entire process related to protein synthesis (including all enzymes) is controlled by genes. Similarly, all other metabolites such as antibiotics, alcohol, and methane are produced by micro-organisms as a result of their genomic activities. Genes are also responsible for the phenotypic and genotypic nature of an organism. With this brief outline it is evident that if the functions of a strain are to be modified, the necessary changes have to be incorporated in the gene itself. Both mutagenesis and rDNA technology are based on the principle of modifying the gene structure of DNA to effect strain improvement. (Refer to *Genes VIII* by Benjamin Lewin for more details.)

Strain Improvement by Mutation

Research has shown that when micro-organisms are exposed to an extreme environment for a prolonged period of time, their survival becomes difficult. This is possibly due to the changes effected in the genetic make-up of the organism. This concept has been utilized to bring about desired changes in the genetic structure of micro-organisms by exposing them to various mutagenic environments for a limited duration. The factors that bring about these changes are called *mutagens* and the resultant effect is known as *mutation*. The various mutagenic agents responsible for mutation are classified as follows.

1. Ionizing radiations such as x-rays, γ-rays, and UV rays
2. Chemicals such as NTG (nitrosoguanidine) and HNO_2 (nitrous acid)
3. Heat

Any of these mutagens may be adopted to effect mutation. It is to be noted that the duration of exposure to mutagenic environment is inversely proportional to the intensity of the mutagenic dose, i.e., if the intensity is high, then less contact time is required to cause mutation.

The combination of time and dose should be so decided that the result must be the death of a vast majority of cells. Through some preliminary experiments, the time versus per cent survival of organisms can be determined. This relation varies from one organism to another. For the convenience of the subsequent steps of screening and selection of mutants, the per cent survival should be around 1%–5%. This will again depend on the initial number of micro-organisms exposed to the mutagenic agents. Among the survivors, however, only some may be mutants, a very small proportion of which may be superior. By the aforementioned methods of mutation it is not possible to predetermine the gene that will be affected by the mutagens. This is because the action of mutagens affects a very large number of genes. Therefore, it is the responsibility of the scientists to differentiate and separate the few superior (desirable) types from the many inferior survivors obtained after the mutation process. During the process of mutation of anaerobic bacteria such as methanogens, which are strictly anaerobes, special care needs to be exercised while conducting the experiment to ensure that strict anaerobic conditions are maintained. In such instances, it is generally helpful to use anaerobic culture jars.

Selection of mutants

The design of the procedure for the isolation of over-producing mutants is as important as the process of mutation. Knowledge of biochemical pathways often helps in the design of procedures that are used for the isolation of mutants that are capable of over-producing primary metabolites. However, to design procedures for the isolation of mutants producing secondary metabolites is more difficult due to the lack of sufficient information. A typical experiment to obtain mutants with improved productivity would be to subject the population of cells to a mutation treatment that would ensure the survival of about 1%–5% cells. The next step is to screen as many survivors as possible for highest productivity. Productivity is usually assessed by culturing the microbes in a shake flask in the case of aerobic organisms and in an anaerobic culture jar in the case of anaerobes. The culture is then assayed for activity or gaseous products.

These procedures are very labour intensive and, therefore, various attempts have been made to miniaturize the system such that the survivors are cultured and assayed for productivity on agar media. Such miniaturized systems are less labour intensive. Although empirical screening for improved secondary metabolite producers has met with considerable success, there has been an increasing tendency to utilize some form of selected culture in the isolation of autotrophs, analogue resistant mutants, and mutants resistant to the autotoxic effect of secondary metabolites.

Recombination

Recombination is a process that helps to generate new combinations of genes that were originally present in different individuals. In comparison to the procedure of mutant induction and selection, the use of recombination for industrial strain improvement is fairly limited. This may be due to the success of the mutation processes, the simplicity with which mutation can be performed, and the non-availability of basic genetic information on industrial strains. However, different techniques of recombination processes for strain improvement are now widely available. The application of the recombination process for industrial strain improvement is very limited in the case of organisms exhibiting sexual reproduction. On the other hand, the recombination procedures are more commonly used for industrially important strains that are not associated with sexual reproduction. Some of the popular recombination methods are conjugation, protoplast fusion, and recombinant DNA technology.

Conjugation

Conjugation is a process, observed in bacteria, in which genetic information is transferred from one cell to another through cell-to-cell contact. Of the two cells involved, one is called the donor cell, designated F^+ (fertility plus), and the other, designated F^- (fertility minus), acts as the recipient. In this process the genetic material is transferred directly from the donor to the recipient cell without being passed through the suspending medium. It has been discovered that F^+ strains contain a plasmid (termed F plasmid) that carries all the genes that encode conjugative genetic transfer. However, the F plasmid is not known to encode any additional function other than to facilitate conjugation and its own replication. That is, the F plasmid encodes its own transfer to cells that lack an F plasmid. Thus, if F^- cells are added to an F^+ culture, all of them rapidly become F^+, i.e., F plasmid gets transmitted very fast. Chromosomal genes of donor cells are also transferred to the recipient cells along with the F plasmid, though at a lower rate.

The amount of genetic information transferred to the recipient cell depends upon the duration of contact between the two participating cells. Thus, longer the cell contact, greater the material transferred. Of late, conjugation as a process of strain improvement has gained enormous industrial significance. The main disadvantage of the conjugation system is that substantial information about the genetic make-up of the organism is required to induce and perform the transfer of genetic material effectively. Lack of sufficient understanding of the genomic structure of the organism, generally, results in incomplete transfer of the genome.

Protoplast fusion

Protoplasts are cells that do not have a cell wall, are capable of complete respiratory activity, and can synthesize proteins and nucleic acids. Protoplasts can be prepared by subjecting cells to the action of lysozymes (i.e., wall-degrading enzymes) in isotonic (i.e., osmotic) solutions. Cell fusion followed by nuclear fusion (i.e., induction of intergenic fusion) occurs efficiently in the presence of PEG (polyethylene glycol) solution. The resulting fused protoplast may regenerate a cell wall and grow as a normal cell. So far, protoplast fusion has been successful in the case of filamentous fungi, yeast, and bacteria. In protoplast fusion the entire genome is subjected to recombination, whereas in conjugation only a portion of it is transferred.

The main advantage of this technique is that through protoplast fusion the desirable characteristics found in different strains can be brought together in one organism. Furthermore, the resultant organism will not only exhibit the desirable properties but will also include other original characteristics of the organism that are important to the fermentation process such as the substrate consumption growth rate. Thus, recombination by protoplast fusion is an excellent tool to improve characteristics other than those pertaining to only product yield.

Recombinant DNA technology

Among the various methods developed for strain improvement, recombinant DNA technology is one of the most sophisticated and specific techniques. rDNA technology is especially useful when the aim is to introduce a gene from an organism into bacteria or yeast. Such manipulation results in the synthesis of important microbial cell properties in microbes, which have a high potential to increase yield.

The various steps involved in rDNA technology for gene cloning are as follows (Fig. 3.12).

1. Isolating the gene of interest from a series of fragments of DNA obtained from different sources.
2. Joining these segments to a DNA molecule known as the vector, which will allow them to be taken up by a bacterium and replicated within it as the cell grows and divides.

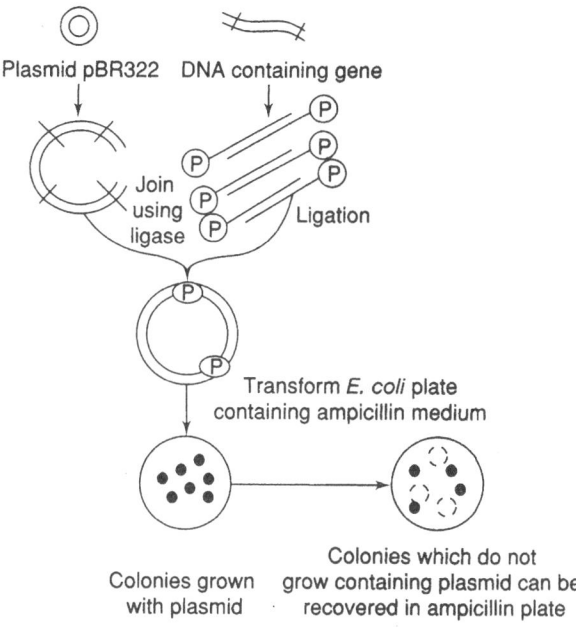

Fig. 3.12 Outline of gene cloning

3. Transforming the recombinant vector into a bacterial cell, so that it first replicates, followed by expression.
4. Selecting the recombinant vectors.

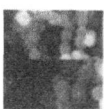

Factors Influencing Anaerobic Digestion Processes

Certain optimum environmental conditions are required for the efficient functioning of an anaerobic digestion process. Some of these system control parameters are temperature, anaerobic conditions, adequate concentration of nutrients, *p*H within tolerable limits, and toxic material either absent or in sufficiently low concentration (Bhattacharyya 1979). The following sections deal with some of the criteria influencing the production of methane.

Inoculum Concentration

The two groups of micro-organisms primarily responsible for effecting biodegradation are normally present in animal wastes such as cow dung and sewage sludge. Non-methanogens are more abundant and fastidious. However, methanogens are less abundantly found, less fastidious, and more sensitive to environmental changes. Therefore, wastes that do not contain dung or sewage sludge will not have naturally occurring methanogens in sufficient numbers.

To initiate an anaerobic digestion process, it is a common practice to seed it with an adequate population of both acid-forming and methane-producing bacteria. Generally, wastes such as an actively digesting sludge from a municipal digester, material from a well-rotted manure pit, or cow dung slurry may be used as the seed to start a new biogas plant. As a guideline, in continuous or semi-continuous operations, the seed material should be at least twice the volume of the fresh manure slurry during the start-up phase. Then the daily addition of the seed can be decreased over a three-week period, which may be considered the initiation period. In the initial stages the gas will contain more carbon dioxide and less methane.

The successful operation of a digester depends on establishing and maintaining a balance between acid-forming and methane-producing bacteria. If the digester accumulates volatile acids as a result of overloading, the situation can be corrected by re-seeding and temporarily suspending the feeding of the digester, or by the addition of lime slurry. It may be noted that in the case of non-availability of pre-digested seeding material, liming is recommended to the extent of 10% on the basis of the volatile solids added, ensuring that a pH of 6.8 to 7.0 is maintained. The temperature needs to be around 35°C. This coupled with mixing will facilitate the start-up operation.

Reports have shown that most of the high-rate digesters are started by seeding them heavily with digested sludge from other units. Laboratory studies, however, show that high-rate units can be started without seeding too. This can be done if proper pH (6.8–7.2) is maintained with the help of liming, the concentration of solids does not exceed 6%, and finally at least a 20-day detention time is made available.

High Rate of Sludge Digestion

High-rate digestion can be characterized as a process involving the complete mixing of the contents on a continuous or intermittent basis to maintain a uniform mixture in all parts of the digestion tank. The digester is fed continuously or intermittently but always with the displacement of the mixed liquor rather than the supernatant. Optimum temperature is maintained by heating. The amount of volatile solids loaded is usually about 1.6 to 6.0 kg/m^3 and the detention time is under 20 days.

Some of the advantages of high-rate digestion are listed here.

1. *Lower capital cost* The capital cost is estimated to be around 25% of that of the conventional digester.
2. *Lesser heat consumption* Heat loss is reduced in direct relation to the surface area of the digester.
3. *Greater process stability* This is due to greater uniformity of the environment and better control.

4. *Quicker solid destruction* A known per cent destruction of volatile solids is achieved in about one-third of the time taken in conventional digestion.

However, there are few disadvantages, as follows.

1. *Dependence upon mixing* If mixing is stopped for a prolonged period, a thick layer of scum forms. This layer is hard to disperse. Even when dispersed it leads to sudden load and as a result might upset the process. The process may even require to be restarted.
2. *Survival of pathogens* Slightly greater numbers of pathogens are likely to be discharged into the digested sludge.
3. *Careful control* Better control of *p*H and temperature is needed.
4. *Post-treatment of sludge* The sludge needs to be elutriated before filtration to get rid of the biochemical end products.

It has been observed that the advantages, however, outweigh the disadvantages.

Significance of Temperature

Temperature has a significant effect on anaerobic digestion of organic material. Studies have shown that methane bacteria of different types grow well in the ranges 15°C–49°C (mesophilic) and 45°C–65°C (thermophilic). The ideal temperature for the digestion process is 30°C–40°C for mesophilic flora, and between 50°C–60°C if thermophilic flora are developed and adapted. The choice between mesophilic or thermophilic bacteria is normally made during the design stage itself and is based upon climatic considerations. For example, if considerable amount of energy is found to be required for maintaining a digester in the thermophilic range, then it may be better to operate the plant at mesophilic temperatures.

The effect of temperature on gas yield for a 4-week detention period is shown in Table 3.4. The data have been obtained with cattle dung as the feedstock.

Table 3.4 Production of gas at various temperatures

Temperature (°C)	Gas produced/kg wet dung (m^3)
15	0.032
20	0.049
25	0.061
30	0.087
35	0.100

In the case of a digester operating in the thermophilic range (45°C–65°C), once the anaerobic digestion system is stabilized, there will be more gas production as compared to that in the mesophilic range of temperature. Further, the retention time at 55°C is quite less (10–20 days). Experimental investigation carried out in Japan has shown that the amount of gas generated per unit weight of urban waste sample under mesophilic conditions (37°C) was 320–340 ml/g/day for a loading of organic substances at the rate of 3.8–3.9 g/L/day, while, under thermophilic conditions (55°C), the value obtained was 340 ml/g/day for 9.0 g/L/day of waste loading.

Methane-producing bacteria are very sensitive to sudden temperature changes. Therefore, any drastic change in temperature needs to be carefully avoided, such that no abrupt decrease in gas production occurs.

Prevention of temperature fluctuations

As already mentioned, the rate of gas production falls if the digester temperature is not maintained within the optimum range. The gas output during the months of winter dips to the minimum value due to the drastic change in ambient temperature. Therefore, for ensuring optimum process stability, the temperature should be controlled carefully within a narrow range (say, 1°C or 2°C) of the selected operating temperature. This can be done by heating the digester and insulating it. The construction of well-functioning and cheap heating systems determines the economical efficiency of the gas plant to a great extent.

The heat requirement of biogas plants comprises of two aspects, namely, the quantity of heat needed to heat the daily charge and the quantity of heat needed to compensate for the heat loss through the sides. The heating system must be planned in conjunction with insulation. The contribution of each arrangement is decided by economic considerations. Heating facilities are technically constructed in the following manner.

1. *Internal heating* By circulating hot water through coils located inside the digester.
2. *External heating* By heating externally all the raw charge to a little above the optimum temperature before releasing it into the digester, or heating a part of the sludge outside and circulating it within the digester.
3. Heating by injecting live steam into the digester or using hot water for making the slurry.

It has been observed that external heat exchangers are more advantageous. External heating gets rid of the problems usually encountered in using internal hot water coils, such as caking and loss of heat transfer efficiency. Heat loss in the gas plant can also be reduced by providing insulation or partially burying the digester tank.

The energy requirement for the heating system can be met by using any of the following.

1. Solar heaters such as collectors, hot houses.
2. Conventional fuels such as oil, coal, wood.
3. Biogas as fuel.
4. Heat recovery from already decomposed material.
5. Waste heat from internal combustion engines.

As we know, biogas is a valuable fuel. Hence, its use for heating purposes should be avoided. Keeping in view Indian conditions, it would be most rational to use solar energy as waste heat.

To minimize the heating requirement, insulation of the digesters with materials such as leaves, sawdust, and straw, all of which are generally available in rural areas, may be considered. These materials may be composted in an annular ring surrounding the digester. However, such composted material needs to be replaced frequently. Insulation bricks can also be used for this purpose. The problem of insulation does not occur only in connection with the material but also with regard to construction and fabrication. Admission of water can nullify the insulation effect.

Effect of Thorough Mixing

The process of methane formation depends on intimate contact between microorganisms and the waste material. Therefore, proper mixing of the digester contents plays an important role. If the sludge is left undisturbed, it leads to stratification in the digester and three distinct zones are formed. These are scum at the top, digested sludge at the bottom (the actively digesting sludge lying on top of well-stabilized sludge), and a thin liquid layer in between the two zones. Methanogens are reported to develop in the sediment and scarcely in the upper layer of the medium. They are present in larger numbers in the sludge than in the liquor. The scum is rich in volatile acids produced by the oxidative bacteria. Scum formation poses many problems. It reduces the effective volume of the digester, causes damage to the cover or piping, it causes sudden increase in organic loading on breaking, prevents proper gas flow, etc. In addition to disrupting the scum layer, mixing provides several other benefits. It has opened up a new area of interest known as 'high-rate digestion'.

One of the most suitable mixing devices is the mechanical mixing device, which is capable of causing high turbulence and thus effecting thorough mixing. Motor-driven mixing devices are troublesome and expensive. For this reason mixing is generally carried out manually in small digester plants. In these small units a plunger is fitted to the inlet pipe or the bell of the gas holder is rotated with pedals (in the case of the floating-dome type) once or twice a day.

If the dilution factor of the incoming material is maintained at an appropriate level, then practically no further stirring is necessary, and this will ensure a reasonable gas yield. Inlet and outlet facilities must be so designed that no short circuit flow occurs. However, in community plants, arrangements for agitation of slurry in the digester should be made to increase the gas yield. For this purpose a part of the gas can be recirculated through the digester sludge.

Impact of Organic Loading

The size of the digester and the quantity of gas produced in a biogas plant per day depend upon the amount of waste material fed per unit volume of the digester capacity. The recommended loading rates are as follows.

Standard municipal digesters 0.48 to 1.6 kg of volatile solids per m^3 of digester capacity per day with detention time varying between 30 and 90 days.

Dairy cattle wastes in India Daily loading rate of 6.7 kg of volatile solids per m^3 of digester capacity; the gas production varies between 0.04 and 0.074 m^3/kg of raw dung fed.

Cow dung Loading rate of about 24 kg raw dung per m^3 per day is reported to be optimal for anaerobic digestion.

Night soil In India, the optimum load ranges from 1.04 to 2.23 kg of volatile solids per m^3 of digester capacity. The comparable volumetric loading ranges from 0.054 to 0.025 m^3 per capita per day and is based on a per capita volatile solids contribution of 0.056 kg per day. A night soil loading rate of 1.6 kg volatile solids per m^3 of digester capacity per day is recommended for temperate climates. An average daily gas production of 0.023 to 0.034 m^3 per capita has been reported.

Influent solid concentration Gas production from a biogas plant also depends upon the concentration of solids in the influent slurry. Investigations indicate that, in the absence of toxic materials, optimum gas production is obtained with slurry made in the ratio of 1:1 (cow dung and water). This means that the concentration of total solids in the slurry should be 10%–12%.

Retention Time

Retention time represents the average time a micro-organism spends in the system and is equal to the hydraulic retention time in a completely mixed digester. The hydraulic retention time is the digester volume divided by the volume of daily feed. The usual retention time is between 20 and 50 days. Lower retention time gives a higher gas yield per unit volume of the digester, but the volatile acid destruction reduces. However, with mixing, volatile acid conversion can be improved. Table 3.5 shows gas production with respect to the retention time

Table 3.5   Retention time versus gas production

Retention time (days)	Volume of gas/vol. of digester/day
30	0.4
20	0.6
15	0.8
12	0.95
10	1.15

Table 3.6   Retention time versus temperature

Digester temperature (°C)	Retention time (days)
15	56
26	30
38	24
50	16

Digestion rate: first week: 37%; second week: 26.5%; third week: 17.5%; fourth week: 10%; fifth week: 5.75%; sixth week: 3.25%.

in the case of a sewage digester. Besides mixing, temperature also affects the retention time. Higher the temperature duration, lower the retention time. The effect of temperature on the hydraulic retention time of an anaerobic digestion process is given in Table 3.6.

Effect of *p*H variation

The *p*H value is the indicator of acidity, alkalinity, or neutrality of the slurry. As we know, a *p*H of 7 indicates that the substance is neutral, as for water; a *p*H lower than 7 indicates acidity; while a *p*H higher than 7 means the substance is alkaline in nature. The *p*H can be determined with the help of a litmus paper.

As mentioned earlier cellulolytic bacteria are usually divided into two classes on the basis of their temperature tolerance. Mesophilic bacteria show optimum activity in the range 30°C–40°C, as in the rumen of cattle, while thermophilic species function optimally between 50°C and 60°C. Both groups show optimum performance in the *p*H range 6.0–7.0. As organic acids are produced during the breakdown of cellulose, the *p*H may fall. Hence, during the initiation of fermentation and the digestive process, it may be necessary to buffer the system with lime in order to stabilize it. When the acid-forming bacteria of stage II and methanogenic bacteria of stage III are present in equal proportions, the pH of the entire system reaches an equilibrium value of about 7, since the organic acids are removed as soon as they are produced.

Methanogenic bacteria are very sensitive to changes in pH. The optimal pH range for methane production is between 7.0 and 7.2, although gas production is satisfactory at a pH of 6.6 to 7.6. When the pH drops below 6.6, there is significant inhibition in the activity of methanogenic bacteria, and the acidic conditions of a pH of 6.2 are toxic to these bacteria. At this pH value, however, acid production will continue, since the acidogenic bacteria will continue to produce acids until the pH drops to 4.5–5.0. Under balanced conditions, the pH is automatically maintained in the desired range due to biochemical reactions. Although the volatile organic acids produced during the first stage of the fermentation process tend to bring down the pH, this effect is counteracted by the destruction of volatile acids and reformation of bicarbonate buffer during the second stage. If imbalance develops, the acid-formers outpace the methane-formers and volatile organic acids build up in the system. If imbalance continues, the buffer capacity may be overcome and a perceptible drop in pH may occur. As has been noted previously, buffering with lime may be necessary. Other agents such as ammonium hydroxide may be used as buffering agents, but care must be exercised, since ammonia, as well as the ammonium ion, can be toxic.

Effect of Nutrient Deficiency

Two of the most important nutrients required for anaerobic digestion are organic carbon and nitrogen. To ensure efficient functioning of the digestion process, the carbon:nitrogen ratio of the substrate should never lie outside the range of 30:1 to 50:1. When sewage sludge or animal wastes are used as substrates, this ratio is usually maintained naturally. However, nitrogen deficiency may occur in systems that utilize food processing or crop residues such as straw as substrates. In such cases, the carbon:nitrogen ratio may be 100:1 or higher, and nitrogen, usually in the form of ammonia, needs to be supplemented.

Disposal of Digested Slurry

Cow dung slurry, in raw as well as in digested condition, is free flowing in nature. In India, the digested slurry is transported through a channel from the digester to a sloping filter bed, where it percolates through a 15-cm layer of a compact bed of dry or green leaves. The slope allows the liquid to be partially decanted. The residue is then handled in a solid or semi-solid form for transportation to the compost pit. The decanted liquid is again made available for mixing with fresh dung to be fed to the digester.

Effect of Unbalanced Digestion

Balanced digestion is that in which the system works with minimum control. This means that the environmental control parameters of the system remain neutrally within their optimum ranges, with only occasional fluctuations that may require intervention. When imbalance does occur, the two main problems are the indication of the commencement of an undesirable change in the control parameters and the cause of the imbalance. The various parameters responsible for causing unbalanced digestion are listed here.

1. Increase in the volatile acid concentration in the slurry and the percentage of CO_2 in the gas.
2. Decrease in the *p*H and total gas production, and hence waste destabilization.

Measurement of Performance

The performance of a biogas plant is measured by the quantity and quality of the methane content in the gas it produces. The quantity of gas produced, however, depends on temperature, loading rate, rate of gas production, retention time, and type of waste material used. It has been found that gas production ranges usually from 0.06 to 1 m^3 per kg of dry volatile solids added. The typical value of dry volatile solids added would be in the range of 0.4 to 0.5 m^3/kg, when primarily animal manure, human waste, and crop residues are used as the feed material with detention between 10 to 20 days. The methane content of the gas can be expected to be about 60%, which means that the rate of methane production is about 0.22–0.3 m^3/kg dry volatile solids digested, over an average detention time of about 15 days.

Cow dung contains 20%–25% dry solids, 10%–15% volatile solids, of which about 25% gets destroyed by digestion in 50 days of retention time. The gas yield is about 0.06 m^3/kg of fresh dung. Since the destruction of volatile solids is proportional to the retention time, gas yield lower than 0.06 m^3/kg of fresh dung, at a given temperature, can be expected when the retention time is less than 50 days. However, mechanical mixing can change the situation, as already discussed before.

Menace Due to Corrosion

Animal and human wastes are corrosive to metal before, during, and after the anaerobic digestion process. Hence, the life of the digestion equipment can be extended if tanks and components are protected against corrosion by fabricating them with corrosion-resistant materials. Digester tanks of wood, fibre glass, concrete, masonry brick, ferrocement, or stone will last longer than steel tanks. The life of metal tanks can be extended by coating the inside surface with rust-resistant paint, epoxy surfacing, or any other similar covering. Pipes, through which hot water is circulated through the digester for maintaining optimum temperature, should be made of stainless steel (or similar corrosion-resistant material). Plastic is resistant to corrosion and lasts long when used for low-pressure gas pipelines, water pipelines operating under conditions of low temperature, and lines conveying waste materials to and from digesters. The use of dissimilar metals in water pipes and various system components should be avoided to eliminate electrolytic corrosion problems.

It is essential that all components of the digestion system be kept free of gas leaks to eliminate gas loss, avoid accumulation of methane in confined areas, and prevent entry of air into the digester. Routine and timely inspection of all pipes and metal components of the digester and gas handling system is necessary to prevent excessive corrosion.

Safety Precautions

'Safety first' is to be kept in mind always. This helps in preventing avoidable hazards. Biogas production mainly involves the risk of fire and explosion.

Potentially lethal situations most commonly occur in confined and poorly ventilated operations. Caution has to be exercised during manure pit agitation. Accidents may also occur during ventilation breakdowns and entry into a storage pit.

It must be borne in mind that methane explodes when mixed with air in proportions ranging from 5% to 15% by volume.

The following is a checklist of safety measures that can be followed to prevent fire and explosion accidents.

1. Biogas should not be discharged into air in a confined place. This suggests that all gas line fittings should be leakproof, and pressure-relief valves should be vented outside buildings and confined spaces.
2. Air should be purged from all delivery lines by allowing gas to flow for some time prior to use.
3. Flame traps should be installed in gas delivery lines if they are located in close proximity to gas burning appliances.
4. Adequate ventilation should be provided around all gas lines.
5. Digester gas carries water vapour and is described as wet gas. Therefore, gas pipes should be installed so as to slope upwards or downwards with a condensation trap at the lower end of the line.
6. All potential sources of sparks or open flames should be cleared from the gas production and storage area.
7. Fire extinguishers should be installed for controlling gas fires at the gas storage location.
8. If the method of pressure storage is used, then storage tanks capable of storing gas safely at pressures up to 170 kg/cm^2 should be used.

Table 3.7 Solubility of CO_2 in water

Pressure (bar)	Solubility (kg CO_2 per 100 kg water)				
	Temperature (°C)				
	0	10	20	30	40
1	0.40	0.25	0.15	0.10	0.10
10	3.15	2.15	1.30	0.90	0.75
50	7.70	6.95	6.00	4.80	3.90
100	8.00	7.20	6.60	6.00	5.40
200	-	7.95	7.20	6.55	6.05

Removal of CO_2 and H_2S

Carbon dioxide is a non-combustible gas, while H_2S causes corrosion. Thus, removal of both these components can improve the gas quality. *Water scrubbing* is the simplest method, but the water requirement for this method is quite large. Table 3.7 shows the approximate solubility of CO_2 at different pressures. *Caustic scrubbing* can also be employed to remove these two gases.

Areas of Improvement

1. Improving the biodegradability of feedstocks (may need pretreatment of raw materials).
2. Reducing the cost of biogas plants.
3. Finding improved techniques of handling, storing, and using digested sludge.
4. Improving the strains for biogas production through the application of genetic engineering.

Some Important Alternative Fuels

The continued development of biosustainable and renewable resource technology is of great importance with respect to environmental concerns. The successful and economical recycling of biomaterials will also assist in slowing down the continued deterioration of the environment. Lignocellulosic materials containing cellulose, hemicellulose, and lignin as the main constituents are the most abundant source of renewable organic material present on earth. With continuous increase in the demand for energy, the natural resources are depleting. Thus fossil fuels such as coal, gas, and oil, which cater to approximately 75%–85% of the total energy demand, are diminishing at an alarming rate. Therefore the need has arisen to scout for alternative fuels to meet the ever-increasing energy requirements. The conversion of both cellulose and hemicellulose for production of the fuel ethanol is being studied intensively, with a view to develop a technically and economically viable bioprocess. The fermentation of glucose, the main constituent of cellulose hydrolyzate, to ethanol can be carried out efficiently. Similarly, the production of other fuels such as biogas, biomass, biodiesel, and biohydrogen is also being given a serious thought.

Bioethanol

To overcome the problem of fuel deficiency, the production of bioethanol from lignocellulosic wastes is being considered as a promising alternative.

Ethanol, an alternative fuel, is replacing petroleum products. Chemically, bioethanol is the same compound as consumable alcohol. Statistically, if worldwide alcohol production is considered, 26 million t of ethanol was produced in 2002, 63% of which was used as biofuel. Europe produced only about 1.6 million t of ethanol in 2002, very little compared to USA and Brazil, where 5.7 and 8.7 million t, respectively, of ethanol was produced to be used as biofuel. This easily makes it the largest fermentation product produced till date.

Presently, ethanol is produced with the help of yeasts that easily convert fermentable carbohydrates obtained from sugar beet, sugarcane, or cereals into ethanol. This process is, perhaps, the oldest fermentation technology known. It has been in use now for at least 4500 years. Egyptian records dating back to 2500 BC refer to the use of grapes for making wine. The present day focus is directed towards producing alcohol from more difficult substrates such as agricultural residues (e.g., corn cobs, straw) or other lignocellulosic biomass. Lignocellulosics and agricultural co-products (preferably not called wastes) such as straw, bran, corn cobs, corn stover, and sawdust, which have either been poorly valued or left to decay, are attracting increasing attention as an abundantly available and cheap renewable source of energy. Estimations from the US Department of Energy have shown that up to 500 million t of such raw materials are available in USA alone each year. As large quantities of these raw materials are available at low prices, a process which would enable to produce an alternative fuel such as bioethanol will not only considerably improve the overall economics of ethanol as a biofuel but would also seriously impact the energy sector in a positive manner. However, because of their complex composition, these substrates are not easily convertible to free sugars.

On the one hand, by genetic modifications, a number of micro-organisms have been modified in such a way that they can convert more complex substrates such as cellulose or lignocellulosic wastes into alcohol. These *superbugs* are then used to convert more complex substrates such as bagasse, straw, and waste paper into ethanol. On the other hand, much effort has been put into biotechnological research for inexpensively producing and improving the required cellulase enzymes. These *supercellulases* must hydrolyse cellulose to glucose, which can then be easily fermented. However, the field application of these processes is yet to be realized.

Bioethanol is also used in chemical, cosmetic, and pharmaceutical industries, and a small part of it is used for human consumption also. Potable alcohol in a certain way can also be considered as a biofuel, as it provides our bodies with 6.5 kcal/g of energy.

Biohydrogen

Hydrogen is envisioned to be used in a very different way to generate energy as compared to conventional biofuels. As hydrogen is one of the smallest known molecules, a number of engineering problems arise with storing and transporting it.

Hydrogen gas is seen as a future energy resource by virtue of the fact that it is renewable and does not emit CO_2, a greenhouse gas. It also liberates large amounts of energy per unit weight in combustion, and is easily converted to electricity by fuel cells. Biological hydrogen production has several advantages over hydrogen production by photoelectrochemical or thermochemical processes. Biological hydrogen production by photosynthetic micro-organisms, for example, requires the use of a simple solar reactor such as a transparent closed box, with low energy requirements.

The basic characteristics of biological hydrogen production and its feasibility at the commercial level, particularly through the use of photosynthetic microorganisms, are of special interest. Progress has also been made in research on anaerobic fermenters. Hydrogenase and nitrogenase enzymes produced by microalgae and cyanobacteria are both capable of producing hydrogen by photolysis of water.

It has been reported that a green alga, *Scenedesmus*, produced molecular hydrogen under light conditions after being kept under anaerobic and dark conditions. Basic studies on the mechanism involved in hydrogen production have determined that the reducing power (electron donation) of hydrogenase does not always come from water; it may sometimes originate intracellularly from organic compounds such as starch. The contribution of the decomposition of organic compounds to hydrogen production depends on the algal species concerned as well as on culture conditions. Even when organic compounds are involved in hydrogen production, an electron source can be derived from water, since organic compounds are synthesized by oxygenic photosynthesis. However, the reason for hydrogenase inactivity in green algae under normal photosynthetic growth conditions is unclear. Hydrogenase is thought to become active in order to excrete excess reducing power under specific conditions such as anaerobic conditions.

The oil crisis of 1973 prompted research on biological hydrogen production, including photosynthetic production, as part of the search for alternative energy technologies. Green algae were known as light-dependent, water-splitting catalysts, but the characteristics of their hydrogen production were not practical for exploitation. Hydrogenase is too oxygen-labile for sustainable hydrogen production. Research has also shown that light-dependent hydrogen production ceases within a few minutes since photosynthetically produced oxygen inhibits or inactivates hydrogenase. A continuous gas flow system designed to maintain low oxygen concentration within the reaction vessel, was employed in basic studies, but has not been found to be practically applicable.

Green algae find their use in another method of hydrogen production. It has already been demonstrated that *Scenedesmus* produced hydrogen gas not only under light conditions, but also fermentation under dark conditions in the absence of oxygen, with intracellular starch as the reducing source. Although the rate of fermentative hydrogen production per unit of dry cell weight was less than that obtained through light-dependent hydrogen production, hydrogen production was sustainable due to the absence of oxygen. Some workers have also proposed that CO_2 is reduced to starch by photosynthesis during the day (under light conditions) and the starch thus formed is decomposed to hydrogen gas and organic acids and/or alcohols under anaerobic conditions during the night (under dark conditions). The technological merits of this proposal include the facts that oxygen inactivation of hydrogenase can be prevented through the maintenance of green algae under anaerobic conditions, the night hours are used effectively, temporal separation of hydrogen and oxygen production does not require gas separation for simultaneous water splitting, and organic acids and alcohols can be converted to hydrogen gas by photosynthetic bacteria under light conditions.

It has been determined that *cyanobacteria* also produce hydrogen gas auto-fermentatively under dark and anaerobic conditions. The *Spirulina* species was demonstrated to have the highest activity among the cyanobacteria tested. The nature of the electron carrier for hydrogenase in cyanobacteria is still unclear. Hydrogenases have been purified and partially characterized in a few cyanobacteria and microalgae.

It was also found that a nitrogen-fixing cyanobacterium, *Anabaena cylindrica*, produced hydrogen and oxygen gas simultaneously in an argon atmosphere for several hours. Nitrogenase is responsible for nitrogen fixation and is distributed mainly among prokaryotes, including cyanobacteria, but does not occur in eukaryotes, under which microalgae are classified. Molecular nitrogen is reduced to ammonium with consumption of reducing power and ATP. The reaction is substantially irreversible and produces ammonia. However, nitrogenase catalyses proton reduction in the absence of nitrogen gas (i.e., in an argon atmosphere).

Hydrogen production catalysed by nitrogenase occurs as a side reaction at a rate of one-third to one-fourth that of nitrogen fixation, even in a 100% nitrogen gas atmosphere. Nitrogenase itself is an extremely oxygen-labile enzyme. Unlike in the case of hydrogenase, however, cyanobacteria have developed mechanisms for protecting nitrogenase from oxygen gas and supplying it with energy (ATP) and reducing power by heterocyst production.

Photosynthetic bacteria undergo a different type of photosynthetic process in the absence of oxygen, with organic compounds or reduced sulphur compounds as electron donors. Some non-sulphur photosynthetic bacteria are potent hydrogen producers, which utilize organic acids such as lactic, succinic, and butyric acids, or alcohols as electron donors. Since light energy is not required for water oxidation, the efficiency of light energy conversion to hydrogen gas by photosynthetic bacteria, is in principle, much better than that by cyanobacteria. Hydrogen production by photosynthetic bacteria is mediated by nitrogenase activity, although hydrogenases may be active for both hydrogen production and hydrogen uptake under some conditions. It has also been demonstrated using *Rhodobacter* sp. in laboratory experiments that the maximum energy conversion efficiency (combustion energy of hydrogen gas produced/incident light energy) is 6% to 8%.

The combined use of photosynthetic and anaerobic bacteria should increase the likelihood of their application in photobiological hydrogen production. Enhancement of the hydrogen-producing capability of micro-organisms by genetic engineering is also being pursued. However, a lot of work has to be done to make it practically feasible for common use.

Biodiesel

Among the different types of fuels, biodiesel is emerging as an important alternative fuel. It is a well-known fact that engines can be operated on biodiesel, which is a transesterified product of vegetable oil or fatty acid methyl ester (FAMES). The vegetable oil fraction (triglycerides) extracted from the oilseeds are treated with alkali (NaOH) in the presence of methanol. During this process a transesterification reaction occurs by which glycerol is released and fatty acid methyl esters are formed. After this reaction (Fig. 3.13), the reaction mixture can be physically separated into an oily biodiesel phase and a watery phase that contains the salts and glycerol. The excess of methanol is recovered by distillation. The biodiesel fraction thus obtained is a slightly yellow, low-viscosity liquid, comparable with normal diesel oil. The mass balance is favourable, since 1 kg of oil reacts with approximately 0.1 kg of methanol, producing 1 kg of biodiesel and 0.1 kg of glycerol. The co-product glycerol is a relatively expensive base chemical that is used in various applications in the cosmetic and chemical industries. Biotechnological alternatives for this type of chemical production, especially by lipases, are being investigated.

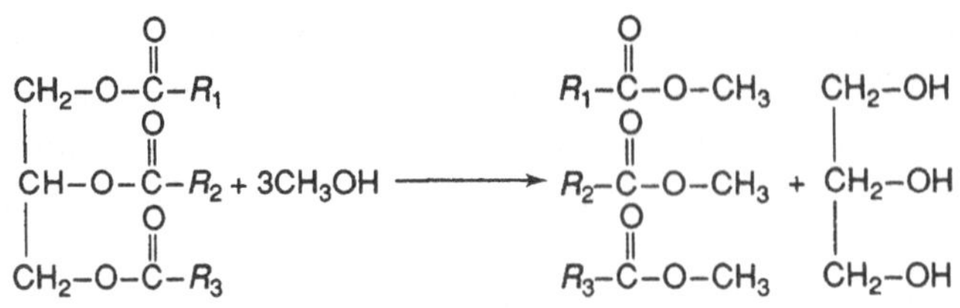

$$
\begin{array}{l}
CH_2-O-\overset{\displaystyle O}{\overset{\|}{C}}-R_1 \\
CH-O-\overset{\displaystyle O}{\overset{\|}{C}}-R_2 + 3CH_3OH \\
CH_2-O-\overset{\displaystyle O}{\overset{\|}{C}}-R_3
\end{array}
\longrightarrow
\begin{array}{l}
R_1-\overset{\displaystyle O}{\overset{\|}{C}}-O-CH_3 \quad CH_2-OH \\
R_2-\overset{\displaystyle O}{\overset{\|}{C}}-O-CH_3 + CH_2-OH \\
R_3-\overset{\displaystyle O}{\overset{\|}{C}}-O-CH_3 \quad CH_2-OH
\end{array}
$$

Fig. 3. 3   Transesterification reaction

Biodiesel contains no nitrogen or aromatics and typically has less than 15 ppm (parts per million) sulphur. It has a specific gravity of 0.88 as compared to 0.85 for diesel fuel. Since biodiesel is slightly heavier than diesel, splash blending biodiesel on top of diesel fuel is a commonly followed mixing procedure. Rack blending is also being considered in some states in our country where B2 blends are being considered. B2 is a high lubricity diesel fuel made with 2% biodiesel.

Biodiesel contains 11% oxygen by weight, which accounts for its slightly lower heating value. It has characteristically low carbon monoxide (–50%), particulate, soot (–30%), and hydrocarbon (–93%) emission. Biodiesel (37–41 MJ/kg) has an energy content comparable to ordinary diesel fuel (36–45 MJ/kg), and can be mixed with normal diesel fuel, usually up to 5%. In France, one in two diesel cars already runs with 2% biodiesel added, without even the awareness of the customer, as it is not indicated at the pump. Pure biodiesel is also used in some countries such as in Germany and Austria. However, this requires certain minor modifications in the engine. Problems can occur during winters because of the crystallization of biodiesel at low temperatures.

Biodiesel can be used in any diesel engine with no need for any modifications. In fact diesel engines run better and last longer with biodiesel. Some of the advantages of the use of biodiesel are listed here.

1. Biodiesel can be used in any diesel engine.
2. Biodiesel is plant-based and hence does not contribute CO_2 to the atmosphere.
3. Biodiesel exhaust is not offensive and does not cause irritation to the eye. (In fact, it smells like French fries!)
4. The fuel is up to 75% cleaner than conventional diesel fuel, which is obtained from fossil fuels.
5. The use of biodiesel leads to substantially reduced unburned hydrocarbons, carbon monoxide, and particulate matter in exhaust fumes.
6. The possibility of sulphur dioxide emissions from exhaust pipes of vehicles is eliminated, as biodiesel contains no sulphur.
7. The ozone-forming potential of biodiesel emissions is nearly 50% less than that of conventional diesel fuel.

8. Nitrous oxide (NOx) emissions may increase or decrease, but can be certainly reduced to well below conventional diesel fuel levels by adjusting the engine timing.
9. Biodiesel is environmental friendly, as it is renewable and said to be 'more biodegradable than sugar and less toxic than table salt'.
10. Biodiesel can be mixed with ordinary diesel fuel in any proportion. Even a small amount of biodiesel means cleaner emissions and better engine lubrication. 1% biodiesel increases lubricity by 65%.
11. Biodiesel is a much better lubricant than conventional diesel fuel and extends engine life. [A German truck won an entry in the *Guinness World Records* by travelling more than 1.25 million km (780,000 miles) on biodiesel with its original engine.]
12. Biodiesel has a high cetane rating, which improves engine performance. 20% biodiesel added to conventional diesel fuel improves the cetane rating by 3 points, thus making it a Premium fuel.
13. Fuel economy is the same as conventional diesel fuel.

Summary

Chapter 3 emphasizes on the production of biofuels such as biogas, biohydrogen, and bioethanol, which are end products of anaerobic fermentation but quite different in nature. Though biogas and biohydrogen are gaseous end products, bioethanol is a liquid fuel that can be used as a petroleum substitute. In this chapter, the biochemical as well as microbiological aspects of the anaerobic fermentation process have been discussed in detail.

With the increased cost of petroleum in the 1970s and the advent of the era of ecological awareness, petroleum microbiology has come into focus. Since then the application of molecular biology for the development of genetically modified organisms for efficient petroleum degradation has been studied and patented. In the 1980s, owing to the energy crisis, studies on micro-organisms were initiated for the quest of alternative fuels. Two different approaches have been taken for solving the energy crisis. One is the use of micro-organisms to increase the efficiency of the recovery of petroleum hydrocarbons from existing wells and aid in the extraction of hydrocarbons from oil sands and shales. The other is the use of micro-organisms to produce fuels such as methane and higher molecular weight hydrocarbons.

This chapter presents in detail the pathways followed by the biochemical reactions occurring in micro-organisms participating in waste degradation. Due emphasis has also been given to the various factors influencing biodegradation processes. The advantages of anaerobic processes, which make them economically viable, as compared to aerobic processes, have been established scientifically.

Review Questions

1. Discuss the structure of bacteria.
2. With a suitable diagram, describe the functions of different organelles in bacteria.
3. How do archaebacteria differ from eubacteria? Give suitable examples.
4. Discuss the metabolism of aerobic and anaerobic bacteria.

5. Cite the biochemical pathways involved in the production of methane from organic wastes.
6. Critically establish the importance of the EMP pathway in aerobic and anaerobic micro-organisms.
7. What are the intermediate and end products of anaerobic fermentation of sugar?
8. Define biogas. What is the mechanism of biogas production?
9. Name the micro-organisms involved in hydrolytic degradation and acetogenesis.
10. What is the importance of *Clostridium* sp. in degradation of organic wastes?
11. Through the biochemical pathway, prove that one molecule of glucose releases three molecules of acetic acid.
12. State the mechanism of the Strickland reaction. What is its importance in anaerobic fermentation?
13. Characterize different methanogenic bacteria.
14. What are the factors governing methane production?
15. Discuss the strategies of strain improvement for anaerobic micro-organisms.
16. What are the different techniques used for improvement of strains by mutation?
17. Define the following.
    (a) Conjugation
    (b) Protoplast fusion
    (c) Recombination
    (d) Fmd
    (e) FTR
18. Briefly describe the different tools of molecular biology involved in the improvement of strains in methanogens.
19. What is the effect of temperature, *p*H, and retention time on methane production?
20. How is biogas enriched? Discuss the importance of biohydrogen and bioethanol as alternative sources of fuel.
21. What is transesterification? How can the fuel value of biodiesel be improved? Identify and logically establish among biohydrogen, biodiesel, bioethanol, and biogas, the fuel that has the most promising future.

References

Atlas, R.M. 1984, *Microbiology Fundamentals and Applications*, 2nd edn, Maxwell Macmillan, New York.

Bache, R. and N. Pfennig 1981, 'Selective isolation of *Acetobacterium woodi* on methoxylated aromatic acids and determination of growth yields', *Arch. Microbiol.*, vol. 130, pp. 255–61.

Beiley, J.E. and F.D. Ollis 1987, *Biochemical Engineering Fundamentals*, McGraw-Hill, New York.

Bhattacharyya, B.C. 1979, 'Design and development of gobar (bio) plant for rural population', *Chemical Age of India*, Monograph.

Daniel, S.L. and R.L. Drake 1998, 'Acetogenesis from methoxylated aromatic acids by *Clostridium thermoautotrophicum*', *Abstr. Ann. Meet. Am. Soc. Microbiol.*, vol. I-105, p. 198.

Ferry, J.G. 1999, 'Enzymology of one-carbon metabolism in methanogenic pathways', *FEMS Microbiol. Rev.*, vol. 23, no. 1, pp. 13–38.

Fuchs, G. 1986, 'CO_2 fixation on acetogenic bacteria variation on a theme', *FEMS Microbiol Rev.*, vol. 39, pp. 181–213.

Gottwald, M. 1973, 'Untersuchungen Zum Pentose Staffwechselbel Homoaetatgarerm', Diplomarbeit, Universitat Gottingen, FRG.

Hanson, R.S. 1980, 'Ecology and diversity of methylotrophic organisms', *Adv. Appl. Microbiol.*, vol. 26, pp. 3–39.

Hao, B., W. Gong, T.K. Ferguson, C.M. James, J.A. Krzycki, and M.K. Chan 2002, 'A new UAG-encoded residue in the structure of a methanogen methyltransferase', *Science*, vol. 296, pp. 1462–6.

Higgins, I.J., D.J. Best, R.C. Hammond, and D. Scott 1981, 'Methane-oxidizing microorganisms', *Microbiol Rev.*, vol. 45, no. 4, pp. 556–90.

Ljungdahl, L.G. 1986, 'The autotrophic pathway of acetate synthesis in acetogenic bacteria', *Ann. Rev. Microbiol.*, vol. 40, pp. 415–50.

MACS-DNES Training Course 1986, 'Microbiological aspects of anaerobic fermentation', Laboratory Manual, Maharastra Association for the Cultivation of Science.

Reeve, J.N. 1992, 'Molecular biology of methanogens', *Ann. Rev. Microbiol.*, vol. 46, pp. 165–9.

Samson, Rejean, L. Bert, Van den Berg, and Kevin J. Kennedy 2004, 'Mixing characteristics and startup of anaerobic downflow stationary fixed film (DSFF) reactors', *Biotechnology and Bioengineering*, vol. 27, no. 1, pp. 10–19.

Shima, S. et al. 2002, 'Structure and function of enzymes involved in the methanogenic pathway utilizing carbon dioxide and molecular hydrogen', *J. Biosci. Bioeng.*, vol. 93, no. 6, pp. 519–30.

Shima, S. and R.K. Thauer 2006, 'Anaerobic ethane oxidation by archaea: A biochemical approach', *Bioscience and Industry*, vol. 64, pp. 23–6.

Stanier, R.Y., J.L. Ingrahm, M.L. Weels, and P.R. Painter 1987, *General Microbiology*, 5th edn, Macmillan, London.

Wiegel, J., M. Braun, and G. Gottschalk 1981, '*Clostridium thermoantorophicum* species novum, a thermopile producing acetate from molecular hydrogen and carbon dioxide', *Curr. Microbiol.*, vol. 5, pp. 255–60.

Wiegel, J. and R. Garrison 1985, 'Utilization of methanol by *Clostridium thermoaceticum*', *Abstr. Ann. Meet. Am. Soc. Microbiol.*, vol. I-115, p. 165.

Winters-Ivey, D., L.G. Ljungdahl, and J. Wiegel 1985, 'Metabolism of methanol in *Clostridium thermoautotrophicum*', *Abstr. Ann. Meet. Am. Soc. Microbiol.*, vol. K-66, p. 182.

Wood, H.G., S.W. Ragsdale, and E. Pzecka 1986, 'The acetyl CoA pathway: A newly discovered pathway of autotrophic growth', *Trends. Biochem. Sci.*, vol. 11, pp. 14–18.

Further Reading

Aasheim, S.E. 1985, *Sludge Stabilization: Manual of Practice No. FD-9*, Water Pollution Control Federation: Task Force on Sludge Stabilization, Washington, DC.

Adams, C.E., W.W. Eckenfelder, and R.M. Stein 1974, 'Modifications to aerobic digester design', *Water Res.*, vol. 8, p. 213.

Ahring, B.K. 2003, 'Perspectives for anaerobic digestion', *Adv. Biochem. Engin./Biotechnol.*, vol. 81, pp. 1–30.

Alexander, A. 1981, 'Biodegradation of chemicals of environmental concern', *Science*, vol. 211, pp. 132–8.

Angelidaki, I., L. Ellegaard, and B.K. Ahring 2003, 'Applications of the anaerobic digestion process', *Adv. Biochem. Engin./Biotechnol.*, vol. 82, pp. 1–33.

APHA, AWWA, WEF 1998, *Standard Methods for the Examination of Water and Wastewater*, 20th edn, Washington, DC.

Borchardt, J.A. 1981, *Sludge and its Ultimate Disposal*, Ann Arbour Science, Collingwood, MI.

Chandler, J.A., W.J. Jewell, J.M. Gosett, P.J. Van Soost, and J.B. Robertson 1980, 'Predicting methane fermentation biodegradability', *Biotechnol. Bioeng. Symp.*, no. 10, pp. 93–107.

Chen, Y.R. and A.G. Hashimoto 1978, 'Kinetics of methane fermentation', *Biotechnol. Bioeng. Symp.*, no. 8, pp. 269–82.

Chin, E.S.K and F.B. De Walle 1977, 'Treatment of high strength acidic wastewater with a completely mixed anaerobic filter', *Water Res.*, vol. 11, no. 3, pp. 295–304.

Colbeau, A., K.L. Kovacs, J. Chabert, and P.M. Vignais 1994, 'Cloning and sequencing of the structural (hupSLC) and accessory (hupDHI) genes for hydrogenase biosynthesis in *Thiocapsa roseopersicina*,' *Gene*, vol. 140, pp. 25–31.

Cook, E.J. 1986, *Anaerobic Sludge Digestion: Manual of Practice No. 16.*, Water Pollution Control Federation: Task Force on Sludge Stabilization, Alexandria, VA.

Das, D. 1978, *Biochemistry*, Academic Publishers, Kolkata.

Dolfing, J. 2001, 'The microbial logic behind the prevalence of incomplete oxidation of organic compounds by acetogenic bacteria in methanogenic environments', *Microb. Ecol.*, vol. 41, pp. 83–9.

Drapal, N. and A. Bock 1998, 'Interaction of the hydrogenase accessory protein HypC with HycE, the large subunit of *Escherichia coli* hydrogenase 3 during enzyme maturation', *Biochemistry*, vol. 37, pp. 2941–48.

Farquar, G.J. and F.A. Rovers 1992, 'Gas production during refuse decomposition', *Water, Air Soil Poll.*, vol. 2, pp. 483–389.

Filipe, C.D.M., G.T. Daigger, and C.P.L. Grady 2001, 'pH as a key factor in the competition between glycogen-accumulating and phosphate-accumulating organisms', *Water Environ. Res.*, vol. 73, pp. 223–32.

Forster, C.F. and D.A.J. Wase (eds) 1987, *Environmental Biotechnology*, Ellis Horwood, Chichester, UK.

Ghosh, S. 1984, 'Microbial production of energy—Gaseous fuel', T.K. Ghose, (ed.) *Biotechnology and Bioprocess Engineering, Proceedings of the 7th International Biotechnology Symposium, New Delhi*, 19–25 February, pp. 187–222.

Hall, E.R., B.E. Jank, and M. Jovanoric 1981, 'Enzyme production from high strength industrial waste water,' *Proc. Bioenergy R&D Seminar*, National Research Council of Canada, Ottawa, pp. 125–29.

Hartman, R.B. 1979, 'Sludge stabilization through aerobic digestion', *J. Water Poll. Control Federation*, vol. 49, p. 2353.

Hofman-Bang, J., D. Zheng, P. Westermann, B.K. Ahring, and L. Raskin 2003, 'Molecular ecology of anaerobic reactor systems', *Adv. Biochem. Eng./Biotechnol.*, vol. 81, pp. 153–203.

Jewell, W.J, S. Dell'Orto, K.J. Fanfoni, T.D. Hayes, A.P. Leuschner, and D.F. Sherman 1980, *Anaerobic Fermentation of Agricultural Residue: Potential for Improvement and Implementation*, Final Report NTIS, vol. 3, US Dept of Commerce, Springfield, VA.

Keeman, J.G. 1974, Paper no. 74-W/A, ASME Winter Annual Meeting, New York.

Lawrence, A.W. 1967, 'Kinetics of methane fermentation in anaerobic waste treatment', PhD thesis, Stanford University.

Lawrence, A.W. 1971, 'Application of process kinetics to design of anaerobic processes', in F.G. Pohland (ed.), *Anaerobic Biological Treatment Processes*, Advances in Chemistry Series no. 105, American Chemical Society, Washington, DC, pp. 163–89.

Lewin, B. 2004, *Genes VIII*, Prentice Hall, Upper Saddle River, NJ.

McCarty, P.L. 1964, 'Anaerobic waste treatment fundamentals', Part I, Chem. Microbiol., vol. 95, no. 9, pp. 107–12.

Monod, J. 1950, 'La technique de culture continue; theorie et applications', *Ann. Inst. Pasture*, vol. 79, pp. 390–410.

Mosey, F.E. 1983, 'Kinetics descriptions of anaerobic digestion', *Proceedings of the 3rd International Symposium on Anaerobic Digestion*, Boston, MA, 14–26 August.

Norman, J. and B. Frosteel 1977, Paper presented at Purdue University, *Industrial Waste Conference, West Lafayette*, IN, 10 May.

Reeve, J.N. 1992, 'Molecular biology of methanogens', *Ann. Rev. Microbiol.*, vol. 46, pp. 165–91.

Renuka, C. 1991, 'High rate stabilization of MSW by using two phase biomethanation process', MTech thesis, IIT Kharagpur.

Sarma, A. 1990, 'Biomethanation of waste biomass', PhD thesis, Gauhati University.

4

Analytical Techniques for Environmental Monitoring

Introduction

In Chapter 3 we have already studied about the various environmental parameters which influence the efficiency of the different types of treatment processes. Furthermore, it is clear that the nature and character of the waste material depends on its source as discussed in Chapter 2. Our learning, as of now, also reveals that the BOD (biological oxygen demand) and the COD (chemical oxygen demand) values of a waste are the indicators of the level of pollution and these need to be monitored. Also, the reduction of BOD during the process of stabilization of wastes results in the yield of some useful products such as biogas, biomanure, SCP (single-cell protein), biohydrogen, and bioethanol, which are produced in proportion to the BOD/COD reduced.

Chapter 3 also discussed that the C/N ratio of the waste materials is an important indicator of microbial growth during the process of waste degradation. Therefore, the content of carbon and nitrogen in the waste material must be known, along with the presence of other trace elements such as sulphur and phosphorous. pH is another parameter that has a significant role in microbial action. In order to effectively monitor environmental pollution, it is essential to know about the various techniques that are employed to determine the contents of waste materials, so that they can be accordingly treated for bioremediation.

This chapter enumerates the analytical techniques that are used to determine the contents of waste materials. Knowledge of these parameters is a prerequisite for designing treatment plants. Some of the parameters are the

Analysing waste

total solid and volatile solid contents, organic substances such as cellulose, hemi-cellulose, starch, lignin, reducing sugars, and so on. These will be discussed in detail in the next few sections.

Estimation of Total Solids

The *total solid* content is defined as the matter that remains as residue after the sample is heated to a temperature of 105°C for 1 hour and then cooled and weighed (W_1). The sample is first thoroughly mixed, put in a petri dish and weighed (W_2). This petri dish is then placed in the oven at 105°C for 1 hour and its weight is measured after cooling the mixture. This process is repeated till a constant weight is obtained (W_3). The percentage of total solid present is calculated as follows.

Calculation

$$\text{Percentage of total solid content in the sample} = \frac{W_3 - W_1}{W_2 - W_1} \times 100$$

Estimation of Volatile Solids

Volatile solids are evolved in the form of gases at 600°C and the inorganic fraction is left behind as ash. A known amount of dry sample is taken in a clean and dry silica crucible. It is kept in the furnace, which has been heated to a temperature range of 550°C ± 50°C for 5 hours. The crucible is then cooled in a desiccator and weighed.

Calculation

Weight of the sample = W_1

Weight of crucible + dry sample = W_2

Weight of crucible + sample after incubation = W_3

Percentage of volatile solids in dry sample = $\dfrac{W_2 - W_3}{W_2 - W_1} \times 100$

The ash is retained for the estimation of phosphorous and potassium.

Estimation of Cellulose

The percentage of cellulose present in a waste material can be determined by any of the two methods explained here. These are the *Updegraff* and the *spectrophotometric* methods.

Estimation of Cellulose—Updegraff Method

This method was given by Updegraff in 1969.

Procedure

1. 0.5 g of dry sample is first taken in a pre-weighed test tube.
2. This sample is then moistened with distilled water and 5 ml of acetic–nitric reagent in 1-ml aliquots is added. It is properly mixed in a cyclomixer.
3. The contents of the test tube are transferred to a boiling water bath and digestion is allowed to take place such that all other organic matter except cellulose are digested. To ensure that there is minimum evaporation of contents, the open end of the test tube is closed with a marble. The digestion process is continued for 30 min and is followed by cooling.
4. Next, the contents are centrifuged at 5000 rpm for 5 min. The supernatant liquid is decanted and the residue is washed several times till it becomes colourless.
5. 10 ml of 67% H_2SO_4 (v/v) in instalments of 1 ml is added with intermittent mixing in a cyclomixer. This mixture is allowed to stand for 1 hour and diluted to 100 ml with distilled water followed by centrifugation at 2000 rpm for 10 min to remove any precipitate.
6. 0.1 ml of this dilute sample is taken and the volume increased to 5 ml by adding 4.9 ml of distilled water.
7. The test tube is next transferred to an ice bath and 10 ml of cold anthrone reagent is added gradually, to avoid spurting.
8. The tube is then transferred to a boiling water bath and heated for about 20 min, ensuring that the open end of the tube is closed with a marble.

9. The tube is immediately cooled by putting it in ice bath and allowing it to stand for 10 min at room temperature.
10. The optical density (OD) of the resultant blue–green coloured mixture is measured at 620 nm. This gives the OD of the unknown.
11. For preparing a standard plot, 0.5, 1.0, 1.5, and 2.0 ml of the sample from the 100 µg/ml standard stock solution of cellulose are taken in a set of test tubes. The volume of each tube is increased to 5 ml by adding 4.5, 4.0, 3.5, and 3.0 ml, respectively, of distilled water.
12. A reagent blank is run along with the experiment. To each of the tubes cold anthrone is added resulting in the appearance of blue–green colour and the optical density is measured at 620 nm.
13. A standard graph is plotted and the unknown cellulose content is calculated.

Calculation

Weight of the test tube $= W_1$
Weight of the test tube + dry sample $= W_2$
Weight of dry sample $= W_2 - W_1 = 5 \times 10^5 \, \mu g$ (i.e., 0.5 g)

$$\text{Per cent cellulose} = \frac{\mu g \text{ of cellulose in 100 ml diluted sample}}{W_2 - W_1} \times 100$$

Estimation of Cellulose—Spectrophotometric Method

The spectrophotometric method involves the use of a series of reagents as listed here.

Reagents

1. *Acetic/nitric reagent* 150 ml of 80% acetic acid with 15 ml of concentrated HNO_3 is mixed.
2. *Anthrone reagent* To prepare this reagent, 200 mg anthrone is added to 100 ml of concentrated H_2SO_4 (prepared fresh and chilled before use).
3. 67% H_2SO_4 solution.
4. *Standard cellulose solution* To 100 mg of cellulose, 10 ml of 67% H_2SO_4 is added and left for an hour. This solution is then further diluted by adding 1 ml of the solution to 100 ml of distilled water (100 µg/ml).

Procedure

1. To about 0.5–1.0 g of the sample, 3 ml of acetic/nitric reagent is added and the two are mixed well, using a vortex mixer.
2. This solution is placed in a water bath at 100°C for 30 min.
3. The mixture is then cooled and centrifuged for 15–20 min. The supernatant is discarded.

4. The residue is washed with water. To this residue 10 ml of 67% H_2SO_4 is added and the mixture is left undisturbed for 1 hour.
5. To 1 ml of this diluted solution, 10 ml of anthrone reagent is added and the solution is mixed well.
6. The tube is next heated in a boiling water bath for 10 min. After properly cooling the solution, its absorbance is measured at 630 nm.
7. A blank with anthrone reagent and water is also prepared.
8. A standard curve is plotted by taking 0.4 to 2 ml of standard cellulose solution (corresponding to 40–200 µg of cellulose), and the volume is equalized. To this, anthrone reagent is added and colour is allowed to develop.
9. The amount of cellulose in the sample is calculated from the standard graph.

Estimation of Starch

The starch content in the waste material can be estimated by using any of the following methods.

Estimation of Starch—Folin–Wu Procedure

It works on the principle of hydrolysis of starch with 10% HCl (v/v), which results in the formation of glucose.

Procedure
1. 1 g of decolourized and solubilized fraction of starch is taken in the flask of a reflux apparatus.
2. To this, 200 ml of 10% HCl (v/v) is added and the mixture is refluxed for 2.5 hours. Then it is cooled to room temperature and with the help of a pH meter neutralized to pH 7 by using $10N$ NaOH.
3. The volume of this solution is increased to 250 ml with distilled water.
4. The starch content is then estimated by following the Folin–Wu procedure (this method is described later in the section on reducing sugar estimation).

Calculation

Weight of the reflux apparatus flask = W_1

Weight of the flask + the decolourized dry residue after deduction of the water-soluble fraction = W_2

Per cent starch content in the residue

$$= \frac{0.9 \times \mu g \text{ of glucose in } 250 \text{ ml of the neutralized solution}}{W_2 - W_1} \times 100$$

Estimation of Starch—Anthrone Reagent Method

Reagents

1. *Anthrone reagent* 200 mg anthrone is dissolved in 100 ml of ice-cold 95% sulphuric acid.
2. 80% ethanol.
3. 52% perchloric acid.
4. *Standard glucose stock* 100 mg in 100 ml distilled water.
5. *Working standard* 10 ml of stock is diluted to 100 ml with water (100 µg/ml).

Procedure

1. 0.1 to 0.5 g of the sample is homogenized in hot 80% ethanol to remove sugars. Then it is centrifuged and the residue is retained. This residue is then washed repeatedly with hot 80% ethanol, till the washings do not give any colour with anthrone reagent. Next the residue is dried well over a water bath.
2. To the residue 5 ml of water and 6.5 ml of 52% perchloric acid are added.
3. The solution is extracted at 0°C for 20 min. It is centrifuged and the supernatant is retained.
4. This extraction process is repeated using fresh perchloric acid. It is repeatedly centrifuged, the supernatants separated, and the volume increased to 100 ml.
5. In the next step, 0.1 or 0.2 ml of the supernatant is pipetted out and the volume is increased to 1 ml with water.
6. The standards are prepared by taking 0.2, 0.4, 0.6, 0.8, and 1 ml of the working standard and the volume is increased to 1 ml in each of the tubes by adding water.
7. 4 ml of anthrone reagent is added to each tube.
8. Each tube is heated for 8 min in a boiling water bath.
9. The test tubes are cooled rapidly and the intensity of green to dark green colour is read at 630 nm.

Calculation

The glucose content in the sample is determined by using the standard graph. The value is multiplied by 0.9 to deduce the content of starch.

Estimation of Hemicellulose

Reagents

1. Acetone.
2. Sodium sulphite.
3. Decahydronaphthalene.
4. *Neutral detergent solution* 18.61 g of disodium ethylenediaminetetraacetate (EDTA-Na salt) and 6.81 g of sodium borate decahydrate are dissolved in

about 200 ml of water by heating. To this, about 100–200 ml of a solution containing 30 g of sodium lauryl sulphate and 10 ml of 2-ethoxy ethanol are added. Then about 100 ml of a solution containing 4.5 g of disodium hydrogen phosphate is added. The *p*H is adjusted to 7.0 and the volume of the solution is increased to 1 L.

Procedure

1. To 1 g of the powdered sample taken in a refluxing flask, 10 ml of cold neutral detergent solution is added.
2. Then 2 ml of decahydronaphthalene and 0.5 g of sodium sulphite are added.
3. The mixture is heated to a boil and then refluxed for an hour.
4. Then it is filtered through a sintered glass crucible (G-2) by suction and washed with hot water.
5. It is washed twice with acetone and the residue transferred to a crucible and dried at 100°C for 8 hours.
6. The crucible is cooled in a desiccator and weighed.

Calculation

Hemicellulose = neutral detergent fibre (NDF)
– acid detergent fibre (ADF)

Estimation of Hemicellulose—Furfural Method

The polymers of pentoses such as mannose and xylose are called hemicelluloses. Hemicelluloses are estimated by converting them into furfural by distillation with 12% HCl (v/v). The mixture is then precipitated with phloroglucinol and quantified gravimetrically, while the hemicellulose content is calculated using Krober's formula (Hurwitz 1960) as described here.

Procedure

1. 4 g of the dry sample is taken in the flask of a distillation assembly.
2. 100 ml of 12% HCl is added to the solution and 30 ml of the distillate is collected.
3. In the next step, 30 ml of 12% HCl is added to the boiling mixture and boiling is continued.
4. During the process of distillation, 30 ml of distillate is collected in each instalment, such that the last 30-ml aliquot would be free of furfural, while the total distillate volume is fixed at 360 ml. The absence of furfural in the last distillate aliquot is checked by using aniline reagent.
5. 80 ml of 12% HCl and 40 ml of 0.7% phloroglucinol (w/v) are added to 360 ml of distillate. The mixture is allowed to stand overnight and the precipitate formed is collected in a watch glass by filtration through a pre-weighed Whatman 1 filter paper. All traces of HCl are removed from

the precipitate by washing it repeatedly with distilled water. Absence of HCl in the filtrate is checked by using 0.1% $AgNO_3$ (w/v).

6. The precipitate in the watch glass is then dried in an oven at $103°C \pm 2°C$ to a constant weight.

Calculation

Weight of the watch glass = W_1
Weight of the watch glass + dry sample = W_2
Weight of the Whatman filter paper = W_3
Weight of the filter paper + precipitate = W_4
Weight of the precipitate $X = W_4 - W_3$

If X is less than 0.03 g, then
Hemicellulose = $(X + 0.0052) \times 0.8949$ g

If X is between 0.03 to 0.3 g, then
Hemicellulose = $(X + 0.0052) \times 0.08866$ g

If X is more than 0.3 g, then
Hemicellulose = $(X + 0.0052) \times 0.8824$ g

$$\text{Percentage of hemicellulose in dry sample} = \frac{\text{hemicellulose}}{W_2 - W_1} \times 100$$

Estimation of Lignin

After the sequential removal of other organic materials such as fats, colouring matter, soluble fraction, cellulose, and hemicellulose, lignin is estimated gravimetrically by igniting it in a muffle furnace. The detailed procedure is as given here.

Procedure

1. After the solubilization of the water-soluble fraction, 2 g of the dry sample, weighed accurately, is taken in a pre-weighed thimble, which is then transferred to the extractor of a Soxhlet unit.

2. The fat is then extracted at 40°C in solvent ether for 5 hours. The thimble is kept in a hot air oven at $103°C \pm 2°C$ till it reaches a constant weight.

3. In the next step the thimble is transferred to the Soxhlet extractor. The colouring matter is extracted completely in an alcohol and benzene mixture (1:2 v/v) at 80°C.

4. The thimble is put in the oven at 103°C to 105°C to achieve constant weight. The residue is then transferred to the 500-ml flask of a reflux apparatus, using 5 ml distilled water.

5. At this stage, 10 ml of 72% H_2SO_4 (v/v) is added to a 2-ml aliquot and kept at room temperature for 2 hours. The sample is diluted to reach a final

concentration of 3% H_2SO_4 by adding distilled water (about 240 ml is found to suffice). It is refluxed for 2 hours and then filtered through a pre-weighed filter paper.

6. The residue obtained is washed repeatedly to remove traces of H_2SO_4. The $BaCl_2$ (1%) test is performed to check for the absence of H_2SO_4 in the filtrate.
7. The H_2SO_4-free filtrate is then transferred to the oven at 105°C to achieve constant weight. The filter paper is placed in a pre-weighed silica crucible and ignited at 550°C ± 50°C in a muffle furnace for 5 hours. It is then cooled in a desiccator and weighed.

Calculation

Weight of the thimble = W_1
Weight of the thimble + dry sample after solubilization of the water-soluble fraction = W_2
Weight of the ash-less filter paper = W_3
Weight of the ash-less filter paper + residue after refluxing = W_4
Weight of the silica crucible = W_5
Weight of the silica crucible + residue after ignition = W_6
Percentage of lignin in the dry matter after solubilization of the water-soluble fraction,

$$Y = \frac{(W_4 - W_3) - (W_6 - W_5)}{W_2 - W_1} \times 100$$

Percentage of water-soluble fraction in the dry sample = X
Percentage of lignin content in the dry matter after solubilization of the water-soluble fraction = Y

Percentage of lignin content in the dry sample $= \dfrac{(100 - X)Y}{100}$

Determination of Cellulose, Hemicellulose, and Lignin—Detergent Extraction Method

Cellulose, hemicellulose, and lignin contents in samples are determined by using the detergent extraction method of Robertson and Van Soest (1981). About 0.5 g of dried sample, powdered to pass through a 20 to 30 mesh (1 mm) is used for analysis. The compositions of the neutral detergent solution and acid detergent solution are as follows:

Neutral detergent (ND) solution

Sodium lauryl sulphate	30 g/L
Disodium EDTA. $2H_2O$	18.61 g/L

$Na_2B_4O_7.10H_2O$	6.81 g/L
Na₂HPO₄ anhydrous	4.56 g/L
2-ethoxy ethanol	10 m/L
pH	6.9–7.1

Acid detergent (AD) solution

| Cetyltrimethyl ammonium bromide | 40.04 g/L |
| H_2SO_4 | $1N$ |

Nitrogen Estimation—Kjeldahl Method

Nitrogen content in sludge may be estimated by the micro-Kjeldahl method. Here, it is assumed that 1 ml of $1N$ H_2SO_4 is equivalent to 0.014 g of nitrogen.

Procedure

1. 20 mg of dried sample is weighed in a micro-Kjeldahl flask.
2. To this, 1 ml of concentrated H_2SO_4 and 20 mg of digestion mixture (K_2SO_4:$CuSO_4$:selenium in the ratio 5:1:0.1) are added.
3. The flask is kept for digestion till its contents become colourless. The digested material is then transferred to the Kjeldahl distillation apparatus.
4. 20 ml of 40% NaOH solution is added to the digested contents and the flask is heated. The open end of the condenser is dipped in a flask containing 10 ml of boric acid mixed with indicators such as bromocresol green and methyl red.
5. The process of distillation is continued for 10 min and then titration is carried out against a standard of $0.1N$ H_2SO_4.

Calculation

Percentage of nitrogen = (sample titre in ml – blank titre in ml)

$$\times \frac{(N)H_2SO_4 \times 0.014}{\text{sample wt (g)}} \times 100$$

Estimation of Inorganic Nitrogen (MACS 1986)

Procedure

1. 10 g of the sample is accurately weighed in a watch glass and transferred into the flask of a steam distillation assembly. The volume is increased to 25 ml with distilled water.
2. 0.1 ml of phenolphthalein indicator and sufficient quantity of sodium hydroxide–sodium thiosulphate reagents are added till the liquid in the flask becomes pale pink when it is distilled.

3. 200 ml of this distillate is collected in a flask containing 25 ml of $0.02N$ H_2SO_4 and two drops of methyl red indicator. The tip of the condenser is submerged in standard H_2SO_4 solution during the collection of the distillate.

4. The quantity of H_2SO_4 that has not been neutralized is determined by titration with $0.02N$ NaOH.

Calculation

Weight of the watch glass = W_1

Weight of the watch glass + sample = W_2

Volume of $0.02N$ NaOH required to neutralize the residual $0.02N$ H_2SO_4 = W_3

Percentage of inorganic N_2 in the sample

$$= \frac{(25 - W_3) \times \text{strength of } H_2SO_4 \ (N) \times 0.014}{(W_2 - W_1)} \times 100$$

Estimation of Organic Nitrogen

The content of organic nitrogen in the sample is calculated by subtracting the value of inorganic nitrogen content from the total nitrogen content.

Estimation of Organic Carbon—Walkely and Black Method (Allison 1965)

Procedure

1. 0.05 g of dry sample is taken in a 500-ml wide-mouthed flask. 10 ml of $1N$ $K_2Cr_2O_7$ is added and the flask is swirled gently. To this 20 ml of concentrated H_2SO_4 is rapidly added; the solution is vigorously mixed and kept aside for 30 min.

2. 200 ml of distilled water is added to the solution. It is titrated against $0.5N$ $FeSO_4.7H_2O$ with ferroin indicator. The colour changes from cast green to dark green. At this point $FeSO_4$ is added very carefully, drop by drop, until the colour changes from blue to maroon red.

3. A blank is also run with the reagents.

4. The percentage of organic carbon content in the raw material is calculated as follows

Calculation

Percentage of organic carbon

$$= \left[\frac{\text{milli equiv. } K_2Cr_2O_7 - \text{milli equiv. } FeSO_4}{\text{sample (g)}} \right] \times 0.003 \times F \times 100$$

where F = correction factor = 1.33.

Estimation of Soluble/Reducing Sugars

Soluble sugars in fermented slurry are found in both reducing and non-reducing forms. Treatment with 10% HCl (v/v) is required to convert non-reducing soluble sugars into reducing sugars, which can then be estimated calorimetrically using the Folin–Wu (Kolmer et al. 1969) method in terms of glucose estimation as explained here.

Procedure

1. To 10 ml of the filtrate, which has been collected during the estimation of the soluble fraction, concentrated HCl is added in 1:9 proportion (i.e., 1 ml concentrated HCl and 9 ml filtrate). This is heated for 1 hour in a boiling water bath so that the non-reducing sugars get converted to reducing sugars.
2. Next the sample is neutralized to pH 7 with $10N$ NaOH using pH indicator paper and the volume is then increased to 100 ml.
3. 2 ml of the diluted sample is taken in a tube, to which 2 ml of copper sulphate reagent is added. This tube is kept in a boiling water bath for 8 min and then cooled to room temperature, without shaking.
4. To this solution 2 ml of phosphomolybdic acid reagent is added. This is then diluted to 25 ml, mixed well, and the optical density value measured at 660 nm.
5. A standard curve of glucose is prepared by taking glucose solution of different concentrations such as 10 µl, 20 µl, 40 µl, 60 µl, 80 µl, and 100 µl from a stock glucose solution of 10 mg/ml.
6. The same procedure (from step 3) is followed taking 2 ml of unknown sample and the value of the unknown sample is deduced from the standard curve.

Calculation

1 µg glucose/ml from the graph is equal to 1 mg of glucose in the aliquot of the dry sample taken for the estimation of the water-soluble fraction.

Estimation of Potassium

Procedure

1. A known amount of sample is taken in a round bottom flask.
2. 1 ml of HNO_3 and 5 ml of trisodium cobalt–nitrate solution are added and the solution is kept for 2 hours. Then it is centrifuged.
3. The precipitate thus formed is washed with about 15 ml of $0.01N$ HNO_3 and further centrifuged.
4. To the precipitate, again, 10 ml of $K_2Cr_2O_7$ solution and 5 ml of concentrated H_2SO_4 are added. The solution is cooled and the volume is increased to 100 ml.

5. The optical density is measured at 425 nm.
6. A standard curve is plotted and the potassium content of the unknown sample is calculated from the curve.

Estimation of Phosphate

Procedure

1. About 0.01 g of the dry sample is taken and 50 ml of di-acid mixture, containing nitric acid and perchloric acid in the ratio of 3:1, is added. Sufficient time is allowed for digestion of the sample.
2. To the digested contents, 2 ml of ammonium molybdate reagent and a few drops of stannous chloride are added and the solution is mixed thoroughly. The mixture is kept aside for 10 min.
3. The optical density is measured photometrically at 690 nm.
4. Distilled water is taken as the blank.
5. A calibration curve for phosphate is prepared and the phosphate content of the unknown sample is calculated therefrom.

Determination of Sodium and Potassium by Atomic Absorption Spectrophotometry

For determination of Na and K by the atomic absorption spectrophotometry (AAS) method, 5 ml of the sample is digested with 25 ml of nitric acid and perchloric acid (9:4) for an hour, followed by filtration through a Whatman 42 filter paper. The final volume is increased to 100 ml with distilled water. The element content of the digested solution can be determined by AAS.

Estimation of Colouring Matter

The extraction of the colouring matter is done in an alcohol–benzene mixture (1:2 v/v) in a Soxhlet extractor.

Procedure

1. A pre-weighed thimble containing 2 g of dry sample is placed in the Soxhlet apparatus.
2. The colouring matter is extracted in alcohol–benzene mixture at 80°C till no further colour appears in the extraction mixture.
3. The thimble is taken out from the Soxhlet apparatus and dried in a hot air oven at 103°C ± 2°C.

Calculation

Weight of the thimble = W_1

Weight of the thimble + dry sample before extraction = W_2

Weight of the thimble + dry sample after extraction = W_3

$$\text{Percentage of colouring matter in the dry sample} = \frac{W_2 - W_3}{W_2 - W_1} \times 100$$

The decolourized sample can be retained for the estimation of sugars and starch.

Estimation of Water-soluble Fraction

Procedure

1. Exactly 3 g of the decolourized sample is taken in a pre-weighed beaker and dissolved in 50 ml of distilled water at room temperature.
2. The sample is filtered through a pre-weighed Whatman 1 filter paper and the residue is washed several times with 10-ml aliquots of distilled water.
3. The volume is then increased to 100 ml.
4. The residue on the filter paper is dried in an oven at $103°C \pm 2°C$.

Calculation

Weight of the beaker = W_1

Weight of the beaker + decolourized sample = W_2

Weight of the filter paper = W_3

Weight of the filter paper + undissolved matter = W_4

Percentage of soluble fraction in the decolourized sample

$$= \frac{(W_2 - W_1) - (W_4 - W_3)}{W_2 - W_1} \times 100$$

Estimation of Volatile Fatty Acid Concentration

This is an indirect method of estimation. In this method, the volatile acid concentration in both the raw feed and the digested residue are required to assay the extent of degradation. The gas chromatographic method is employed for detection and estimation of the different gases, alcohol, and volatile fatty acids produced during the process of anaerobic digestion. In the event of non-availability of a gas chromatograph, the volatile acid concentration is estimated by the following procedure, where it is assumed that 70% of the volatile acids will be found in the distillate.

Procedure

1. 50 ml of leachate is taken in a 500-ml round bottom flask. 50 ml of distilled water is added to it followed by 2.5 ml of concentrated H_2SO_4.
2. The contents are mixed properly by gently shaking the flask.
3. The flask is heated and about 70 to 75 ml of the distillate is collected at the rate of 5 ml/min.
4. 25 ml of this distillate is then taken in a conical flask. 2 to 3 drops of phenolphthalein indicator are added to the distillate and it is titrated with $0.1N$ NaOH solution. (In the case of lower concentrations of volatile fatty acid in the distillate, lower concentrations of NaOH solutions are used.) The end point is indicated by a light pink colouration.

Calculation

The volatile acid concentration is calculated by the following equation:

$$\text{mg/ml of volatile acid} = \frac{\text{NaOH titre (ml)} \times \text{strength of NaOH (N)} \times 60,000}{\text{volume of sample (ml)} \times 0.7}$$

Gas Analysis

The analysis of gas produced from the biogas plant is necessary to know the constituents present in the gas and, in particular, the concentration of CO_2 and CH_4. Besides chromatographic analysis, gas can be analysed in an Orsat gas analyser.

Estimation of Gas Using Orsat Gas Analyser

With the Orsat gas analyser, a gas mixture containing CO_2, CO, and O_2 can be analysed. In this apparatus, there are three cylinders containing 50% potassium hydroxide (KOH) solution, alkaline pyrogallol, and ammoniacal cuprous chloride for absorbing CO_2, O_2, and CO, respectively. The gas produced in the biomethanation process contains CO_2 in large quantities, apart from CH_4. In the Orsat gas analyser CH_4 is estimated by subtracting the CO_2 content from the total gas mixture of CO_2 and CH_4.

Procedure

1. The levels of KOH, alkaline pyrogallol, and ammoniacal cuprous chloride are adjusted in the three cylinders. The glass stoppers over the pipettes are closed. The indicator liquid in the burette is taken up to 100 ml by expelling the air out of the burette. The gas sampling bottle is connected to the analyser and suction is created with the help of the suction bottle for easy entry of the gas, so that the indicator liquid is displayed from the zero mark.

2. To begin with, the glass stopper of the connecting bottle of the first pipette containing 50% KOH solution is opened, while the stopper connecting the sample bottle with the analyser is closed. The gas is allowed to go in and out of the pipette using the levelling bottle and the burette reading is recorded about 12 times. The CO_2 present in the gas gets absorbed in the solution with this method.
3. Now the levelling bottle is lowered, the liquid is allowed to go inside the pipette and the initial level is adjusted. The stopper is closed. The gas volume is measured by adjusting the liquid level inside the burette. The rise of indicator solution to this level gives the volume fraction of CO_2 absorbed.
4. The same procedure is followed for the estimation of O_2 and CO contents in the gas by using the other pipettes.

Calculation

The methane content of the gas is calculated by using the following formula:

$$\text{Percentage of } CH_4 = \frac{\text{total gas taken in burette} - (CO_2 + O_2 + CO) \times 100}{\text{total gas taken in the burette}}$$

Gas Chromatography (GC)

Gas chromatography (GC) is an analytical technique employed for separating compounds, based primarily on their volatilities. This technique can also be used to quantify methane and carbon dioxide in biogas, and volatile fatty acids in leachate. Gas chromatography provides both qualitative and quantitative information for individual compounds present in a sample.

Each compound moves through a GC column in the gaseous state, either because the compound is normally a gas or it can be heated and vapourized

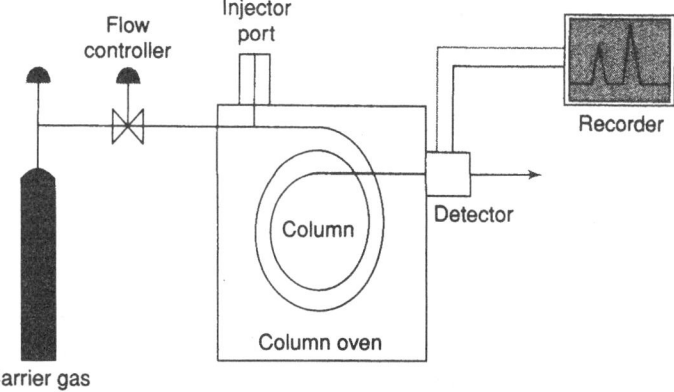

Schematic diagram of gas chromatograph

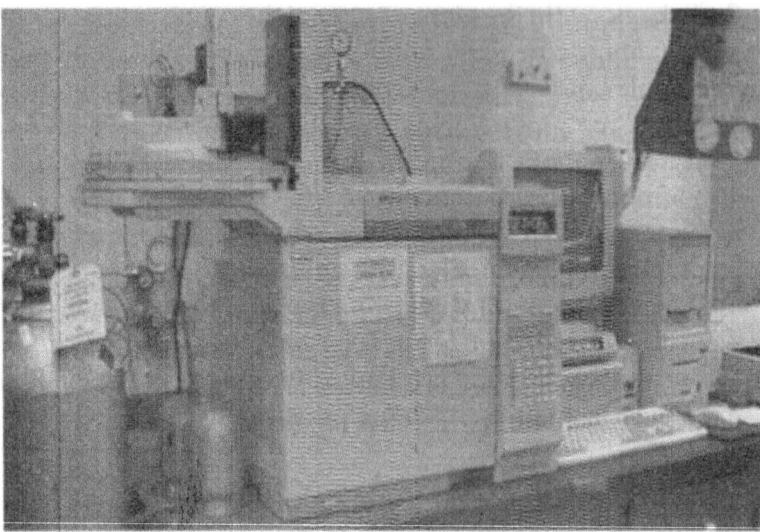

Gas chromatograph

into the gaseous state. The different compounds get separated into a stationary phase, which can be either solid or liquid, and a mobile phase (gas). The differential partitioning into the stationary phase allows the compounds to be separated in time and space. The carrier gas is usually inert, e.g., helium, argon or nitrogen. It serves as the mobile phase that moves the sample through the column. The flow of the carrier gas can be quantified by either linear velocity, expressed in cm/sec, or volumetric flow rate, expressed in ml/min.

The sample is introduced into the GC through the injector, which is a hollow, heated, glass-lined cylinder. The temperature of the injector is controlled in such a manner that all the components present in the sample are vapourized. The GC column is the heart of the system. It is coated with a stationary phase, which greatly influences the separation of the compounds. Some typical stationary phases are large molecular weight compounds such as polysiloxane, polyethylene glycol, or polyester polymers of 0.1 to 2.5 µm film thickness. As the compounds come off the column, they enter a detector. The compound and the detector interact to generate a signal. There are several different types of detectors (mainly electron capture detector, flame ionization detector, thermal capture detector) that can be employed, depending on the nature of compounds being analysed. These detectors can measure from 10^{-15} to 10^{-6} g of a single component. The data recorder plots the signal from the detector over a period of time. This plot is called a *chromatogram*. The retention time, which is when the component elutes from the GC system, is qualitatively indicative of the type of compound. The data recorder also has an integrator component to calculate the area under the peaks or the height of the peak. This area or height is indicative of the amount of each component.

Estimation of Chemical Oxygen Demand

The chemical oxygen demand (COD) is a measure of the oxygen, equivalent of the organic content of a sample, that is susceptible to oxidation by a strong chemical oxidant. COD can be related empirically to the BOD (biological oxygen demand), organic carbon, or organic matter (MACS-DNES 1986), in the case of samples obtained from a known and specific source.

One of the common methods used for COD estimation is the *dichromate open reflux* method. It has a wide application because of its superior oxidizing ability and ease of operation. The extent of oxidation of most organic compounds is 95% to 100% of the theoretical value. Most organic matter is oxidized by a boiling mixture of chromic and sulphuric acid. Then the sample is refluxed in excess amount of potassium dichromate ($K_2Cr_2O_7$). After digestion, the residual non-reduced $K_2Cr_2O_7$ is titrated with ferrous ammonium sulphate [$Fe(NH_4)_2(SO_4)_2$] and the amount of $K_2Cr_2O_7$ consumed is calculated. The oxidizable organic matter is calculated in terms of its oxygen equivalent.

It has been observed that halides and nitrites may interfere in the estimation of COD. Interference by halides can be largely overcome by complexing the sample with mercurous sulphate ($HgSO_4$) before refluxing, whereas by adding sulphamic acid before refluxing, nitrite (NO_2^-) interference can be eliminated. Generally, addition of 10 mg of $HgSO_4$ per 1 mg Cl^- and 10 mg of sulphamic acid per 1 mg NO_2^- are recommended.

Apparatus

    500-ml flask with ground glass joints
    Condenser with ground glass joints
    Hot plate/gas burner

Procedure

1. 25 ml of standard $K_2Cr_2O_7$ solution is placed in a reflux flask.
2. To this, 50 ml of the sample, 1 g of $HgSO_4$, and several glass beads are added. 75 ml of concentrated H_2SO_4 is added to this mixture with constant mixing. (Samples having a COD of more than 900 mg/L are diluted appropriately to bring the COD down to less than 900 mg/L.)
3. The flask is then attached to the condenser and the mixture is refluxed for 2 hours and cooled.
4. The condenser is washed with distilled water and the entire volume is titrated against FAS (ferrous ammonium sulphate) using 2 to 3 drops of

ferroin indicator. The end point is indicated by the sharp change of colour from blue-green to reddish-brown.

5. The blank titrate value is obtained from $K_2Cr_2O_7$ solution, H_2SO_4, and a volume of distilled water equal to the sample volume (without the addition of sample).

Calculation

$$\text{COD (mg/L)} = \frac{(x-y)A \times 8000}{\text{sample (ml)}}$$

where x is the amount of FAS used for the sample (ml), y is the amount of FAS used for the blank (ml), and A is the molarity of FAS.

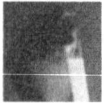

Estimation of Biological Oxygen Demand

Biological oxygen demand (BOD) is the measure of the process that determines the relative oxygen requirement of wastewater, effluents, and pollutant water. The test measures the oxygen required for biological degradation of organic material.

The sample is placed in a completely airtight bottle and incubated under specified conditions for a specific duration. This is followed by the measurement of the amount of dissolved oxygen (DO) in the water. The difference in the DO values between the initial sample, before incubation, and the after incubation sample, helps to calculate the BOD value.

As most wastewaters contain more oxygen-demanding material compared to air-saturated water, it is necessary to appropriately balance the O_2 demand and its supply by dilution. Since micro-organisms require other ingredients such as nitrogen, phosphorous, and trace metals as nutrients, these are added to the diluted water, which is buffered to ensure that the pH stability of the sample is maintained for the proper growth of bacteria. A 5-day period of incubation has been accepted as the standard incubation period for all practical purposes.

Apparatus

Incubation bottles of 250–300 ml capacity with ground glass stoppers and water seal.
Incubator adjusted at $20°C \pm 1°C$.

Procedure

Sample preparation The desired amount of distilled water is taken in a bottle and from the different nutrient solutions, 1 ml each of phosphate buffer, $MgSO_4$, $CaCl_2$, and $FeCl_3$ solution/L are added. This mixture is vigorously aerated for 1 hour. The pH of the sample is adjusted between 6.5 and 7.5 by the addition of concentrated H_2SO_4 or NaOH solutions. The dilution factor of the sample is maintained between 10^{-1} and 10^{-4}. During dilution bubbling is avoided. Four BOD bottles are filled as follows:

Aerated distilled water	B_1
Aerated distilled water	B_5
Dilute sample	E_1
Dilute sample	E_5

The open ends of bottles B_5 and E_5 are tightly closed and water sealed, and the bottles are incubated for 5 days at 20°C. The initial DO value in bottles B_1 and E_1 is determined by the following method.

DO estimation   1 ml of $MnSO_4$ solution is added to the bottle followed by the addition of 1 ml alkali-iodide-azide reagent. The open end is properly closed with the stopper to eliminate air bubbles and its contents are mixed by inverting the bottle a few times. The precipitate is allowed to settle. 1 ml of concentrated H_2SO_4 is added and mixed thoroughly. Exactly 200 ml of the contents from the bottle is titrated with $0.002M$ $Na_2S_2O_3$ solution until a pale yellow colour appears when a few drops of starch solution are added, and further titration is carried out till the blue colour starts to disappear.

Calculation

DO for titration of 200-ml sample:

1 ml of $0.002M$ $Na_2S_2O_3$ = 1 mg DO/L

BOD mg/L = $[(E_1 - E_5) - (B_1 - B_5)]f$

where E_1 is the DO in bottle E_1, E_5 is the DO in bottle E_5, B_1 is the DO in bottle B_1, B_5 is the DO in bottle B_5, and f is the dilution factor.

Estimation of Alkalinity

Alkalinity is determined following the standard method (APHA-AWWA-WPCF 1976). In this method, 50 ml of the sample is titrated potentiometrically using a commercial pH meter with standard $0.1N$ H_2SO_4 to an end point of pH 3.7. The alkalinity is expressed as mg $CaCO_3$ per litre.

High-pressure Liquid Chromatography

High-pressure liquid chromatography (HPLC) was developed in the mid-1970s and quickly improved with the development of column packing materials and the additional convenience of on-line detectors. In the late 1970s, new methods including reverse phase liquid chromatography allowed for better separation between very simple compounds. By the 1980s HPLC was commonly used for the separation of chemical and biochemical compounds. HPLC is mainly used as an analytical technique in biotechnological, biomedical, and biochemical research, and pharmaceutical industries. However, currently HPLC is also being used in a variety of other fields such as cosmetics, energy, food, and environmental sectors.

Different compounds, based on their chemical or molecular properties, can be separated through the specific stationary and mobile phases of the HPLC.

The *mobile phase* in HPLC refers to the solvent that is continuously applied to the column or the stationary phase. The mobile phase acts as a carrier of the sample solution. A sample solution is injected into the mobile phase of an assay through the injector port. As the sample solution flows through the column along with the mobile phase, the components of the solution migrate according to the non-covalent interactions of the compound taking place with the column. The chemical interactions occurring between the column and the mobile-sample phase determine the degree of migration and separation of components contained in the sample. There are several types of mobile phases such as isocratic, gradient, and polytypic.

The *stationary phase* in HPLC refers to the solid support contained within the column over which the mobile phase continuously flows. As the sample solution flows with the mobile phase through the stationary phase, the components of the sample will migrate according to their non-covalent interactions and the stationary phase. The differential distribution of components between the stationary-sample phase and the mobile phase determine the degree of migration and separation of the components contained in the sample. Columns containing various types of stationary phases are commercially available.

Procedure

1. The sample is injected into the HPLC via an injection port. The injection port of an HPLC commonly consists of an injection valve and the sample loop. The sample is typically dissolved in the mobile phase before injection into the sample loop. The sample is then drawn into a syringe and injected into the loop via the injection valve. A rotation of the valve rotor closes the valve and opens the loop in order to inject the sample into the stream of the mobile phase. Loop volumes can range between 10 µl to over 500 µl. In modern HPLC systems, the sample injection is typically automated.

2. There are several types of pumps available for use with HPLC analysis. Some of these are the reciprocating piston pumps, syringe-type pumps, and constant pressure pumps. The *detector* for an HPLC is a component that emits a response due to the eluting sample compound and subsequently signals a peak on the chromatogram. It is positioned immediately posterior to the stationary phase, in order to detect the compounds as they elute from the column. There are many types of detectors that can be used with HPLC. Some of the more common detectors include refractive index (RI), ultraviolet (UV), fluorescent, radiochemical, electrochemical, near-infrared (near-IR), mass spectroscopy (MS), nuclear magnetic resonance (NMR), and light scattering (LS).

Detection of Bacteria

Different techniques are employed to detect varied types of bacteria. Some of them are enumerated here.

Negative Staining of Bacteria

In this technique the background takes the stain, hence the name *negative staining*. Nigrosine or Indian ink is used for this type of analysis. The advantage of negative staining is that heat fixing can be avoided and the capsulated bacteria, which are difficult to stain, can be observed.

Material required

> Nigrosine solution
> Cover slips
> Slides
> Microscope
> Inoculation loop
> Bunsen burner
> Biological sample: bacteria

Procedure

1. A clean and dry slide is taken.
2. One drop of nigrosine dye is placed at one end of the slide.
3. To this a drop of water is added.
4. Then a loopful of bacteria are placed on top of this mixture and mixed well. Precaution is exercised to ensure that the slide is not heated to fix the bacteria when the technique of negative staining is used.
5. With the help of another slide, a drop of nigrosine is spread in such a way that the dye gets uniformly distributed on the slide.
6. The smear is allowed to air dry.
7. After this the slide is observed under the microscope.

Inference   Transparent bacteria are seen against a blue background.

Simple Staining

In simple staining, bacteria are directly stained. This technique helps to determine the shape, size, and arrangement of bacterial cells. In this method crystal violet, carbol fuchsin, or methylene blue stains are used.

Material required

> Glass slides
> Bunsen burner/spirit lamp

Blotting paper
Tissue paper
Inoculation needle
Staining tray
Methylene blue
Biological sample: 24-hour-old grown bacterial culture

Procedure

1. A clean and dry slide is taken.
2. A loopful of bacterial smear is fixed onto the slide by heating it.
3. Five drops of methylene blue dye are added to the bacterial smear and the preparation is kept aside for 50–120 sec.
4. Excess stain is poured off and the slide, containing the bacterial smear, is very slowly washed in running tap water to remove the excess stain.
5. The slide is dried by using blotting paper (it is not wrapped as that might disturb the smear).
6. The top side of the slide is soaked with tissue paper without hampering the smear.
7. The prepared slide is observed under the microscope.

Inference   Deep blue colour stained bacteria are seen under the microscope.

Bacterial Spore Staining

Some bacteria are capable of forming dormant structures inside the cells, called *endospores*, which are remarkably resistant to heat, radiation, chemicals, and other agents, having all metabolic activities. Such spores have a high calcium and dipicoline acid levels. A single bacterium forms a single spore by the process of sporulation. It has a very thick cell spore coat because of which it can withstand the adverse conditions. The endospore has a characteristic position within the cell, i.e., either central, sub-terminal, or terminal. Endospores can remain dormant for long periods of time and sometimes for a few thousand years.

Different staining procedures are used to detect the presence of an endospore in bacteria. One of the most popular techniques is the staining of the spore wall with the help of dyes such as aqueous primary stain (malachite green), which is applied and steamed to enhance penetration into the impermeable spore coat. Once stained, the endospores do not readily decolourize and appear green within the red bacterial cells.

Material required

Spore-forming bacterial species such as *Bacillus cereus*
Glass slides
Inoculating loop

Blotting paper
Spirit lamp
Malachite green (5% aqueous)
Safranin (0.5% aqueous)
Microscope

Procedure

1. A heat-fixed smear of *B. subtilis* is made on a clean and dry slide.
2. To this smear malachite green is applied in sufficient amount.
3. The slide is steamed for 5 min by adding more stain to the smear.
4. The slide is then washed slowly under running tap water.
5. The material is counter-stained with safranin for 30 sec.
6. The smear is again washed with distilled water.
7. The slide is blotted dry with absorbent/blotting paper.

Inference   The slide is examined through a microscope under an oil immersion objective. The position and size of endospores within the individual cells as well as the size and shape of free spores are clearly visible. In *Bacillus*, the endospores stain green and the vegetative cells stain red.

Viability Staining Technique for Bacteria

Viability staining is carried out in order to quantify the actual number of living and dead cells in a culture medium. The composition of the living cells can be distinguished from that of dead cells due to the fact that some reactions keep taking place in living cells, which results in these cells giving a different stain from that of dead cells.

Material required

Bacillus culture
Glass slides
Bunsen burner
Blotting paper
Inoculation needle
Loeffler's methylene blue solution
Dilute carbol fuchsin solution
Microscope

Procedure

1. A bacterial smear is made by taking a loopful of bacteria on a clean and dry slide and heat-fixed.
2. The smear is flooded with excess quantity of methylene blue for a duration of 10 min.

3. The excess stain is washed off thoroughly with tap water until the smear appears pale blue in appearance.
4. Carbol fuchsin is added to the slide and retained for few seconds. This is again followed by washing the slide under tap water to remove excess stain and dry the slide.
5. The smear is examined under the microscope.

Inference   Living bacteria appear purple, while dead bacteria stain red or pink. Living spores stain faintly pink and dead spores stain blue.

IMViC Tests

IMViC tests are in fact a combination of the following four different tests.

(i) Indole production
(ii) Methyl red
(iii) Voges–Proskauer
(iv) Citrate utilization

The name IMViC stands for the first letter of the name of each test in the series, with the lower case 'i' included for ease of pronunciation. The IMViC tests

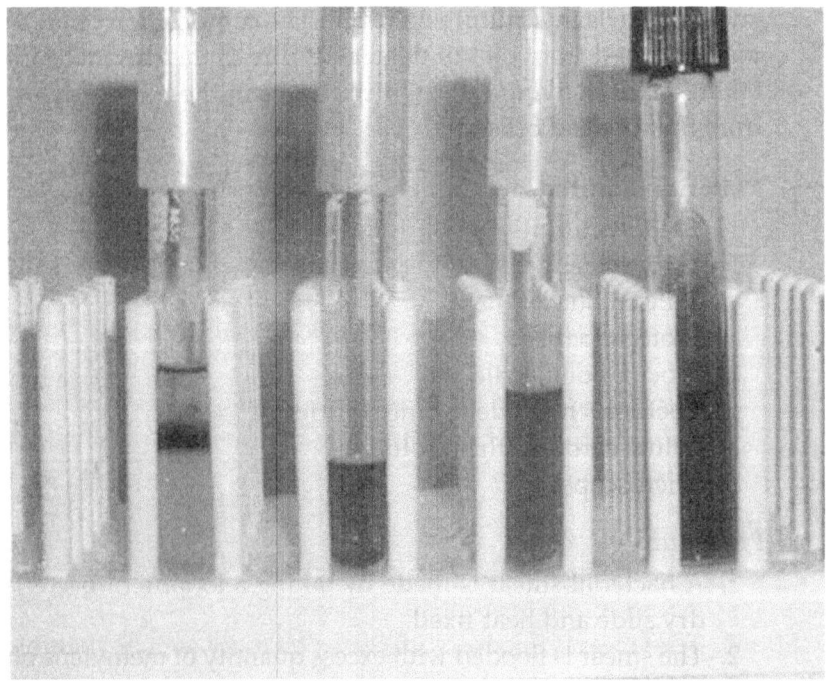

IMViC test

were designed to differentiate gram-negative intestinal *Bacilli* (family *Entero-bacteriaceae*), particularly *Escherichia coil* and the *Enterobacter-Klebsiella* group, on the basis of their biochemical properties and enzymatic reactions in the presence of specific substrates.

Indole production test

Indole is produced when bacteria are inoculated into tryptone broth. During the reaction, Kovac's reagent (dimethylaminobenzaldehyde) is added, which gives a cherry red colour. Some bacteria oxidize tryptophan, an essential amino acid, by using the enzyme trytophanase, resulting in the formation of indole, pyruvic acid, and ammonia.

The aim of this test is to determine the amount of indole produced as a result of microbial catabolism of tryptophan.

$$\text{Tryptophan} \xrightarrow{\text{Tryptophanase}} \text{indole} + \text{pyruvic acid} + NH_3$$

$$\text{Indole} + \text{Kovac's reagent} \xrightarrow{\text{HCl}} \text{rosindole dye} + H_2O$$
$$\text{(cherry red compound)}$$

Material required

Nutrient broth
Bacterial strains
Inoculating needle
Tubes containing 1% tryptone broth
5-ml test tube

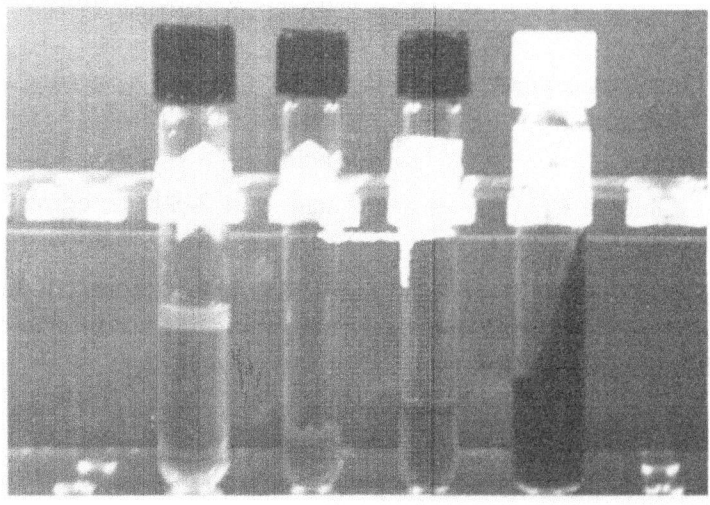

IMViC test

Kovac's reagent
Dropper bottle
1-ml pipette

Procedure

1. *Tryptone broth preparation* 10 g of peptone is dissolved in 1 L of distilled water and autoclaved.
2. The tube containing tryptone broth is inoculated with *E. coli*, another with *E. aerogenes*, and the third tube is kept uninoculated and treated as a control (tryptone should always be freshly prepared before use).
3. The three tubes are now kept for incubation at 35°C for 48 hours (the control tube is not incubated for more than 48 hours, as indole is auto-degraded resulting in erroneous results).
4. After incubation, 1 ml of Kovac's reagent is added to each tube including the control.
5. The tubes are gently shaken at regular intervals for 10–15 min for the development of cherry red colour.
6. The tubes are allowed to stand to permit the regent to come to the surface.

Inference A deep cherry red colour appears in the top layer of the tubes, indicating that the sample is indole positive. The colouration proves that *E. coli* is indole positive, while *E. aerogenes* is an indole negative bacterium.

Methyl red test

The methyl red test is a qualitative test for the acid produced from the oxidation of glucose. It is used primarily to differentiate *E. coli* from *E. aerogenes*—*E. coli* produces large amounts of acid; *E. aerogenes* produces neutral or non-acidic end products.

$$\text{Glucose} + H_2O \longrightarrow \begin{array}{l} \text{lactic acid} \\ \text{acetic acid} + CO_2 + H_2 \\ \text{formic acid} \end{array}$$

$$\text{Glucose} + H_2O \longrightarrow \text{acetic acid} \longrightarrow \begin{array}{l} \text{2, 3-butanediol} \\ \text{acetylmethylcarbinol} \end{array} + CO_2 + H_2$$

To perform the test, the micro-organism is grown in an MR-VP (methyl red/ Voges–Proskauer) broth and methyl red indicator is added to culture to identify the amount of acid present.

Observation When the organism is *E. coli*, the *p*H of the medium, which turns red (positive), is 4. When the organism is *E. aerogenes*, i.e., when a much lower concentration of acid is present, the *p*H of the medium (which turns yellow) is 6.

Voges–Proskauer test

This test is performed simultaneously with the methyl red test. It determines the ability of a micro-organism to produce non-acid or neutral end products from the organic acids present, following glucose metabolism. The organism is grown in an MR-VP broth and Barritt's reagent is added to the culture.

Observation A pink/red colour indicates the presence of acetylmethylcarbinol (positive reaction).

Citrate test

Some micro-organisms use citrate as a carbon source for energy when glucose or lactose is not present. These organisms have two primary enzymes—citrate permease and citrase. Simmon's agar slants are used to perform this test, wherein bromothymol blue indicator is incorporated into the medium.

Observation If the culture grows (citrate positive, *p*H 7.6), the medium turns blue. If there is no growth, the medium is green (*p*H 6.9) and the culture is citrate negative.

Examination of Fungi

One of the most common and quick methods of studying any type of fungus is the lactophenol cotton blue staining technique. This method can be used for studying any type of fungal hyphae. To study the vesicular arbuscular mycorrhiza (VAM) structures in fungi, a different staining technique is employed. Both these methods are discussed here.

Lactophenol Cotton Blue Mounting of Fungi

Lactophenol cotton blue is the most popular technique for staining any fungal hyphae. It stains the fungal cytoplasm and provides a light blue background, against which the walls of hyphae can readily be seen. It contains four constituents: phenol, which serves as a fungicide; lactic acid, which acts as a clearing agent; cotton blue, which stains the cytoplasm of the fungus; and glycerin, which gives a semi-permanent preparation. For a quick and routine examination of almost all types of fungi, a clean slide with a drop of mounting fluid (lactophenol cotton blue), the unknown specimen, and a cover slip placed over the preparation are sufficient for microscopic examination.

Staining of Vesicular Arbuscular Mycorrhiza

Mycorrhiza is the symbiotic association of a fungus with roots and/or rhizomes of a plant. The fungi are called mycorrhizal fungi, which are found either as a

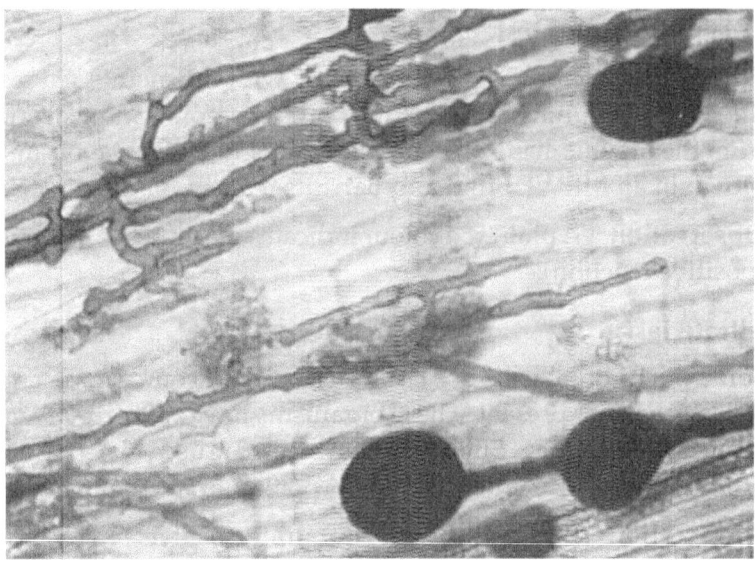

Arbuscular mycorrhiza

sheet of mycelium around the root or as intercellular penetrations into the root. Mycorrhiza can be classified as ectomycorrhizae, endomycorrhizae, and ectendomycorrhizae. The endomycorrhizae, most popularly known as VAM, are caused mainly by aseptate hyphae. These are members of endogonales, which form dichotomously branched haustoria within the host cells known as *arbuscles*. Besides this, fungal hyphae also bear a second type of swollen thick-walled structures called *vesicles*. Lactoglycerine or lactophenol staining spores are the standard mounting media for vesicular arbuscular mycorrhizal fungi (endogonales). Water melzer reagent is used for this study, zygospores (zygosporangia) appear deep orange whereas the spore and gametangial walls stain reddish-brown. Vesicles and internal hyphae can be seen clearly, particularly in squashed roots, while arbuscles can be seen clearly by adding one drop of lactofuchsin to the roots mounted on the slide.

Material required

    VAM-infected roots
    FAA (formalin–acetic acid–alcohol)
    10% KOH
    18% lactic acid
    Glycerin
    95% ethyl alcohol
    Chlorazol black E
    Chloral hydrate
    Basic fuchsin

Mounting fluid [chloral hydrate 20 g, gum arabic 20 g, glycerine 20 g, glucose syrup 3 ml, basic fuchsin 10 drops (0.3 g, 10 ml 95% ethyl alcohol), water 35 ml]

Autoclaved jars

Clean glass slides

Water bath at 90°C

Procedure

1. The VAM-infected roots are washed thoroughly and fixed overnight in FAA.
2. FAA is removed from the roots by washing it with tap water.
3. The sample is kept in KOH solution taken in autoclave-resistant jars.
4. The sample is made clear by autoclaving it at 15 psi for 15 min.
5. It is rinsed with tap water followed by deionized water.
6. The roots are stained by keeping them in the staining solution for 1 hour or a little longer at 90°C [made of equal volumes of 80% lactic acid, glycerin, and distilled water with 0.1% (w/v) chlorazol black E]. The staining solution is prepared several hours before use and is kept undisturbed to allow for the settling of undissolved particles.
7. The stained roots are kept in glycerin overnight for destaining.
8. The roots are mounted on clean glass slides using mounting fluid.

Inference The stained slide is observed under a bright-field microscope for the VAM structures.

Streak-plate Method

The streak-plate method offers the most practical method of obtaining discrete colonies and pure cultures. In this method, a loop of a suitably diluted suspension of organisms is streaked on the surface of an already solidified agar plate. The basic objective of this method is to continuously dilute the suspension present at the tip of the needle till it finally results in a single colony.

Material required

Biological sample

Nutrient agar plates

Inoculating loop

Bunsen burner

Procedure

1. The inoculating loop is sterilized by heating it over a flame after removing the cotton wool plug. As the plug is removed, the mouth of the tube is set aflame immediately.

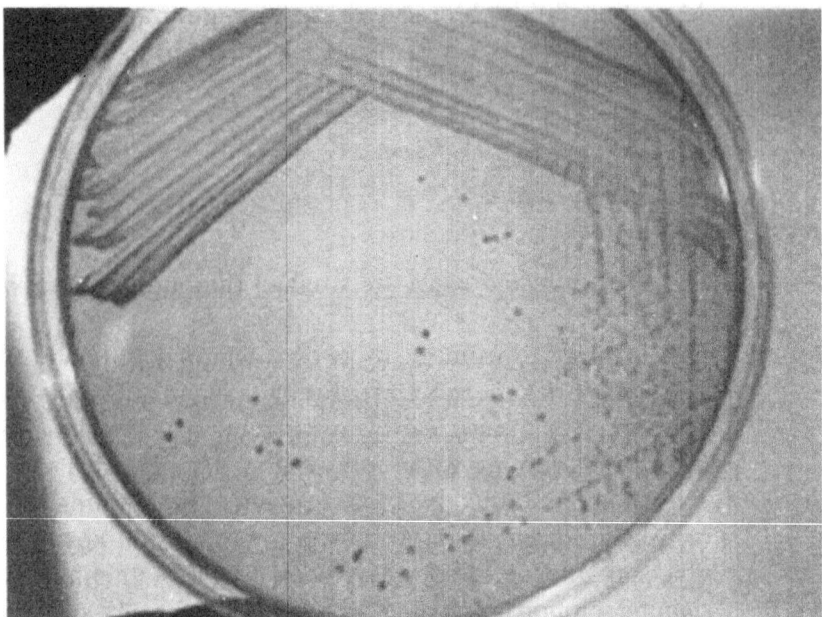

Streak-plate method

2. The inoculating loop is introduced into a broth that has a rich growth of the organism and one loopful of culture is withdrawn.
3. The mouth of the tube is immediately set aflame and plugged with cotton wool.
4. The cover of the petri plate is removed and held at an angle of 60°.
5. The end of the inoculation needle is touched on the agar surface at the edge farthest from the surface and the inoculum is streaked from side to side in parallel lines across the surface area.
6. The inoculation loop is re-ignited and cooled. The loop is touched to a corner of the culture and the inoculum is streaked across the agar. The rest of the agar surface is quickly covered by zigzag movements of the hand and the streaking is completed (the entire process is carried out very fast to avoid contamination).
7. The inoculated plate is kept for incubation for 48–72 hours to ensure proper growth of the micro-organisms.

Inference   A dense growth of bacterial colony is observed where the streak was initially made. The growth is less dense in the area away from the streak, and discrete colonies develop farthest away from the streak (i.e., towards the end of the streak). The single colony formed farthest from the initial inoculation is considered the purest colony and is further studied under the microscope.

Pour Plate Method

The starter culture is serially diluted with the liquid medium such as buffer, distilled, or deionized water. This is followed by the application of the culture into a sterile petri plate; it is poured with melted agar medium and cooled for solidification (42–45°C). After incubation, the plates are examined for the presence of individual colonies growing in the medium. Pour plates are also used as a means of determining the number of viable organisms in a liquid medium.

Material required

  Biological sample
  20 ml nutrient agar (or Czapek–Dox agar for fungi) deep tubes
  Empty sterile culture tube (with cotton plug)
  Sterile 9-ml water blanks
  Sterile 1-ml pipettes
  Test tube rack
  Water rack
  Bunsen burner
  Wax-marking pencil

Procedure

1. A clean and dry test tube is taken.
2. The spores of filamentous fungi are scraped to prepare a cells/spores suspension.
3. After mixing, the cotton plug is removed and 1 ml of the bacterial suspension is aseptically transferred from tube 1 to water blank tube 2 and the pipette returned to tube 1.
4. Tube 2 is shaken and 1 ml is transferred with a fresh sterile pipette from it to tube 3. The pipette is returned to tube 2.
5. Serial dilutions are made till the six water blanks (2 to 7) use fresh sterile pipettes each time.
6. One millilitre of the bacterial suspension from each of tubes 1 to 7 is transferred to petri plates 1 to 7 using respective pipettes.
7. A nutrient agar tube is removed from the water bath (at 45°C), the medium is poured into plate 1, and the plate is gently rotated to ensure uniform distribution of cells in the medium.
8. Step 7 is repeated for the addition of medium to plates 2 to 7.
9. The medium is allowed to solidify.
10. The inoculated plates are incubated for 24–48 hours at 37°C in an inverted position (lid at the bottom).

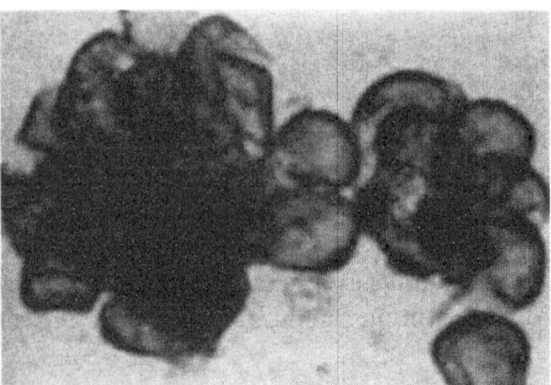

VAM fungi spores

11. The plates are examined for the appearance of individual colonies growing throughout the agar medium.

Inference It is observed that with progressing dilution, the poured plates have lesser and lesser number of colonies, which are more or less sparsely distributed in the plates. These may be transferred (subcultured) to other media (fresh plates) or agar slants (tubes containing solid media prepared by keeping the tubes tilted as the agar solidifies) for further study.

Precautions
1. Only freshly sterilized pipettes should be used for each dilution.
2. The medium poured in the petri plates should have a temperature of 45°C.
3. The suspension should be in continuous motion when it is being transferred from one tube to another to ensure uniform distribution of cells.
4. The plates need to be incubated in an inverted position in order to prevent condensation on the agar surface. Discrete surface colonies cannot be obtained unless the surface is dry.

Spread-plate Technique

In the spread-plate technique micro-organisms are spread over the solidified agar medium with a sterile L-shaped glass rod. The petri dish is placed on a turntable. As the bent glass rod moves on the rotating petri dish, cells are singly spread over the solid agar plate. Some of these cells get separated from each other and this allows for the colonies to develop.

Material required
Fully grown bacterial sample
Nutrient agar plates

L-shaped bent glass rod
95% alcohol
Beaker (50 ml)
Bunsen burner

Procedure

1. The L-shaped bent glass rod is surface sterilized by dipping it in 95% alcohol, taken in a coupling jar.
2. Bacteria from the matured culture are taken on an inoculating loop and aseptically transferred onto a nutrient agar plate.
3. The inoculated plate is transferred onto the turntable.
4. The L-shaped rod is surface sterilized in the Bunsen burner flame.
5. The rod is cooled for 10–15 sec.
6. The petri dish is uncovered and kept on the turntable.
7. The tip of the sterile bent rod is touched lightly to the agar surface and this movement is continued throughout the surface of the agar plate, maintaining proper sterilization.
8. The inoculated plates are incubated by placing them in an inverted position for 24–48 hours.

Inference It is observed that some discrete colonies appear on the surface of the agar plate. The morphology, form, elevation, pigmentation, and size of the formed colonies are studied.

Subculturing

The isolated colonies grown on the agar plate by the aforementioned streak-plate, pour plate, or spread-plate techniques are examined for the next step, i.e., subculture. Subculturing is a term used to describe the procedure of transferring micro-organisms from their parent growth source to another medium. This technique is also used routinely for preparing and maintaining stock cultures.

Material required

Biological material such as fully grown bacterial species
Nutrient agar slants or nutrient agar plates
Inoculating loop
Microscope

Procedure

1. The inoculating loop is surface sterilized by dipping it in 95% alcohol (taken in a coupling jar) and holding it to the Bunsen burner flame until the entire wire becomes red hot.
2. The loop is cooled for a few seconds by dipping it into a fresh plate (uninoculated).

3. The tip of the loop is touched to the discrete colony, agar streak-plate, or pour plate where the fully grown organisms are present.
4. The plug of the agar slants is removed and the neck of the tube is placed quickly over the Bunsen burner flame.
5. The plate is subcultured aseptically over the agar surface in a straight or zigzag line.
6. The inoculating needle is re-ignited by dipping it into alcohol and sterilizing it after use.
7. The inoculated plates are incubated by placing them in an inverted position for 24–48 hours.
8. After incubation, the plates or slants are studied for the growth of pure colonies.

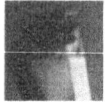

Anaerobic Culture Procedure

Micro-organisms can grow in a culture medium, provided the medium contains all the ingredients and environmental factors suitable for their growth. The medium generally contains carbon, nitrogen, vitamins, and some trace amount of mineral salts essential for the proper growth and metabolism of bacteria. The composition of the medium varies from one organism to another. The simplest culture procedure is the batch culture. In this culture, the micro-organisms grow on a limited quantity of medium, until either the nutrient component is exhausted or enough toxic by-products accumulate to inhibit growth. The techniques employed for the isolation and culture of the anaerobic bacteria are different from those used for aerobic bacteria. The basic difference lies in the removal of the last trace of oxygen from the medium. The most common method used for creating anaerobic conditions is the purging of an inert gas or nitrogen to eliminate oxygen contamination. Besides this, anaerobic conditions can be created by the following methods (MACS-DNES 1986).

Alkaline pyrogallol
Thioglycollic acid
Cysteine hydrochloride
Sodium sulphide
Semi-solid agar
Cooked meat medium
Gas mixture
Anaerobic jar
Gas-pack anaerobic jar
Anaerobic glove boxes

Some of these processes are discussed next.

Alkaline pyrogallol Alkaline solution of pyrogallol has the property of absorbing oxygen. The main advantage of this method is that it does not require any special equipment and the reagents are easily available. To create anaerobic conditions, a test tube or petri dish containing NaOH solution with pyrogallol is placed inside the culture jar and immediately sealed. During the process of O_2 absorption, CO is also formed, causing harm to micro-organisms. Further, excess amount of NaOH absorbs the CO_2 formed during anaerobic digestion, thereby giving erroneous results. This can, however, be overcome by replacing NaOH with CO_2.

Sodium sulphide This method is especially suitable for the cultivation of rumen anaerobes. Na_2S of 0.025% concentration is added to the culture medium to eliminate oxygen. However, in the case of many anaerobic bacteria, sodium sulphide with cystein chloride is also used.

Anaerobic jar Anaerobic jar is a cylindrical vessel made of materials such as stainless steel, good quality glass, or special grade plastic. The mouth of the vessel is generally flanged to provide for a large contact surface area for the lid, to ensure an airtight joint. To remove O_2, pure gases such as CO_2, H_2, N_2, or their mixtures are used. The residual O_2 is removed by a catalytic reaction between O_2 and H_2 to form H_2O. This reaction is accomplished in the presence of hot palladium as the catalyst. Before starting the experiment the capsulated palladium is activated by heating it over a flame and H_2 is supplied from outside. In a semi-solid agar medium, methylene blue is used as the redox indicator.

Cultivation of Anaerobes

Based on shape, micro-organisms have been classified into rod, round/cocci, or spiral bacteria. Similarly, based on the temperature tolerance capability, bacteria are grouped as psychrophilic, mesophilic, and thermophilic. Micro-organisms are also grouped based on their molecular oxygen uptake capabilities for cellular respiration. Respiration involves the oxidation of substrates to produce the energy necessary to survive. As we already know, organisms which require the presence of molecular oxygen for metabolism are aerobes and those that function both in the absence and in the presence of oxygen are facultative anaerobes. A third group appear to require oxygen in a less quantity, or too much of oxygen is detrimental to them. There is a fourth group of micro-organisms that utilize no molecular oxygen, and some may even get killed in its presence, called *obligate anaerobes*.

Some of the simple and commonly used techniques used to achieve the different types of oxygen-dependent and non-dependent conditions are listed here.

(a) Pyrogallic acid and sodium hydroxide are kept in the container for the absorption of oxygen, thereby creating the anaerobic environment in which the micro-organisms are incubated.

(b) In the shake-culture technique the molten and cooled (45°C) nutrient agar is inoculated with a loopful of culture. It has been observed that strict anaerobes grow in the deeper portions of the culture, facultative anaerobes grow throughout the tube, microaerophiles grow near the surface, and aerobes grow at the surface.

(c) The anaerobic jar is a special jar in which some kits are kept to absorb the oxygen present in the jar. Hydrogen or a mixture of other gasses ($95\% N_2 + 5\% CO_2$) can also be introduced after the jar is sealed.

Shake-culture Technique

Material required

24- to 48-hour thioglycollate broth culture of *Clostridium sporogenes* and *Escherichia coli*
Nutrient agar deep tubes
Inoculating loop
Bunsen burner

Procedure

1. The nutrient agar plate is poured and allowed to cool to room temperature.
2. The tubes are aseptically inoculated with each type of culture.
3. Each tube is gently shaken by striking it with fingers.
4. The tubes are cooled rapidly under running cold water to solidify the medium.
5. The tubes are incubated for 24 to 48 hours at 37°C.

Inference The growth of *C. sporogenes* is observed in the deeper portions of the tube and *E. coli* are found at the surface of agar.

Pyrogallic Acid–Sodium Hydroxide Method

Material required

24- to 48-hour thioglycollate broth cultures of *Clostridium sporogenes* and *E. coli*
Cotton plugged nutrient agar slants
Pyrogallic acid crystals
4% NaOH
Rubber stopper
Pasteur pipette
Glass rod
Inoculating rod
Bunsen burner
Glass-marking pencil

Procedure

1. The marked slants are inoculated aseptically, by streaking the agar surface.
2. Pyrogallic acid crystals are added to the space between the cotton and the top of the slant tube.
3. 2 ml of sodium hydroxide (4%) is added to the tube.
4. The rubber stopper is fitted tightly to the tube.
5. The tubes are incubated in an inverted position at 37°C for 24 to 48 hours.

Inference   The growth of *Clostridium sporogenes* is observed on the nutrient agar slant, while no growth is noticed in the *E. coli* inoculated slant. Thus it can be inferred that *C. sporogenes* is an anaerobic bacterium and *E. coli* is an aerobic bacterium.

Maintenance of Pure Cultures

Once a micro-organism is isolated in a pure form, it is subcultured on plates or agar slants (a tube containing solid medium prepared by keeping tube tilted as agar solidifies). The prepared cultures can be stored in a refrigerator, at a temperature of 4°C, to slow down the rate of growth and protect the culture from damage due to evaporation of the medium. The method also preserves the culture, which can then be subcultured routinely. To preserve micro-organisms, it is necessary to reduce their rate of metabolism to minimum.

Storage of Micro-organisms

Cultures can be successfully stored in refrigerators or cold rooms if the temperature is maintained at 4°C. Cultures can also be preserved for several years in glycerol at –40°C in a deep freezer, by transferring to a 0.5-ml ampoule, which can be placed in a mixture of industrial methylated spirit and carbon dioxide and frozen rapidly at –70°C, or by lyophilization.

The liquid nitrogen method is a technique in which micro-organisms are stored at low temperatures such as –196°C; these micro-organisms survive unchanged for long periods. In this method the cell suspension is preserved in the presence of a stabilizing agent such as glycerol or dimethyl sulphoxide (DMSO), which prevents the formation of ice crystals. Lyophilized cultures retain their viability for several years.

Control of Mites in Fungal Cultures by Fumigation

Mites feed on moulds and contaminate cultures by spreading from one culture tube to another tube, regardless of the presence of cotton plugs. Mite infection

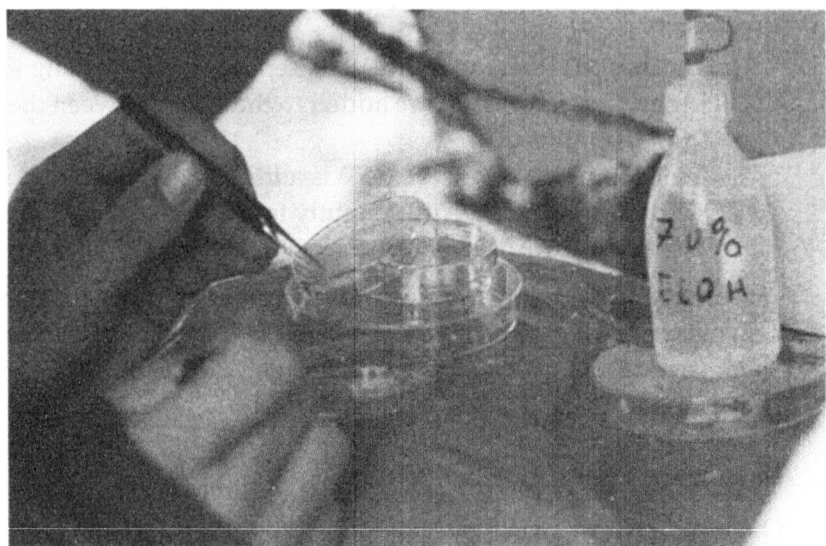

Fungal culture laboratory

can be prevented by adding 0.01% lindane powder, an insecticide, to the Sabouraud dextrose agar medium. This results in killing the mites within 3 min after they start crawling on the agar medium. Dry fumigation of the cultures with para-dichloro-benzene also results in the elimination of mites without adversely affecting the cultures.

Material required

Airtight box
Para-dichloro-benzene crystals
Fungal culture infested with mites
Transfer needle
Sabouraud agar plate
Bunsen burner

Procedure

1. The infested tubes are kept in an airtight box.
2. An insecticidal agent such as para-dichloro-benzene is added in sufficient amount, followed by the sealing of the box for at least 4 hours.
3. Para-dichloro-benzene crystals are added once again to the box after one week.

Inference The fumigated cultures are examined for the destruction of mites and the fertility of their eggs is checked from the culture by subculturing them on a fresh agar plate.

Bacteriological Examination of Water by Multiple Tube Fermentation

Coliform bacteria are indicators of faecal contamination in water. It is detected by using the multiple tube fermentation technique, which has three sequential stages, namely, presumptive, confirmed, and completed.

Coliform bacteria are an aerobic/facultative anaerobic, gram-negative, rod-shaped, and non-endospore-forming group of microorganisms. They ferment lactose, leading to the production of acid. They are gas-producers, and the gas is produced 24 hours after the start of the period of incubation, which is done at 37°C.

Presumptive Coliform Test

When a known volume of water is added to lactose fermentation tubes, it results in the production of acid and gas from coliforms. The lactose broth used has bile and lauryl sulphate or brilliant green added to it. Hence it is selected for the isolation of coliforms. A *p*H indicator is also added to the broth. The statistical method is used to estimate the population of coliforms and the result obtained is expressed as the *most probable number* (MPN) of coliforms. A count of the number of lactose fermentation tubes showing production of gas is matched with the respective value from the statistical table. This test is termed presumptive because the MacConkey broth used here may show positive results due to the presence of non-coliform organisms also.

Material required

Water sample
Six 9-ml single-strength lactose fermentation broth tubes
Three 20-ml double-strength lactose fermentation broth tubes
Sterile pipettes of 10 ml, 1 ml, and 0.1 ml capacity
Bunsen burner or spirit lamp

Procedure

1. Water sample from a pond or a sewage plant is taken and thoroughly shaken.
2. Three single-strength lactose broth tubes are labelled '0.1', the next set of three tubes as '1', and the last three double-strength broth tubes as '10'.
3. 10 ml of sterile water is added aseptically into each of the tubes marked '10'.
4. 0.1 ml of sterile water is added in '0.1' tubes.
5. 1 ml of sterile water is added to the tubes marked '1'.
6. These nine inoculated tubes are incubated at 37°C for 24–48 hours.
7. All the lactose fermentation tubes are examined for the production of yellow coloured acid and gas after 24 to 48 hours of incubation.

Coliform test

Inference

1. The production of acid and gas after 24 hours of incubation indicates a positive presumptive test for coliform bacteria.
2. If gas is produced even after 48 hours of incubation, then the presumptive test is doubtful.
3. If no gas is produced, then it shows a negative presumptive test (i.e., coliforms are absent).
4. The tubes showing a positive presumptive test are used for the confirmed test.

Confirmed Coliform Test

This test is done to confirm the presence of coliforms in water samples that give a positive or doubtful presumptive test. The medium used here is EMB (eosin methylene blue), which is selective in nature because of the presence of the dye methylene blue, which inhibits the growth of gram-positive bacteria while allowing for the growth of gram-negative bacteria. Lactose fermen-tation bacteria form coloured colonies due to the complex form of the EMB that is precipitated. Bacteria that do not ferment lactose produce colourless colonies.

Material required

EMB agar plate
24-hour-old positive lactose broth culture (for presumptive test)
Spirit lamp
Inoculating loop

Procedure

1. The EMB agar plate is inoculated with the 24-hour-old positive lactose broth culture with a sterile inoculating needle.
2. The plates are incubated at 37°C for 24–48 hours in an inverted position.
3. The inoculated plates are examined for the presence or absence of *E. coli* colonies.

Inference The appearance of typical coliform colonies (*E. coli*) with dark centres and metallic sheen indicates a positive coliform test.

Completed Coliform Test

This is the confirmatory test for the presence of *E. coli* in the water sample. Lactose-positive colonies are isolated and inoculated into a lactose broth tube and streaked onto a nutrient agar plate to perform Gram's staining. The production of acid and gas in the inoculated lactose broth and the presence of rod-shaped bacteria giving a gram-negative reaction confirm the presence of *E. coli* in the water sample.

Material required

Lactose fermentation broth tubes
Nutrient agar slant
24-hour-old coliform positive EMB agar culture
Gram's stain reagent
Inoculating loop
Bunsen burner or spirit lamp

Procedure

1. The lactose fermentation broth tubes are inoculated on an EMB agar plate using the inoculating loop.
2. Nutrient agar slants are streaked with colonies from the EMB agar plate with the help of the inoculating loop.
3. The inoculated broth and slants are incubated at 37°C for 24 hours.
4. The organisms on the slants are stained with Gram's stain.
5. The lactose fermentation tubes are examined for the production of acid and gas.
6. The slides are observed for positive or negative Gram reaction and cell morphology.

Inference The production of acid and gas from the inoculated lactose broth and the presence of gram-negative rods indicate a positive completed test. Hence the presence of coliform bacteria in the sample is confirmed.

Biomarkers

Biomarkers are defined as quantitative measures of changes in the biological system that respond to either (or both) exposure to, and/or doses of, xenobiotic substances that lead to biological effects. In most definitions, 'biomarker' or 'biomarker response' is often restricted to the changes that are measured at the molecular, biochemical, or physiological level in cells, body fluids, tissues, or organs and which indicate exposure to xenobiotic compounds. Biomarkers not only help to understand a bioprocess but also enable to predict the cause and outcome of any process.

The monitoring of biological effects has recently become an integral component of environmental assessment. Till date, the harmful effects of contamination in biological systems have been efficiently monitored by biomarkers. The purpose of biomarkers can be broadly divided into two major groups based on their performance.

1. Risk prediction
2. Disease prediction

The use of biomarkers in risk prediction, screening, as well as diagnostics is already well established. Despite the fact that they offer very prominent advantages, they do have some limitations.

Biomarkers are generally of three types—biochemical, immunochemical, and genetic indicators.

In any biological process detoxification of contaminants takes place through the enzymatic process, for which some genes are responsible at the molecular level. This enzymatic detoxification of xenobiotics is carried out by a group of hydrolytic enzymes such as esterases and amidases. However, the major reactions are catalysed by the oxido-reductase group of enzymes, which sometimes catalyse different complex reactions. For example, xenobiotic detoxification takes place through oxidation by Cytochrome P450 oxidase, resulting in the formation of Cytochrome P450 complex and hydroxylated products. Similarly, glutathione S transferase, in the presence of glutathione, forms a conjugant with xenobiotic compounds.

There are different types of antibodies developed against different antigens to trace the presence of xenobiotic compounds present as contaminants. The antibodies against some herbicides, insecticides, PCBs, and PCDFs (polychlorinated dibenzofurans) have already been reported, which are used for determining the presence of xenobiotic compounds in trace quantities through immunochemical techniques.

For genetic indicators, the *lux* gene can be considered as an appropriate example. With the development of genetically modified (GM) organisms, some efficient techniques have been applied to use them as indicator organisms for determining the level of contaminants. Several techniques are used for screening engineered organisms for gene expression. The most popular technique for the screening of gene expression is to fuse the relevant sequence of the promoter to the gene sequences having characteristics that can be used for identification either through some antibiotic, through heavy metal resistance, or by emitting light. The luciferase enzymes (lux AB) emit light through light-producing proteins and are used as markers when the *lux* gene is placed under the control of a promoter, associated with the response to toxic compounds.

Biosensors

A biosensor is an analytical device which converts a biological response into an electrical signal. Biosensors are often used to determine the concentration of substances and/or other predefined parameters. The two main elements of a biosensor are the biological recognition element and the signal transducer. Biosensors play an important role in pollution monitoring. The environmental applications of biosensors include groundwater monitoring, effluent monitoring at water treatment facilities, drinking water analysis, and the rapid analysis of extracts of soils and sediments at hazardous waste sites. The easy availability of enzymes, antibodies, and genetically engineered micro-organisms with specific interactions coupled with greater durability and sensitivity, and low cost of signal transducers for handling environmental pollutants has contributed to the recent interest of the scientific community in applying biosensors to environmental monitoring.

Biological recognition mainly covers three basic mechanisms: biocatalytic, bioaffinity, and microbe-based systems. For environmental application the enzyme-based reactions involve either the catalytic transformation of a pollutant (say, in very minor quantities), or the detection of pollutants that either inhibit or activate enzyme activity. Enzyme-based inhibition reactions mainly detect a large number of environmental pollutants, usually from a particular group of chemicals, even at very low concentrations. However, in some cases, biosensors need to be reactivated or the sensing element needs to be disposed of because of some enzyme–substrate reaction.

Environmental biosensors are currently being used to monitor chemical and organic pesticides. This is because pesticides typically function by means of interacting with specific biochemical targets as substrates (e.g., organophosphorus insecticides/organophosphate hydrolase), inhibitors (e.g., dithiocarbamate fungicides/aldehyde dehydrogenase, organophosphorus insecticides/acetylcholinesterase), or herbicides. Moreover, a wide variety of antibodies have been developed to counteract various classes of insecticides, herbicides, and fungicides such as triazines, alachlor, aldicarb, 2,4-D, and paraquat, which are commercially available.

For the detection of pesticides, two enzyme systems based on organophosphorus hydrolase and acetylcholinesterase have been reported; some advantages and limitations of these systems have been noted while applying these systems for field screening. Biosensors have also been developed for non-agricultural organics that fall under the broad spectrum of chemical classes for which the principle applied was the same as discussed earlier, i.e., it was either enzymes, antibodies, or micro-organisms based. With respect to environmental monitoring, one of the major challenges is that a large number of contaminants may be involved, from a variety of related and unrelated chemical classes. Among polyaromatic hydrocarbons, phenols have received particular attention, as they are often found at hazardous waste sites but are difficult to measure, even under laboratory conditions. There are other techniques available for identifying phenols in multianalyte mixtures including groundwater, soil, and sludge. Some researchers are working on the bioluminescence approach to detect phenols present in industrial wastewater. Fibre-optic biosensors for measuring multiple organic contaminants in groundwater have also been developed. The device uses a two-layer detection element immobilized on the tip of an optical fibre.

Biosensors meant for the detection of environmentally significant metal ions primarily use enzymes or genetically modified organisms as recognition elements. Some biosensors are composed of biological assays interfaced with various signal transducers to measure parameters such as micro-organism toxicity, enzyme inhibition, biological oxygen demand, and DNA damage, and identify and enumerate micro-organisms of environmental concern. In a number of cases, the interface of a biological assay with a signal transducer has been shown to reduce the time and complexity involved with these assays. Research is being continued for the development of biosensors for monitoring different pollutants efficiently, with a minimum time period and the formation of minimum amount of intermediates.

Summary

To effectively monitor the environment as well as devise pollution abatement methods, it is necessary to know the quality of the environment quantitatively. It is essential to be aware of the components responsible for environmental pollution and their concentration in wastewater and solid waste. In addition, the environment required for the growth of eco-friendly micro-organisms, which can be utilized to reduce the amount of pollutants in waste, is also required to be evaluated.

Waste material contains biological oxygen demand, chemical oxygen demand, total solids, volatile solids, biomass, cellulose, hemicellulose, lignin, carbon, nitrogen, phosphorus, sulphur, chromium, glucose, sugar, dissolved oxygen, etc. This chapter presents analytical methods to quantitatively estimate all these components. It also explains aerobic and anaerobic culture procedures as well as isolation and culture of micro-organisms.

Using this information, the level of residual pollutants can be evaluated vis-à-vis the prescribed minimum. The results of such quantitative analysis are further used for mass and energy balance, design kinetics of biodegradation processes, design of bioreactors and fermenters, etc., which will be elaborated upon in the subsequent chapters.

Review Questions

1. How can the total volatile solids, per cent moisture, and ash content of any organic agroresidue be determined?
2. What is the importance of the quantification of macromolecules in organic wastes?
3. Describe the method to estimate starch content in agroresidues.
4. What is the importance of lignin in agroresidues? Which group of micro-organisms can degrade such complex molecules?
5. Elaborately discuss the procedures employed to estimate the content of organic carbon, nitrogen, potassium, and phosphorus in a waste material.
6. How can you estimate volatile fatty acids and methane in anaerobic fermentation process?
7. Discuss the principle of gas chromatography.
8. How is HPLC different from gas chromatography?
9. Write notes on the following.
   Negative staining of bacteria
   Spore staining
   Viability staining
   IMViC tests
   Streak-plating
   Coliform test
10. Describe the methods used to create anaerobic conditions. What are the special precautions to be exercised for anaerobic cultivation?
11. Briefly discuss the importance of biosensors and biomarkers in pollution monitoring.
12. How will you quantify BOD? How is it different from COD?

References

Allison, A.E. 1965, 'Organic carbon', in C.A. Black (ed.-in-chief), *Methods of Soil Analysis*, Agronomy Series no. 9, part 2, American Society of Agronomy, Madison, WI, pp. 1367–78.

APHA, AWWA, WPCF 1976, *Standard Methods for the Examination of Water and Wastewater*, 14th edn, John D. Lucas, Baltimore.

Breemer, J.M. 1965, 'Total nitrogen', in C.A. Black (ed.-in-chief), *Methods of Soil Analysis*, Agronomy Series no. 9, part 2, American Society of Agronomy, Madison, WI, pp. 1171–5.

Hurwitz, W. (ed.) 1960, *Official Method of Analysis of AOAC*, AOAC, Washington, DC.

Kolmer, J.A., E.H. Spalnding, and H.W. Robinson 1969, *Approved Laboratory Techniques*, 5th edn, Scientific Book Agency, Kolkata, India.

MACS-DNES Training Course 1986, 'Microbiological aspects of anaerobic fermentation', Laboratory Manual, Maharastra Association for the Cultivation of Science.

Robertson, J.B. and P.J. Van Soest 1981, 'The detergent system of analysis and its application to human foods', in W.P.T. James and O. Thiandev (eds), *The Analysis of Dietary Fibers in Food*, Marcel Dekker, New York, pp. 123–58.

Thimmaiah, S.R. 2004, *Standard Methods of Biochemical Analysis*, Kalyani Publishers, Ludhiana.

Updegraff, D.M. 1969, 'Semimicro determination of cellulose in biological materials', *Anal. Biochem.*, vol. 32, pp. 420–424.

CHAPTER

5

Stoichiometry and Design Kinetics for Waste Treatment Processes

Introduction

The first four chapters of the book have provided the basic information required for the design of waste treatment processes. Chapter 1 dealt with the different aspects of waste treatment systems. While, Chapter 2 enumerated the characteristics of waste materials obtained from different sources, the knowledge of which is essential for the design of any bioremediation process. The scientific aspects including the biological mechanism of biodegradation and the basic science behind pollution control and product formation from waste materials were presented in Chapter 3. Another important requirement for successful monitoring of pollution is to know the nature and characteristics of the contents present in waste material. Chapter 4 presented the different analytical methods used to determine the characteristic features of waste that causes pollution.

In a nutshell, to design pollution control devices, knowledge of the nature of pollutants, the micro-organisms involved in the biodegradation of waste material, biochemical pathways, etc. is a prerequisite. To utilize this information it is equally necessary to know about the stoichiometry (material and energy balances) as well as the reaction kinetics of the processes involved. Stoichiometry and reaction kinetics together provide the quantitative approach essential for the design of bioreactors and any design calculations. In this chapter these facets will be discussed in detail.

Basic Mass Balance

As already mentioned, the knowledge of mass balance and energy balance is essential for the design of any system to be used for chemical or biological processes including waste treatment. For example, consider the case of the biological oxidation of a sample of wastewater containing casein, where casein is oxidized in the presence of oxygen to stabilize the wastewater by reducing the BOD (biological oxygen demand) level. The equation for this oxidation reaction is as follows (the values given below the equation in parentheses are the corresponding molecular weights):

$$C_8H_{12}O_3N_2 + 3O_2 \longrightarrow C_5H_7O_2N + NH_3 + 3CO_2 + H_2O$$

(Casein)   (Cell)

(184)   + (96) =   (113)   + (17) + (132) + (18)

(280) =   (280)   (5.1)

Equation (5.1) is a simple account of mass balance, where it can be seen that the input is equal to the output. The generalized balanced equation is

Input – output = accumulation   (5.2)

Let us now examine the design information that we can get from Eqn (5.1). The equation shows that to effect biological degradation of, say, 850 kg of casein present in a sample of wastewater in a day, 444 kg/d of oxygen needs to be supplied in the dissolved form by aerating the wastewater. Accordingly, it also implies that 522 kg/d of biomass solids (i.e., dry sludge) must be dewatered and disposed of. By knowing the concentration of casein in wastewater, the volume of wastewater to be treated per day can be determined. From kinetic data, the retention time required for digestion can be calculated. The basic volume of the digester is calculated by multiplying the volume of wastewater to be treated per day by the retention time. Finally, the size and shape of the reactor is decided on the basis of its volume. The aeration system should be appropriately designed, as it is important to effect adequate transfer of oxygen into the system. Besides C, H, O, and N, which are the major constituents required for microbial cell mass growth, wastewater must also contain elements such as phosphorus, sulphur, and iron in trace amounts. For example, phosphorus normally represents about 2% of the cell mass on a dry weight basis.

To develop a mass balance equation similar to Eqn (5.1), one must know at least the empirical, if not exact, chemical formulae of the cells involved. From the literature it has been observed that the empirical formula for calculating cell mass, on an average, can be taken as follows, with a few exceptions:

N:O:C:H::1:2:5:7

Two per cent of the molecular weight can be taken as phosphate–phosphorus.

The general formula for cell mass is given as $C_nH_aO_bN_c$. Theoretically, the COD (chemical or calculated oxygen demand) required for complete oxidation of cellular carbon is calculated using the following correlation:

$$C_nH_aO_bN_c + (2n + 0.5a - b - 1.5c)\left(\frac{O_2}{2}\right) \longrightarrow nCO_2 + (0.5a - 1.5c)H_2O$$

$$+ cNH_3 \qquad (5.3)$$

From Eqn (5.3), we get

$$\frac{COD}{\text{weight of cell mass}} = \frac{(2n + 0.5a - b - 1.5c) \times 16}{(12n + a + 16b + 14c)} \qquad (5.4)$$

If mass distribution, per cent, of the four main organic elements C, H, O, and N is taken as X_C, X_H, X_O, and X_N, respectively, then

$$n = \frac{X_C}{12X}$$

$$a = \frac{X_H}{X}$$

$$b = \frac{X_O}{16X}$$

$$c = \frac{X_N}{14X}$$

where

$$X = \frac{X_C}{12} + X_H + \frac{X_O}{16} + \frac{X_N}{14} \qquad (5.5)$$

Illustration 5.1

The Microbial Biotechnology laboratory of AGFE department of IIT Kharagpur receives a sample of a biological culture from a wastewater treatment plant of a starch-manufacturing industry for analysis. Students in the laboratory perform step-by-step experiments to determine the percentage, by weight, of the four major organic elements present in the cell culture. First of all, the sample is bone-dried by keeping it in the oven for about 24 hours at 150°C. The organic portion of the dried cell mass is analysed and thereafter the sample is burnt at 550°C in a muffle furnace to determine the weight of the ash content of the cell mass, which contains all inorganic elements such as phosphorus, sulphur, and iron. The composition of the biomass by weight per cent is found to be $X_C = 47.2\%$, $X_H = 5\%$, $X_O = 24.3\%$, $X_N = 10.0\%$, and ash = 11.5%. Then, the students derive the empirical formula for the cells assuming $c = 1$. Calculate the COD/organic weight ratio for the cells.

Solution

From Eqn (5.5), we have

$$X = \frac{47.2}{12} + 5.0 + \frac{24.3}{16} + \frac{10.0}{14} = 11.17$$

Now,

$$n = \frac{47.2}{(12 \times 11.17)} = 0.352$$

$$a = \frac{5.0}{11.17} = 0.448$$

$$b = \frac{24.3}{(16 \times 11.17)} = 0.136$$

$$c = \frac{10.0}{(14 \times 11.17)} = 0.064$$

For normalization with $c = 1$, all the values are divided by 0.064 and the empirical formula for cell mass is obtained as $C_{5.5} H_7 O_{2.1} N$. From Eqn (5.4),

$$\frac{COD}{\text{organic weight}} = \frac{(2 \times 5.5 + 0.5 \times 7 - 2.1 - 1.5) \times 16}{(12 \times 5.5 + 7 + 16 \times 2.1 + 14)}$$

$$= \frac{174.4}{120.6}$$

$$= 1.45 \text{ g COD/g cells}$$

Oxidation–Reduction Reactions

Oxidation–reduction reactions in the fermenter supply the energy needed for the growth, maintenance, and functioning of micro-organisms. An electron donor and an electron acceptor are always involved in oxidation–reduction reactions. In most cases, organic substrates (i.e., the food of the micro-organisms) are considered as electron donors. As we know, fermentation systems can function under both aerobic and anaerobic conditions. In the aerobic system, molecular oxygen (O_2) is a common and most effective electron acceptor, generating maximum possible free energy for the system. As an example, the aerobic oxidation of glucose is shown here:

$$C_6H_{12}O_6 + 6O_2 \longrightarrow 6CO_2 + 6H_2O \tag{5.6}$$

(free energy = –2880 kJ/g per mol of glucose)

It can be inferred from this reaction that aerobic micro-organisms would prefer oxygen as the electron acceptor as it provides the maximum amount of energy possible from a reaction. However, the anaerobic group of organisms cannot tolerate the presence of molecular oxygen. Therefore, it has been observed that rate of growth of anaerobic micro-organisms is much lower compared to the rate of growth of aerobes. Accordingly, the time period required for the digestion of organic waste matter under anaerobic conditions is much more than that required for aerobic digestion.

After oxygen, nitrate as an electron acceptor (denitrification) generates the maximum quantity of energy in a bioreactor. For example,

$$5C_6H_{12}O_6 + 24NO_3^- + 24H^+ \longrightarrow 30CO_2 + 42H_2O + 12N_2 \qquad (5.7)$$

(free energy = −2720 kJ/g per mol of glucose)

In the sulphate reduction process, sulphate acts as the electron acceptor. The process is not as energy efficient as aerobic oxidation or denitrification. A typical sulphate reduction process reaction is as follows:

$$2C_6H_{12}O_6 + 6SO_4^{--} + 9H^+ \longrightarrow 12CO_2 + 12H_2O + 3H_2S + 3HS^- \qquad (5.8)$$

(free energy = − 492 kJ/g per mol of glucose)

In methanogenesis, CO_2 acts as the electron acceptor. Therefore, the presence of CO_2 is essential for methanogenesis under anaerobic conditions. The reaction is as follows:

$$C_6H_{12}O_6 \longrightarrow 3CO_2 + 3CH_4 \qquad (5.9)$$

(free energy = − 428 kJ/g per mol of glucose)

Only methanogens can carry out methane-forming reactions, and these organisms cannot tolerate the presence of oxygen. There are certain types of prokaryotes that use organic matter both as an electron acceptor and electron donor under anaerobic conditions. These reactions are specifically termed *fermentation* reactions, for example, fermentation of ethanol:

$$C_6H_{12}O_6 \longrightarrow 2CO_2 + 2CH_3CH_2OH \qquad (5.10)$$

(free energy = −244 kJ/g per mol of glucose)

It is clear that micro-organisms obtain energy from oxidation–reduction reactions, which involve the transport of electrons between the donor and the acceptor. Of the total energy involved in the microbial process, which can be expressed as the electron equivalent (e^- eq) of the electron–donor substrate, a few electrons (E_1) are first transferred to the electron acceptor to provide energy for the conversion of the remainder of the electrons (E_2) into the new cell mass. E_1 and E_2 represent fractional values of the total, while the sum of E_1 and E_2 is always unity.

From this discussion of reaction energetics, an interesting conclusion can be drawn. When oxygen is the electron acceptor, only a small number of electrons from the donor substrate prove to be sufficient to provide the necessary energy to the system for the synthesis of a given number of new cells. That is, when E_1 is less, E_2 will accordingly be more, as $E_1 + E_2$ is always unity. The yield coefficient Y (of the cell mass), which is proportional to E_2, will also be high. On the contrary, when the reaction is carried out in the absence of oxygen, where the same substrate acts as the electron acceptor, E_1 will be more, as it needs to provide more energy to synthesize a particular amount of cell mass. This explains the higher yield in the case of aerobic micro-organisms as compared to anaerobic micro-organisms.

Illustration 5.2

In a tannery, toxic chemicals such as sulphide are used to effect unhairing of skin. In this process, a large quantity of hair gets dissolved and amino acids are formed, which are removed through tannery effluents. However, this also increases the COD of the tannery wastewater. These effluents are treated anaerobically in the presence of methanogens. Let us assume the different kinds of amino acids present in the wastewater to be alanine (CH_3CHNH_2COOH) equivalent; then the effluent can be said to contain 125 mM alanine. During the treatment process, if all the alanine is used for the supply of energy for the process of biomethanation, at a pH of about 7, determine (a) the composition of the gas produced and (b) the other products produced in the solution and their concentrations.

Solution

(a) The overall oxidation–reduction reaction that occurs during anaerobic methanogenic fermentation of alanine is

$$2CH_3CHNH_2COOH + 4H_2O \longrightarrow 3CH_4 + CO_2 + 2NH_4^+ + 2HCO_3^-$$

From this correlation it is clear that the gas contains 75% CH_4 and 25% CO_2 by volume.

(b) The reaction also shows that the other products formed in the solution are ammonium (NH_4^+) and bicarbonate (HCO_3^-). Generally, after the oxidation of an organic compound at neutral pH, positively charged NH_4^+ are released. To balance this charge, a negatively charged species is also formed, and normally at neutral pH organic substrates undergo oxidation to produce HCO_3^-. The equation also shows that 1 mol of alanine forms 1 mol each of NH_4^+ and HCO_3^-. Hence, from 125 mM of alanine, 125 mM each of NH_4^+ and HCO_3^- will be formed in the solution. Taking the molecular weight of nitrogen as 14, the ammonium nitrogen concentration in the solution will be increased by 1750 mg/L. Alkalinity in wastewater is expressed as equivalent $CaCO_3$ (equivalent wt = 50). Therefore, bicarbonate alkalinity in the solution will be increased by 6250 mg/L. The formation of bicarbonate in the solution produces a neutral pH buffer and helps to maintain a near-neutral pH. It is interesting to note that if the pH of the solution is well above 9.3, then NH_3 would be expected to dominate instead of allowing the formation of NH_4^+ during the oxidation of the substrate.

Degree of Reduction (γ) and Mass Balance

For organic compounds, *degree of reduction* is defined as the number of equivalents of available electrons (e^- eq) per gram atom of carbon. The available electrons are those that can be transferred upon oxidation of the compound to CO_2, H_2O, and NH_3. The degrees of reduction for some key elements are $C = 4$, $H = 1$, $N = -3$, $O = -2$, $P = 5$, and $S = 6$. The degree of reduction of any element in a compound is equal to the valence of that element. For example, 4 is the valence of carbon in CO_2 and -3 is the valence of nitrogen in NH_3. Table 5.1 shows the calculation of equivalents of available electrons (e^- eq) and the degree of reduction (γ) of a few compounds. If a compound has a high degree of reduction, it indicates that it has a correspondingly low degree of oxidation. In oxidation–reduction reactions, the degree of reduction is significant. It directly relates to the quantity of oxygen consumed and the amount of energy released. For example,

$$CH_4 + 2O_2 \longrightarrow CO_2 + 2H_2O \tag{5.11}$$

In Eqn (5.11), the oxygen consumption ratio per unit weight of CH_4 oxidized is 4. Let us consider another example of glucose:

$$C_6H_{12}O_6 + 6O_2 \longrightarrow 6CO_2 + 6H_2O \tag{5.12}$$

Table 5.1 Electron equivalents and degrees of reduction of some selected organic compounds

Compound	e^- equivalent	γ (degree of reduction)
Methane (CH_4)	$4 \times 1 + 1 \times 4 = 8$	$8 \div 1$ (C) $= 8$
Glucose ($C_6H_{12}O_6$)	$4 \times 6 + 1 \times 12 + (-2) \times 6 = 24$	$24 \div 6$ (C) $= 4$
Ethanol (C_2H_5OH)	$4 \times 2 + 1 \times 6 + (-2) \times 1 = 12$	$12 \div 2$ (C) $= 6$
Oxalic acid ($C_2H_2O_4$)	$4 \times 2 + 1 \times 2 + (-2) \times 4 = 2$	$2 \div 2$ (C) $= 1$
Succinic acid ($C_4H_6O_4$)	$4 \times 4 + 1 \times 6 + (-2) \times 4 = 14$	$14 \div 4$ (C) $= 3.5$
Glycerol ($C_3H_8O_3$)	$4 \times 3 + 1 \times 8 + (-2) \times 3 = 14$	$14 \div 3$ (C) $= 4.67$
Carbon dioxide (CO_2)	$4 \times 1 + (-2) \times 2 = 0$	$0 \div 1$ (C) $= 0$

In this case, the oxygen consumption ratio per unit weight of glucose oxidized is $(6 \times 32 \div 180)$ or 1.07. Therefore, on oxidation methane will release more energy than glucose, per unit weight.

Two simplified biological conversion equations have been formulated based on the biological material defined by empirical formulae containing 1 gram atom of carbon. In the first case, except for cellular biomass, no extracellular products, other than CO_2 and H_2O, are produced. However, in the second case an additional extracellular product is formed. Both the processes occur, however, under aerobic conditions.

$$CH_mO_n + aO_2 + bNH_3 \longrightarrow cCH_\alpha O_\beta N_\delta + dH_2O + eCO_2 \qquad (5.13)$$
Substrate $\qquad\qquad\qquad\qquad$ Biomass

$$CH_mO_n + aO_2 + bNH_3 \longrightarrow cCH_\alpha O_\beta N_\delta + dCH_xO_yN_z + eH_2O + fCO_2$$
Substrate $\qquad\qquad\qquad\qquad$ Biomass $\quad$ Product $\qquad\qquad (5.14)$

In both the equations CH_mO_n represents 1 mol of carbohydrate, $CH_\alpha O_\beta N_\delta$ stands for 1 mol of cellular biomass, and $CH_xO_yN_z$ for 1 mol of bioproduct.

In Eqn (5.13) there are five unknown coefficients (a, b, c, d, and e) and from elemental mass balance on C, H, O, and N, four equations will be obtained. One more equation is required to determine the values of these coefficients. If the respiratory quotient (RQ) is experimentally measured, then this will give the fifth equation as $RQ = e/a$.

There are six unknown coefficients (a, b, c, d, e, and f) in Eqn (5.14). From elemental mass balance on C, H, O, and N, we can get four equations, electron balance gives one more equation, and from RQ the sixth equation can be obtained, and all these can be used together to determine the unknown coefficients. From the energy balance formula of aerobic growth, one more equation can be obtained, though this will not be independent of the electron balance equation. Accordingly, we have the following equations:

$$\text{Electron balance: } E_s - 4a = cE_b + dE_p \qquad (5.15)$$

$$\text{Heat balance: } Q_0E_s - Q_04a = Q_0cE_b + Q_0dE_p \qquad (5.16)$$

$$\text{or} \quad 1 = \frac{cE_b}{E_s} + \frac{dE_p}{E_s} + \frac{4a}{E_s} = F_b + F_p + F_o \qquad (5.17)$$

where E_s is the e^- equivalent of the substrate, E_b is the e^- eq of the biomass, E_p is the e^- eq of the product, Q_0 is a constant denoting the heat evolved per equivalent of available electrons transferred to oxygen $= 26.95$ kcal/g equivalent of available electrons transferred to oxygen (this helps to predict the amount of heat evolved, based on estimates of oxygen consumption), F_o is the fraction of available electrons of the organic substrate transferred to oxygen, F_b is the fraction of available electrons incorporated into biomass, and F_p is the fraction of available electrons incorporated into extracellular products. In the case of anaerobic processes, F_o is absent.

As already discussed, in order to solve the mass balance equations for unknown coefficients, various formulae such as the elemental and mass balance equations, measured values of respiratory quotient, yield coefficients, and energy balances can be employed. It may also happen that the number of equations formulated may be more than the number of unknown quantities. In such instances the less significant correlations are generally omitted. For example, in elemental mass balance equations, the relations pertaining to C and N are more significant than those related to H and O. Similarly, compared to energy balance equations, electron balance equations are more significant. If data are available, then RQ and yield coefficient formulae are very important. The various yield coefficients are

$Y_{b/s}$ = amount of biomass produced per gram or mole of substrate consumed

Y_{b/O_2} = amount of biomass produced per mole of oxygen consumed = c/a

$Y_{p/s}$, Y_{b/O_2}, Y_{b/e^-}, $Y_{b/ATP}$, etc. are other similar yield coefficients. It is interesting to note that the value of $Y_{b/ATP}$ is approximately constant for many substrates. It is 10 to 11 g dry wt (dw)/mol of ATP for heterotrophic growth under anaerobic conditions and about 6.5 g biomass synthesized/mol of ATP generated for many autotrophic organisms that fix CO_2, but the same value is greater than 10.5 under aerobic conditions.

Depending on the substrate and organisms, Y_{b/O_2} varies from 0.17 to 1.5. It may also be noted that $Y_{b/e^-} = 3.14 \pm 0.11$ g dry wt/g equivalent of electrons in oxygen molecules when ammonia is used as the nitrogen source in aerobic fermentation. When the number of oxygen molecules consumed per mol of substrate is known, the growth yield coefficient $Y_{b/s}$ can be calculated easily.

For example, consider the aerobic catabolism of glucose, which can be represented by the following equation:

$$C_6H_{12}O_6 + 6O_2 \longrightarrow 6CO_2 + 6H_2O \qquad\qquad (5.18)$$

The total number of available electrons per mole of glucose is 24. The cellular yield per available electron is given as

$$Y_{b/e^-} = 3.14$$

Therefore, the cellular yield per mole of glucose consumed is

$$Y_b/\text{mol substrate} = 24 \times 3.14 \approx 76 \text{ g dw cells/mol substrate}$$

The theoretically predicted growth yield coefficient is equal to

$$Y_{b/s} = \frac{76}{180} = \frac{0.42 \text{ g dw cells}}{\text{glucose consumed (g)}}$$

Most of the experimental values of $Y_{b/s}$ for aerobic growth in the case of glucose lie between 0.38 and 0.51 g/g.

The ATP yield ($Y_{b/ATP}$) in most of the anaerobic fermentation cases is approximately equal to 10.5 ± 2 g dw cells/mol ATP. In aerobic fermentation this yield varies approximately between 6 and 29 g dw cells/mol ATP. When the energy yield of a metabolic pathway is known (i.e., we know that, say, n mols of ATP are produced per gram of substrate consumed), the growth yield $Y_{b/s}$ can be calculated using the following equation:

$$Y_{b/s} = Y_{b/ATP} n \tag{5.19}$$

and the rate of growth of microbial cells can be expressed as

$$\frac{dX_a}{dt} = Y\left(\frac{-dS}{dt}\right) - bX_a \tag{5.20}$$

where dX_a/dt is the net growth rate of active cells (mg $L^{-1} d^{-1}$), X_a is the active cells concentration (mg L^{-1}), $-dS/dt$ is the rate of consumption of the substrate (mg $L^{-1} d^{-1}$), S is the substrate concentration (mg L^{-1}), Y is the true yield of micro-organisms (g/g), and b is the decay rate of organisms (d^{-1}). It is also known that

Net growth rate = growth from substrate consumption
– decay due to self/endogenous respiration or predation

From Eqn (5.20), we have

$$Y_n \text{ (net yield)} = \frac{\left(\dfrac{dX_a}{dt}\right)}{\left(\dfrac{-dS}{dt}\right)} = Y - \frac{bX_a}{\left(\dfrac{-dS}{dt}\right)} \tag{5.21}$$

Y_n is always less than Y because a portion of the substrate is consumed as energy required for the maintenance of cells. From Eqn (5.21) it is clear that the decay portion, $(bX_a)/(-dS/dt)$, becomes large if the decay rate b or cell concentration X_a increases, or if the rate of substrate consumption reduces. Equation (5.21) also reveals that the right side value will be near zero when the substrate utilization rate per unit mass of cells is significantly low. Hence it can be said that $Y_n = 0$. In such situations the substrate utilization rate is found to be just sufficient to maintain the cells and there is no net growth of active cells. Under these conditions, from Eqn (5.21), we get

$$\frac{\left(\dfrac{-dS}{dt}\right)}{X_a} = \frac{b}{Y} = m \tag{5.22}$$

Here m (gg^{-1}d^{-1}) is known as the maintenance energy, which indicates the substrate utilization rate per unit mass of cells. From Eqn (5.22), it is clear that

$$m \, \alpha \, b, \, Y^{-1}$$

If $(-dS/dt)/X_a < m$, then it represents a kind of starvation indicating that the substrate available is insufficient to satisfy the metabolic requirements of the micro-organisms.

Illustration 5.3

In a batch culture reactor system acetate is used as the source of carbon for the cultivation of a particular micro-organism. The data available are as follows. $X_a = 600$ mg/L, $-dS/dt = 800$ mg L^{-1}d^{-1}, $Y = 0.6$ g cells per gram substrate, and $b = 0.15$ d^{-1}. Using the data provided determine the following.

(i) Specific growth rate of micro-organisms
(ii) Specific rate of substrate utilization
(iii) Net yield of cells

Solution

(i) The specific growth rate μ of micro-organisms is calculated as the net growth rate of active cells (dX_a/dt) divided by the active cells concentration (X_a) in the broth. Thus, we have

$$\mu = \frac{\left(\dfrac{dX_a}{dt} \right)}{X_a}$$

From Eqn (5.20), we know

$$\frac{dX_a}{dt} = Y \left(\frac{-dS}{dt} \right) - bX_a$$

or
$$\frac{\left(\dfrac{dX_a}{dt} \right)}{X_a} = \left(\frac{Y}{X_a} \right) \left(\frac{-dS}{dt} \right) - b$$

$$= \left(\frac{0.6}{600} \right) (800) - 0.15 = 0.65^{-1}$$

This value of specific growth rate is indicative of the increase of the cell mass population with respect to the existing cell mass concentration under the prescribed situation. The rate of increase here is 65% per day.

(ii) The specific rate of substrate utilization is given as

$$= \frac{\left(\dfrac{-dS}{dt} \right)}{X_a}$$

$$= \frac{800}{600} \text{ (from the data given)}$$

$$= 1.33 \text{ g acetate per g cells per day}$$

The result indicates that the micro-organisms are consuming substrates weighing about 1.33 times their own weight, per day.

(iii) From Eqn (5.21), we get

$$Y_n = Y - \frac{b}{\left[\dfrac{\left(\dfrac{-dS}{dt} \right)}{X_a} \right]}$$

$$= 0.6 - \frac{0.15}{1.33}$$

$$= 0.49 \text{ g cells produced per g acetate consumed per day}$$

This is $0.49/0.6$ or 82% of the actual yield Y.

Part-reactions in Mass Balance

As we already know, in environmental biotechnology, micro-organisms play a very important role in the degradation of pollutants present in wastewater or solid wastes. The overall mass balance correlation is formulated on the basis of three types of standard part-reactions. These are related to the generation of new cell mass, product formation, and maintenance of microbial growth environment. From these three part-reactions, two basic reactions can be proposed. These are

$$R_E = R_A + (-R_D) \tag{5.23}$$
$$R_S = R_C + (-R_D) \tag{5.24}$$

where R_E is the basic balanced reaction for energy generation, R_S is the basic balanced reaction for cell synthesis, R_A is the balanced part-reaction for electron acceptors for energy supply, R_C is the balanced part-reaction for cell synthesis, and R_D is the balanced part-reaction for the substrate acting as an electron donor. In Eqn (5.23), R_A depends on the type of electron acceptor, such as CO_2, O_2, NO_3, SO_4, and Fe^{3+}. Similarly, R_C in Eqn (5.24) depends on the nature of the source of nitrogen (e.g., NH_4^+, NO_3, NO_2^-, N_2). Any of them can be used as a source of nitrogen for the synthesis of cells. R_D, naturally, depends on the nature of the substrate. As already mentioned, all the equations must be written in different combinations.

Theoretically, when Eqns (5.23) and (5.24) are considered independently, the electron donor substrate is fully utilized. However, for the overall reaction (R_o), in which both energy generation (R_E) and cell synthesis (R_S) are involved simultaneously, the expression for R_o will be as follows:

$$R_o = E_1 R_A + E_2 R_C + (E_1 + E_2)(-R_D)$$
$$= E_1 R_A + E_2 R_C + (-R_D) \tag{5.25}$$

Here $E_1 + E_2 = 1$ and the magnitudes of E_1 and E_2 need to be decided judiciously. The negative sign before R_D is used to differentiate the donor from the acceptor.

The generalized part-reaction (R_D) for organic substrates as available in the literature is presented here. If the organic matter formula is known or has been determined empirically, the following correlation can be used to solve mass balance problems, as has been demonstrated in Illustration 5.4:

$$R_D = \left[\frac{(n-c)}{d} \right] CO_2 + \left(\frac{c}{d} \right) NH_4^+ + \left(\frac{c}{d} \right) HCO_3^- + H^+ + e^-$$
$$= \left(\frac{1}{d} \right) C_n H_a O_b N_c + \left[\frac{(2n-b+c)}{d} \right] H_2 O \tag{5.26}$$

In this equation, $d = 4n + a - 2b - 3c$ and $C_n H_a O_b N_c$ is the formula for the organic substrate.

Illustration 5.4

Consider the organic matter of an industrial effluent from a food processing unit. After analysing it for its organic carbon, hydrogen, oxygen, and nitrogen contents, the empirical formula $C_9 H_{19} O_3 N$ was proposed and the concentration of the organic matter was found to be 25 g/L. The effluent was treated anaerobically for BOD reduction and methane production. If the effluent flow rate is 180 m^3/d, calculate how many m^3/d of methane, at 35°C and 1 atm pressure, approximately are produced and what would be the percentage of methane? The process efficiency may be assumed to be 90% and $E_2 = 0.1$.

Solution

This problem can be solved by two methods. However, there can be some minor difference between the two results. The first method is based on mass balance and some logical assumptions and the second method is based on oxidation–reduction equations.

Method I We know that methane fermentation results in the formation of products such as CH_4, CO_2, new cell mass having the empirical formula $C_5 H_7 O_2 N$, and some ions as buffering agents (HCO_3^- and NH_4^+). With this information the following overall balanced equation can be proposed:

$$C_9H_{19}O_3N + aH_2O \longrightarrow bC_5H_7O_2N + cCH_4 + dCO_2 + eNH_4^+ + eHCO_3^-$$

By elemental mass balance the following equations are obtained:

1. By C balance

$$9 = 5b + c + d + e \qquad\qquad\qquad\qquad (i)$$

2. By H balance

$$19 + 2a = 7b + 4c + 5e \qquad\qquad\qquad\qquad (ii)$$

3. By O balance

$$3 + a = 2b + 2d + 3e \qquad\qquad\qquad\qquad (iii)$$

4. By N balance

$$1 = b + e \qquad\qquad\qquad\qquad (iv)$$

There are four equations and five unknown coefficients. The magnitude of one coefficient needs to be selected judicially to determine the magnitudes of other coefficients. From experimental knowledge it can be said that the value of the coefficient b of the cell mass will be very small, as the growth rate of methanogens is low. From the data given it is known that 10% of the e^- eq ($E_2 = 0.1$) donated by the organic substrate is utilized as energy for the synthesis of new cell mass. With this in mind, it can be assumed that 10% of the 90% organic matter gets converted into new cell mass. The molecular weight of organic matter is 189 and that of cell mass is 113. Therefore,

$$b = \frac{189 \times 0.90 \times 0.1}{113}$$

$$= 0.15$$

Another check will be the molal ratio of methane to carbon dioxide, i.e., c/d, which should lie between 1 and 3. From Eqn (iv), we have

$$e = 1 - b = 0.85$$

Now, by solving equations (i) to (iii) simultaneously, the magnitudes of other coefficients are obtained. These are $a = 3.9$, $b = 0.15$, $c = 5.375$, $d = 2.025$, and $e = 0.85$. Thus the mass balance equation becomes

$$C_9H_{19}O_3N + 3.9H_2O \longrightarrow 0.15C_5H_7O_2N + 5.375CH_4 + 2.025CO_2$$
$$+ 0.85NH_4^+ + 0.85HCO_3^-$$

The molal ratio of CH_4 to CO_2 is found to be 2.65. This is reasonable because, depending on the nature of substrate, this ratio varies between 1 and 3. The number of moles of organic matter in the effluent digested per day at 90% efficiency

$$= \frac{180 \times 1000 \times 25 \times 0.9}{189} = 21{,}428.6 \text{ g mol}$$

From the mass balance equation, g mol of CH_4 produced $= 21{,}428.6 \times 5.375 = 115{,}178.7$. The volumetric CH_4 production rate

$$= 115{,}178.7 \times 0.0224 \times \left[\frac{(273 + 35)}{273} \right] = 2911 \, \frac{m^3}{d}$$

$$\text{mol \% of methane} = \left[\frac{5.375}{(5.375 + 2.025)} \right] \times 100$$

$$= \left(\frac{5.375}{7.400} \right) \times 100$$

$$= 72.639$$

In determining the coefficient b of the mass balance equation, it was assumed that 10% of the 90% organic matter was converted into new cell mass. Instead of this, as per the problem statement, if we consider that 10% of the available electrons in the organic matter are utilized for new cell mass generation, then the value of the coefficient b is estimated as follows:

e^- eq of organic matter $= 46$

Available e^- eq of organic matter $= 46 \times 0.9$ or 41.4

e^- eq of cell mass $= 20$

Thus,

$$41.4E_2 = 20b$$

or

$$b = \frac{(41.4 \times 0.1)}{20} = 0.207 \text{ (approximated as 0.2)}$$

If this value is used to find the value of the other four coefficients of the mass balance equation, then we have

$$a = 3.7, \ b = 0.2, \ c = 5.25, \ d = 1.95, \text{ and } e = 0.8$$

If these values are compared with the earlier values of the coefficients, it is clear that the difference is not much. Moreover, the molal ratio of CH_4 to CO_2, i.e., $c/d = 2.69$, is also very close to the earlier value of 2.65. This indicates that the assumption made initially was reasonably correct.

Method II To solve the problem by the method of part-reactions, the standard reactions of electron acceptors and electron donors for product formation and cell synthesis should be known. According to the empirical formula for organic matter given in the problem, we have

$$n = 9, \ a = 19, \ b = 3, \text{ and } c = 1$$

Substituting these values, we get $d = 46$ and $-R_D$ becomes the following [Eqn (5.26)]:

$-R_D$:

$$0.0217C_9H_{19}O_3N + 0.3472H_2O = 0.1736CO_2 + 0.0217NH_4^+ \\ + 0.0217HCO_3^- + H^+ + e^- \qquad \text{(i)}$$

It is to be noted carefully that all the equations are, individually, balanced. The standard forms of the electron acceptor part-reaction (R_A) for CO_2 to CH_4 and the cell synthesis part-reaction (R_C) are the following.

R_A:

$$0.125CO_2 + H^+ + e^- = 0.125CH_4 + 0.25H_2O$$

R_C:

$$0.2CO_2 + 0.05NH_4^+ + 0.05HCO_3^- + H^+ + e^- = 0.05C_5H_7O_2N + 0.45H_2O$$

To get the expression for R_o, first R_A and R_C should be multiplied by E_1 and E_2, respectively, to get expressions for E_1R_A and E_2R_C:

E_1R_A:

$$0.1125CO_2 + 0.9H^+ + 0.9e^- = 0.1125CH_4 + 0.225H_2O \qquad \text{(ii)}$$

E_2R_C:

$$0.02CO_2 + 0.005NH_4^+ + 0.005HCO_3^- + 0.1H^+ + 0.1e^- = 0.005C_5H_7O_2N \\ + 0.045H_2O \qquad \text{(iii)}$$

By adding Eqns (i), (ii), and (iii) we get

R_o:

$$0.0217C_9H_{19}O_3N + 0.0772H_2O = 0.005C_5H_7O_2N + 0.1125CH_4 \\ + 0.0411CO_2 + 0.0167NH_4^+ \\ + 0.0167HCO_3^-$$

Dividing by 0.0217,

R_o:

$$C_9H_{19}O_3N + 3.5576H_2O = 0.2304C_5H_7O_2N + 5.1843CH_4 + 1.894CO_2 \\ + 0.7696NH_4^+ + 0.7696HCO_3^-$$

Here 1 g mol of organic matter gives 5.1843 g mol of CH_4. So, CH_4 (m^3/d) produced with 90% conversion efficiency is calculated as

$$\left[\frac{(180 \times 1000 \times 25 \times 0.9)}{189} \right] \times 5.1843 \times 0.0224 \times \left[\frac{(273 + 35)}{273} \right] = 2807.5$$

According to method I, CH_4 (m^3/d) is 2911 and the per cent difference $= 3.7$

$$\text{Per cent of } CH_4 \text{ in biogas} = \frac{5.1843}{(5.1843 + 1.894)} = 73\%$$

$$\text{Molal ratio of } CH_4 \text{ to } CO_2 = \frac{5.1843}{1.894} = 2.74$$

By method I it is 2.65.

Fermentation of ethanol

It has already been stated that in a fermentation process an organic substrate can serve as both an electron acceptor and an electron donor. Consider the example of the fermentation of glucose to ethanol. Here the most important point is to develop the part-reactions of both the electron donor and the electron acceptor. The part-reaction for ethanol is

$$\frac{1}{6}CO_2 + H^+ + e^- = \frac{1}{12}CH_3CH_2OH + \frac{1}{4}H_2O \qquad (5.27)$$

From this equation it is clear that ethanol has 12 electron equivalents (see also Table 5.1). The part-reaction for glucose is as follows:

$$\frac{1}{4}CO_2 + H^+ + e^- = \frac{1}{24}C_6H_{12}O_6 + \frac{1}{4}H_2O \qquad (5.28)$$

In this equation we find that glucose has got 24 electron equivalents (same as shown in Table 5.1). Equations (5.27) and (5.28) show the reduction of CO_2 to ethanol and glucose, respectively, for which energy is required. A similar part-reaction will also occur when ethanol and glucose undergo oxidation and release energy. Such oxidation–reduction equations are carried out by micro-organisms to obtain the energy required for their growth and cell maintenance. However, it is to be noted that the amount of energy released per electron equivalent of an electron donor oxidized varies considerably from reaction to reaction. The availability of electron–donor substrates in limited or unlimited quantity plays an important role in cell growth. If electron–donor substrate is available in unlimited concentration along with the other required factors, then cell growth will be fast. On the other hand, if the electron–donor substrate is limited in concentration, then a large portion of the energy obtained from the substrate oxidation is used for cell maintenance, with little left for cell growth. With further decrease in substrate concentration, the energy released will not even suffice for cell maintenance. As a result micro-organisms go into net decay.

The value of the electron equivalent of organic matter in waste provides a lot of useful information. As already studied, the strength of waste is expressed by its oxygen demand, which is more popularly known as BOD or COD. Theoretically, when the oxygen demand (OD) of any waste stream is calculated, it is expressed as calculated oxygen demand. The OD value can be calculated from a balanced oxidation reaction. We have already seen this from Eqn (5.6), where it was found that 1 mol of glucose consumes 6 mols of oxygen to complete the oxidation reaction. From the calculation of the electron equivalent of the substrate we get a similar result. One electron equivalent of oxygen is 8 g of O_2. Consider the example of glucose whose electron equivalent is 24 (see Table 5.1). It will require 192 g of O_2 as its OD. The same result may be obtained from Eqn (5.6), which shows that per mol of glucose, 6 mols of O_2 or 192 g of O_2 are required.

The analysis done so far reveals that if the number of electron equivalents per litre or per kg of any waste material is known, then the OD value can be calculated directly. This makes it very easy to compute the value of Y, as it contains the widely used units of COD or BOD. Illustration 5.5 exemplifies this in a better way.

Illustration 5.5

An industrial effluent contains 15 g/L of glucose. Calculate the e^- eq/L and the COD (g/L) for this effluent.

Solution

From Table 5.1, it is clear that glucose can donate 24 e^- eq/mol, since the molecular weight of glucose is 180; so the equivalent weight is 7.5 g/e^- eq. The glucose concentration in the effluent is, therefore, 15/7.5 or 2e^- eq/L. From this, the COD value of the effluent can be calculated as 2×8 or 16 g/L.

Ethanol fermentation involves two steps. The first step is aerobic, by which yeast cells are produced, and in the second step alcohol is produced under anaerobic conditions. Two sets of part-reactions are involved and in both the cases glucose can be considered as the electron donor.

Step I

$E_{11}R_A$:

$$(0.25O_2 + H^+ + e^-)E_{11} = (0.5H_2O)E_{11} \tag{5.29}$$

$E_{21}R_C$:

$$(0.2CO_2 + 0.05HCO_3^- + 0.05NH_4^+ + H^+ + e^-)E_{21} = (0.05C_5H_7O_2N + 0.45H_2O)E_{21} \tag{5.30}$$

$-R_D$:

$$0.0417C_6H_{12}O_6 + 0.25H_2O = 0.25CO_2 + H^+ + e^- \tag{5.31}$$

$(R_o)_I$:

$$0.0417C_6H_{12}O_6 + 0.05E_{21}(HCO_3^- = 0.05E_{21}(C_5H_7O_2N) + (0.25 - 0.2E_{21})CO_2$$
$$+ NH_4^+) + 0.25E_{11}O_2 \qquad + (0.45E_{21} + 0.5E_{11} - 0.25)H_2O \tag{5.32}$$

Here,

$$E_{11} + E_{21} = 1$$

Step II

$E_{12}R_A$:

$$E_{12}(0.1667CO_2 + H^+ + e^-) = E_{12}(0.0833CH_3CH_2OH + 0.25H_2O) \qquad (5.33)$$

$E_{22}R_C$:

$$E_{22}(0.2CO_2 + 0.05NH_4^+ + 0.05HCO_3^- + H^+ + e^-) = E_{22}(0.05C_5H_7O_2N$$
$$+ 0.45H_2O) \qquad (5.34)$$

$-R_D$:

$$0.0417C_6H_{12}O_6 + 0.25H_2O = 0.25CO_2 + H^+ + e^- \qquad (5.35)$$

$(R_o)_{II}$:

$$0.0417C_6H_{12}O_6 = 0.0833E_{12}CH_3CH_2OH + 0.05E_{22}C_5H_7O_2N$$
$$+ 0.05E_{22}(NH_4^+ \quad + (0.25 - 0.1667E_{12} - 0.2E_{22})\ CO_2$$
$$+ HCO_3^-) \qquad + (0.25E_{12} + 0.45E_{22} - 0.25)\ H_2O \qquad (5.36)$$

Here,

$$E_{12} + E_{22} = 1$$

It is to be noted that in step I donor electrons will be mostly utilized for the synthesis of cells. Therefore, E_{21} will be much larger than E_{11}. In step II product formation is predominant and some energy is spent in the maintenance of the cells. The cells produced in step I will be utilized in step II. For this reason, in step II, E_{12} is much larger than E_{22}. These two steps occur independently, as the first one is aerobic and the second step requires anaerobic conditions.

Illustration 5.6

From the literature it is known that ethanol production from glucose by *Saccharomyces cerevisiae* is a two-stage process. The first stage is aerobic, where the cell mass is produced, which converts glucose into ethanol in the second stage. If $E_{11} = 0.28$ and $E_{12} = 0.76$, formulate the overall biological reactions for each stage of ethanol fermentation. Also comment on the results.

Solution

The overall biological reaction for the first stage of ethanol production is given by Eqn (5.32). We have E_{11} is 0.28 and $E_{21} = 1 - 0.28 = 0.72$. Substituting these values, we get

$(R_o)_I$:

$$0.0417C_6H_{12}O_6 + 0.07O_2 + 0.036HCO_3^- + 0.036NH_4^+ = 0.036C_5H_7O_2N$$
$$+ 0.106CO_2$$
$$+ 0.214H_2O \quad \text{(i)}$$

For the second stage of the biological reaction, Eqn (5.36) is applied, where $E_{12} = 0.76$ and $E_{22} = 0.24$. Substituting these values, we get
$(R_o)_{II}$:

$$0.0417C_6H_{12}O_6 + 0.012HCO_3^- + 0.012NH_4^+ = 0.0633CH_3CH_2OH$$
$$+ 0.012C_5H_7O_2N$$
$$+ 0.0753CO_2 + 0.048H_2O \quad \text{(ii)}$$

From solution (i), we can conclude that the molar yield of biomass per mol of substrate utilized is

$$Y_{b/s} = \frac{0.036}{0.0417} = 0.863$$

On dry weight basis, this will become

$$Y_{b/s} = 0.863 \times \frac{113}{180} = 0.542 \text{ g dw biomass per g of glucose consumed}$$

The RQ can also be obtained from step I and is given as

$$RQ = \frac{0.106}{0.07} = 1.5$$

From solution (ii), we have

$$Y_{p/s} = 1.52 \text{ mol ethanol per mol of glucose consumed} = 0.388 \text{ g ethanol}$$
$$\text{per g of glucose consumed}$$

$$Y_{b/s} = 0.18 \text{ g dry biomass per g of glucose consumed}$$

$$Y_{p/e^-} = 0.388 \times \frac{180}{24} = 2.91 \text{ g ethanol produced per g equivalent of}$$
$$\text{available electrons}$$

Design Kinetics

The knowledge of the concepts of reaction kinetics is essential for the design of reactors. The study of reaction kinetics involves different aspects such as types and rates of reactions and rate constants, which will be dealt with in detail in the next few sections.

Types and Rates of Reactions

Homogeneous and *heterogeneous* are the two principal types of reactions that occur during waste treatment. In homogeneous reactions, the reactants are uniformly distributed throughout the effluent, whereas heterogeneous reactions occur in the presence of multiple phases such as solid, liquid, and gaseous. Compared to homogeneous reactions, heterogeneous reactions are more complex because they involve different types of interrelated steps.

The *rate of reaction* describes the change in substrate or product concentration with time. The unit will be the change in the number of moles per unit time per unit volume for homogeneous reactions and per unit surface area or mass for heterogeneous reactions. Again the reaction can be either reversible (A $\longleftrightarrow$ B) or irreversible (A $\longrightarrow$ B).

For homogeneous reactions, the rate of reaction, r, is given as

$$r = \left(\frac{1}{V}\right)\left(\frac{dN}{dt}\right) = \frac{\text{mol}}{\text{(volume) (time)}} \tag{5.38}$$

If N is replaced by VC, where V is the volume and C is the concentration, then

$$r = \left(\frac{1}{V}\right)\left[d\,\frac{(VC)}{dt}\right] = \left(\frac{1}{V}\right)\left(\frac{VdC}{dt} + \frac{CdV}{dt}\right) \tag{5.39}$$

If V is constant, then

$$r = \frac{\pm dC}{dt} \tag{5.40}$$

Here the '+' sign indicates increase or accumulation of product and the '–' sign indicates decrease or consumption of substrate.

For heterogeneous reactions, where S is the surface area, the corresponding expression is

$$r = \left(\frac{1}{S}\right)\left(\frac{dN}{dt}\right) = \frac{\text{mol of product}}{\text{(area) (time)}} \tag{5.41}$$

Reaction Rate Constant

This is also termed *specific reaction rate*. From Eqn (5.40), it can be seen that the rate of reaction is proportional to the residual concentration of the reactants. Thus, for a reaction involving a single component A,

$$r = \pm kC_A \tag{5.42}$$

where k is a constant of proportionality defined as the specific reaction rate or reaction rate constant. Substituting r from Eqn (5.38), we have

$$k = \frac{r}{C_A} = \left(\frac{1}{C_A}\right)\left(\frac{1}{V}\right)\left(\frac{dN}{dt}\right) \quad \text{(time)}^{-1} \tag{5.43}$$

where C_A is the residual or remaining concentration of reactants in the reactor (mol/L), V is the effective volume of the reactor (L), N is the number of moles of the reactant, and t is the time.

Temperature Effect on Specific Rate Constant

The temperature dependence of rate constants is given by the Van't Hoff–Arrhenius equation:

$$\frac{d\,(\ln k)}{dT} = \frac{E}{RT^2} \tag{5.44}$$

where T is the temperature (K), R is the ideal gas constant = 8.314 J/mol/K (1.987 cal/K mol), and E is a constant characteristic of the reaction called the activation energy.

Integration of Eqn (5.44) between the limits T_1 and T_2 gives

$$\ln\left(\frac{k_1}{k_2}\right) = \frac{E\,(T_1 - T_2)}{RT_1T_2} \tag{5.45}$$

From known values of k_1, T_1, and E, k_2 can be calculated by using Eqn (5.45) at T_2. Again, using the same equation, the activation energy can be calculated by determining k at two different temperatures. The values of E for waste treatment processes are usually fall in the range 8400–84,000 J/mol (2000 to 20,000 cal/mol).

The quantity $E/(RT_1T_2)$ may be assumed to be a constant for all practical purposes, because most wastewater treatment operations and processes are carried out at or near ambient temperatures. If this quantity is designated by C, then Eqn (5.45) can be rewritten as

$$\ln\left(\frac{k_2}{k_1}\right) = C(T_2 - T_1) \tag{5.46}$$

or

$$\frac{k_2}{k_1} = e^{C(T_2-T_1)} \tag{5.47}$$

$$= \theta^{(T_2-T_1)} \tag{5.48}$$

where $\theta = e^C$

Order of Reactions

For irreversible reactions, in which the rate of product formation is found to be independent of the concentration, the reaction is said to be of zero order and is defined as

$$r_A = \frac{d(A)}{dt} = -k_0 \tag{5.49}$$

Integrating for the initial condition $A = A_0$ gives

$$A_t = A_0 - k_0 t \tag{5.50}$$

For a first-order reaction, the rate is defined as

$$r_A = \frac{d(A)}{dt} = -k(A) \tag{5.51}$$

Integrating for the initial condition $A = A_0$, gives

$$\ln \frac{A_t}{A_0} = -k_1 t \quad \text{or} \quad A_t = (A_0)\, e^{-k_1 t} \tag{5.52}$$

For a second-order reaction $(A + A \longrightarrow P)$, the rate is defined as

$$\frac{d(A)}{dt} = -k(A)^2 \tag{5.53}$$

Integrating, we get

$$\frac{1}{A} - \frac{1}{A_0} = k_2 t \tag{5.54}$$

For a reversible reaction of the form

$$A \underset{k_2}{\overset{k_1}{\rightleftharpoons}} B \tag{5.55}$$

the rate of reaction is given by

$$r_A = \frac{d(A)}{dt} = -k_1 A + k_2 B \tag{5.56}$$

Biological Growth Kinetics

Information regarding the kinetics of biological growth is required while mass balancing micro-organisms and the substrate. In batch culture, bacteria increase in proportion to their mass during the log-growth phase. The rate of growth of this phase is defined by the following relationship:

$$r_g = \mu X \tag{5.57}$$

where r_g is the rate of bacterial growth[mass/(unit volume) (time)], $\mu = (dX/dt)/X$ is the specific growth rate (1/time), X is the concentration of micro-organisms (mass/volume), and t is the time. For batch culture,

$$\left(\frac{dX}{dt}\right) = r_g = \mu X \tag{5.58}$$

The following expression proposed by Monod (1950) is valid in the case of a limiting substrate or nutrient that is essential for microbial growth:

$$\mu = \frac{\mu_m S}{K_s + S} \tag{5.59}$$

where μ_m is the maximum specific growth rate (1/time), S is the concentration of growth-limiting substrates in media (mass/volume), and K_s is the half-velocity constant, i.e., the substrate concentration at one-half the maximum growth rate (mass/volume).

If μ from Eqn (5.59) is substituted in Eqn (5.57), then the expression for the rate of bacterial growth becomes

$$r_g = \frac{\mu_m XS}{K_s + S} \tag{5.60}$$

Cell Growth and Substrate Utilization

In the batch culture growth system, as described earlier, a portion of the substrate is converted into new cells and a portion is oxidized to inorganic and organic end products (recall part-reactions). This process is represented in general terms by using the following equations for bacterial oxidation and synthesis:

$$\underset{\text{(Organic matter)}}{COHNS + O_2 + \text{bacteria}} \longrightarrow \underset{+ \text{ energy}}{CO_2 + NH_3 + \text{other end products}} \tag{5.61}$$

Synthesis (assimilating):

$$\underset{\text{(Organic matter)}}{COHNS + O_2 + \text{bacteria} + \text{energy}} \longrightarrow C_5H_7O_2N \text{ (new bacterial cells)} \tag{5.62}$$

Endogenous respiration (anti-oxidation):

$$C_5H_7O_2N + 5O_2 \longrightarrow 5CO_2 + NH_3 + 2H_2O + \text{energy} \tag{5.63}$$

The following relationship has been developed between the rate of substrate utilization and the rate of growth.

$$r_g = Yr_{sc} \tag{5.64}$$

where r_g is the rate of bacterial growth [mass/(volume) (time)], Y is the maximum yield coefficient measured during any finite period of logarithmic growth consumed (mass cells/mass substrate), and r_{sc} is the substrate consumption rate [mass/(volume) (time)]. The yield depends on the various factors listed here.

1. Oxidation–reduction state of the carbon source and nutrient elements
2. Degree of polymerization of the substrate
3. Metabolism pathways
4. Specific growth rate
5. Various physical parameters of cultivation

If the value of r_g from Eqn (5.60) is substituted in Eqn (5.64), the rate of substrate consumption is modified as follows:

$$r_{sc} = \frac{-\mu_m XS}{(K_s + S) Y} \tag{5.65}$$

The term (μ_m/Y) is often replaced by k and is defined as the maximum rate of substrate consumption per unit mass of micro-organisms. So,

$$k = \frac{\mu_m}{Y} \tag{5.66}$$

By substituting this value in Eqn (5.65), the resulting expression is

$$r_{sc} = \frac{-kXS}{K_s + S} \tag{5.67}$$

Effects of Endogenous Metabolism

In any biological process, including wastewater treatment, all the cells are not at the same time, in the log-growth phase. Thus, a portion of the energy is utilized for cell maintenance as well. Therefore, the expression for the rate of growth should be modified such that it includes other factors such as death and predation. For convenience, these factors are clubbed together and it is assumed that the decrease in cell mass caused by these factors is proportional to the concentration of organisms present. This decrease is often identified as the endogenous decay, which can be formulated as follows:

$$r_d \text{ (endogenous decay rate)} = -k_d X \tag{5.68}$$

where k_d is the endogenous decay coefficient (d^{-1}) and X is the concentration of cells (mass/volume).

When Eqn (5.68) is combined with Eqns (5.60) and (5.64), the following expressions are obtained for the net rate of growth:

$$r_g' = \frac{\mu_m XS}{K_s + S} - k_d X \tag{5.69}$$

$$r_g' = -Y r_{sc} - k_d X \tag{5.70}$$

where r_g' is the net rate of bacterial growth [mass/(volume) (time)]. The corresponding expression for the net specific growth rate is

$$\mu' = \frac{\mu_m S}{K_s + S} - k_d \tag{5.71}$$

where μ' is the net specific growth rate (1/time).

The effects of endogenous respiration on the net bacterial yield are accounted for by defining an observed yield as follows:

$$Y_{obs} = \frac{r_g'}{r_{sc}} \tag{5.72}$$

Mass Balance and Application of Kinetics

Mass is neither created nor destroyed, it only undergoes some changes in its nature. The analysis of mass balance, as shown in Fig. 5.1, reveals that

1. the volumetric flow rate into and out of the container is the same,
2. the contents inside the reactor are uniformly distributed,
3. all chemical and biological changes occur within the reactor vessel, and
4. the rate of change within the reactor is governed by the rate of the, first-order reaction $r_c = -kC$.

From the stated assumptions, the mass balance equations can be formulated as follows:

$$\text{Rate of accumulation of reactants} = \text{rate of inflow} - \text{rate of outflow} + \text{rate of utilization} \tag{5.73}$$

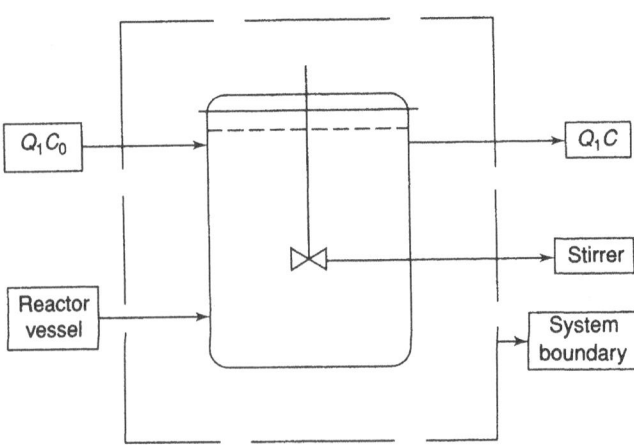

Fig. .   Mass balance system

$$V \frac{dC}{dt} = QC_o - QC + Vr_c \tag{5.74}$$

$$V \frac{dC}{dt} = QC_o - QC + V(-kC) \tag{5.75}$$

where V is the volume of the reactor, dC/dt is the rate of change of reactant concentration in the reactor [mass/(volume)(time)], Q is the volumetric rate of inflow and outflow of liquid (volume/time), C_o is the reactant concentration in the inflow (mass/volume), C is the concentration of the reactant in the reactor and outflow (mass/volume), and k is the first-order reaction rate constant (1/time). The solution of Eqn (5.75) is

$$C = \frac{QC_o}{V\beta} + Ke^{-\beta t} \tag{5.76}$$

where $\beta = (k + Q/V)$ and K is the integration constant. When $t = 0$, $C = C_o$, then K is given by

$$K = C_o - \frac{QC_o}{V\beta} \tag{5.77}$$

Substituting this value of K in Eqn (5.76), we get

$$C = \frac{QC_o}{V\beta}(1 - e^{-\beta t}) + C_o e^{-\beta t} \tag{5.78}$$

When $t \longrightarrow \infty$, Eqn (5.78) gives

$$C = \frac{QC_o}{V\beta} = \frac{C_o}{\left[1 + k\left(\dfrac{V}{Q}\right)\right]} \tag{5.79}$$

Under steady-state conditions, $dC/dt = 0$; then the value of C from Eqn (5.75) is given as

$$C = \frac{C_o}{\left[1 + k\left(\dfrac{V}{Q}\right)\right]} \tag{5.80}$$

We find that both Eqns (5.79) and (5.80) give the same expression for C.

Figure 5.1 represents a continuous stirred tank reactor (CSTR) system, for which mass balance equations have been developed without the use of micro-organisms. We will now study the application of kinetics in mass balances for different reactor systems including CSTR (Fig. 5.2), where the micro-organisms are active agents. For the CSTR system, the rate of accumulation of microbes is given by the following expression:

Rate of accumulation = rate of inflow of micro-organisms
of micro-organisms   – rate of outflow of micro-organisms
                       + net rate of growth of micro-organisms
                       within the system

Mathematically, we have

$$\left(\frac{dX}{dt}\right)V = QX_o - QX + V(r_g')$$ (5.81)

where dX/dt is the rate of change of micro-organism concentration in the reactor measured in terms of mass, i.e., volatile suspended solids [mass VSS/(volume)(time)], V is the reactor volume, Q is the flow rate (volume/time), X_o is the concentration of micro-organisms in the influent [mass VSS/volume), X is the concentration of micro-organisms in the reactor [mass VSS/(volume) (time)], and r_g' is the net rate of micro-organism growth [mass VSS/(volume) (time)]. Substituting the value of r_g' from Eqn (5.69) into Eqn (5.81):

$$\left(\frac{dX}{dt}\right)V = QX_o - QX + V\left[\frac{\mu_m XS}{K_s + S} - k_d X\right]$$ (5.82)

If it is reasonably assumed that $X_o = 0$ and we know that for the steady-state condition $dX/dt = 0$, Eqn (5.82) can be simplified to give

$$\frac{Q}{V} = \frac{1}{\theta} = \frac{\mu_m S}{K_s + S} - k_d$$ (5.83)

where θ is the hydraulic detention time (V/Q). The substrate balance (Fig. 5.2) corresponding to the micro-organism's mass balance, given by Eqn (5.82), is as follows:

$$\left(\frac{dS}{dt}\right)V = QS_o - QS + V\left[\frac{-kXS}{K_s + S}\right]$$ (5.84)

At steady state, $dS/dt = 0$, therefore the resulting equation will be

$$S_0 - S - \theta\left[\frac{kXS}{K_s + S}\right] = 0$$ (5.85)

where $\theta = V/Q$.

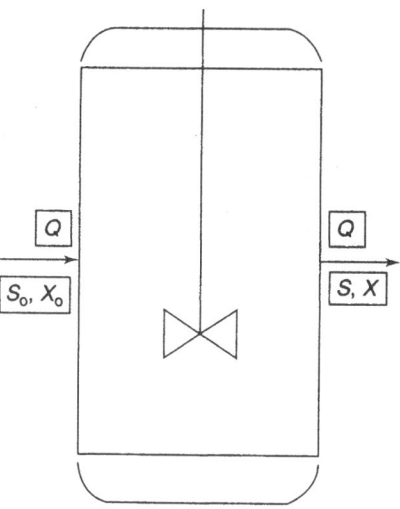

Fig. .   Continuous stirred tank reactor

Prediction of Micro-organism and Substrate Concentrations in CSTR

From Eqn (5.83) it can be seen that

$$\frac{S}{K_s + S} = \frac{1 + \theta k_d}{\mu_m \theta} \tag{5.86}$$

If this equation is substituted in Eqn (5.85) and rearranged, we will get the following expression for cell mass concentration in the CSTR and effluent:

$$X = \frac{\mu_m (S_o - S)}{k (1 + \theta k_d)} \tag{5.87}$$

Again, from Eqn (5.66), we have $Y = \mu_m / k$, and after substituting this in Eqn (5.87), we get

$$X = \frac{Y (S_o - S)}{1 + \theta k_d} \tag{5.88}$$

Rearranging Eqn (5.86) and substituting $\mu_m = Yk$, we can obtain the following expression for substrate concentration in the CSTR under steady-state conditions:

$$S = \frac{K_s (1 + \theta k_d)}{\theta (Yk - k_d) - 1} \tag{5.89}$$

Thus, if the kinetic coefficients are known, then with the help of Eqns (5.88) and (5.89), the concentration of micro-organisms and the substrate can be predicted in the CSTR or the outflow.

Process Design Correlations

If r'_g in Eqn (5.81) is substituted with its value from Eqn (5.70), the resulting equation becomes

$$\left(\frac{dX}{dt}\right)V = QX_o - QX + V(-Yr_{sc} - k_d X) \qquad (5.90)$$

Assuming $X_o = 0$ and steady-state conditions, Eqn (5.90) reduces to

$$\frac{Q}{V} = \frac{1}{\theta} = \frac{-Yr_{sc}}{X} - k_d \qquad (5.91)$$

From Eqns (5.71) and (5.83) it is clear that the net specific growth rate (μ') can be defined as

$$\mu' = \frac{Q}{V} \qquad (5.92)$$

If Q and V are multiplied by X (cell concentration), the reciprocal of Eqn (5.92) gives the mean cell residence time (θ_c):

$$\theta_c = \frac{VX}{QX} \qquad (5.93)$$

where VX/QX is the mass of cells in the reactor/mass of cells wasted per day. Substituting θ_c for θ in Eqn (5.91), we get

$$\frac{1}{\theta_c} = \frac{-Yr_{sc}}{X} - k_d \qquad (5.94)$$

The term (r_{sc}/X) is known as the *specific substrate consumption rate* and the quantity r_{sc} is determined by using the following expression:

$$r_{sc} = -\left(\frac{Q}{V}\right)(S_o - S) = \frac{-(S_o - S)}{\theta} \qquad (5.95)$$

where ($S_o - S$) is the mass concentration of the substrate utilized (mass/volume or mg/L) and θ is the hydraulic detention time (d).

A term, closely related to the specific rate of consumption and also useful in practice, is the food–micro-organism ratio (F/M), which is defined as follows:

$$\frac{F}{M} = \frac{S}{\theta X} \qquad (5.96)$$

If E is the process efficiency, then

$$E = \left[\frac{(S_o - S)}{S_o}\right] \times 100 \qquad (5.97)$$

Then,

$$-\frac{r_{sc}}{X} = \left(\frac{F}{M}\right)\frac{E}{100}$$

(5.98)

where E is the process efficiency (%), S_o is the influent substrate concentration (mg/L), S is the effluent substrate concentration (mg/L), X is the mass of VSS in the reactor or effluent (mg/L), and r_{sc} is the substrate consumption rate (mg/L d).

Aerobic Treatment Process

Among the various biological treatment processes such as aerated lagoons, high-rate oxidation ponds, and the nitrification process, the activated sludge process is considered to be most effective in terms of its application and will be discussed under this section. The schematic diagram given in Fig. 5.3 represents the continuous flow stirred tank activated sludge system (reactor with cellular recycle).

To begin with, the contents of the reactor are thoroughly mixed and, for the purpose of analysis of the system, it is assumed that there are no micro-organisms in the influent wastewater. As shown in Fig. 5.3, the cells from the reactor are first made to settle in the settling tank and then returned to the reactor to increase the cell concentration or maintain the same at a higher level. To develop the kinetic model the following assumptions are made.

1. Biological reactions occur only in the reactor and no microbial action is assumed to be occurring in the settling tank.
2. For the calculation of the mean cell residence time, only the reactor volume is considered.

If θ_h denotes the mean hydraulic retention time for the total system, then

$$\theta_h = \frac{V_{rs}}{Q}$$

(5.99)

where V_{rs} is the volume of reactor + the volume of settling tank and Q is the flow rate of the influent wastewater. Again,

$$\theta_r = \frac{V_r}{Q}$$

(5.100)

where θ_r is the mean hydraulic retention time for the reactor and V_r is the volume of the reactor. Considering the flow diagram given in Fig. 5.3, the mean cell residence time θ_c can be calculated as follows:

$$\theta_c = \frac{V_r X}{Q_w X + (Q - Q_w) X_e}$$

(5.101)

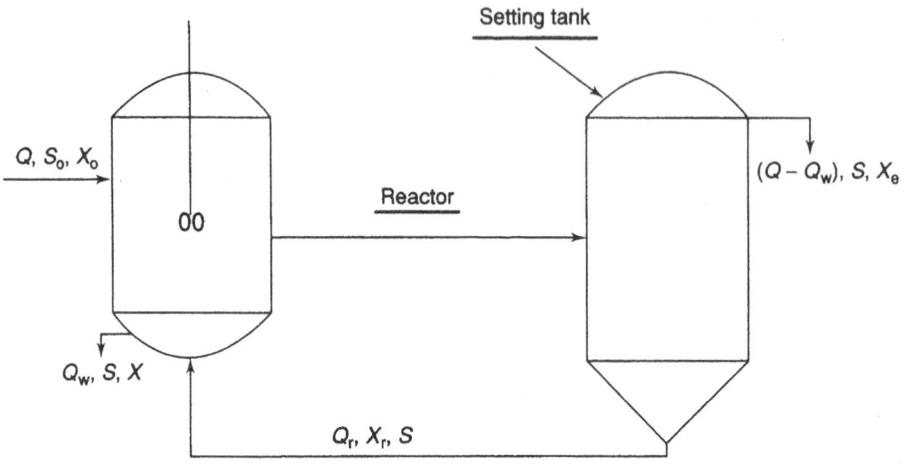

Reactor with cellular recycle

where Q_w is the volumetric flow rate of the liquid containing biological cells to be removed from the reactor/system and X_e is the concentration of micro-organisms in the effluent from the settling tank. For efficient operation of the settling unit, X_e can be assumed to be zero for all practical purposes. Then the expression for mean cell residence time becomes, approximately,

$$\theta_c = \frac{V_r}{Q_w} \tag{5.102}$$

If Eqn (5.102) is compared with Eqns (5.99) and (5.100), it can be seen that for a given reactor volume, θ_c is theoretically independent of both θ_r and θ_h, although practically it cannot be so. A mass balance relationship for the micro-organisms for the system is the following:

> Rate of accumulation = rate of inflow of micro-organisms
> of micro-organisms   – rate of outflow of micro-organisms
> + net growth of micro-organisms

i.e.,

$$\left(\frac{dX}{dt}\right)V_r = QX_o - [Q_w X + (Q - Q_w)X_e] + V_r\,(r_g') \tag{5.103}$$

Substituting r_g' from Eqn (5.70) and assuming $X_o = 0$ and steady-state condition (i.e., $dX/dt = 0$), Eqn (5.103) is modified as

$$\frac{[Q_w X + (Q - Q_w)X_e]}{V_r X} = \frac{-Y r_{sc}}{X - k_d} \tag{5.104}$$

Table 5.2 Kinetic coefficients

Coefficient	Unit	Value at 25°C	
		Range	Typical
k	d^{-1}	2–10	5.0
K_s	mg/L, BOD_5	25–100	60
Y	mg VSS/mg substrate, BOD_5	0.4–0.8	0.6
	mg VSS/mg substrate, COD	0.25–0.6	0.4
k_d	d^{-1}	0.04–0.075	0.06

With the use of Eqn (5.101), Eqn (5.104) can be simplified as

$$\frac{1}{\theta_c} = \frac{-Yr_{sc}}{X} - k_d \tag{5.105}$$

Table 5.2 gives the typical kinetic coefficients for the activated sludge process and these values can be used to solve Eqn (5.105) and other rate-dependent equations. The mass concentration of micro-organisms, X, in the reactor can be obtained by using Eqns (5.95) and (5.105) and solving for X, which gives

$$X = \left(\frac{\theta_c}{\theta}\right)\left[\frac{Y(S_o - S)}{1 + k_d\theta_c}\right] \tag{5.106}$$

In this expression $(S_o - S)$ can be determined from the corresponding values of BOD_5 or COD. However, determination of X is difficult, because the VSS of the effluent may not comprise only active cellular material. As θ_c can be determined easily from Eqn (5.102), this will be a better parameter to assess wastewater treatment efficiency.

Cell wastage can take place either from the reactor or from the sludge recycle line. However, in most of the biological treatment processes, cell wastage from the sludge recycle line is preferred. For this condition Eqn (5.101) is modified as

$$\theta_c = \frac{V_r X}{[Q'_w X_r + (Q - Q'_w) X_e]} \tag{5.107}$$

where X_r is the micro-organism concentration in the return sludge line (mg/L) and Q'_w is the cell wastage rate from the recycle line (L/d). Assuming X_e is negligible

$$\theta_c = \frac{V_r X}{Q'_w X_r} \tag{5.108}$$

Here, both X and X_r must be known in order to solve for θ_c.

Kinetics of Two-stage Bioremediation Process of Solid Wastes

Kinetic expressions for bioconversion processes are essential for the design of bioreactors. Various kinetic relationships have been proposed for anaerobic digestion of waste biomass and methane production. Most of the earlier models were based on the Monod equation. However, the validity of this equation for the treatment of municipal solid waste (MSW) is questionable. This is because, in the Monod equation, the specific growth rate is expressed only as a function of the concentration of the limiting substrate. It does not take into consideration the complex nature of the feed material such as of MSW. Another deficiency of the Monod equation is that it does not consider the influence of the influent substrate concentration on the effluent substrate concentration or vice versa. The Monod equation was, however, found to be true in the case of pure cultures (and not for heterogeneous cultures), when the substrate was measured as, say, glucose.

Some critical reviews on kinetic models of substrate utilization and methane production in the anaerobic digestion of complex organic wastes are available as literature. Two complimentary kinetic equations, one each for substrate utilization and methane production, based on the Contois and Monod models for continuous digestion in steady state in a completely mixed reactor without recycling are presented here for study.

First-stage Kinetics of Solid Waste Digestion

In the first stage, the complex solid wastes are first hydrolysed by extracellular enzymes into soluble and assimilable product such as glucose, which is easily transported into the cells and utilized for cell multiplication and the production of acetic acid and other metabolites. Though the first-stage operation is not strictly anaerobic, the system is assumed to be an anaerobic batch process for kinetic analysis. Therefore, the first-order decay model proposed by Jewell et al. (1980) is still followed but with some modifications. According to the first-order reaction kinetics, the change in the substrate concentration in a constant volume batch reactor can be expressed as follows on the basis of mass balance:

$$-\frac{dS_b}{dt}\,V = KS_bV \tag{5.109}$$

where S_b is the biodegradable VS concentration (kg m^{-3}) at time t, V is the digester volume (m^3), and K is the biodegradable VS decay coefficient (kg kg^{-1} d^{-1}). If Eqn (5.109) is integrated within the limits $S_b = S_{b0}$ at $t = 0$ and $S_b = S_b$ at $t = t$, it will give the following equation:

$$\ln\left(\frac{S_{b0}}{S_b}\right) = Kt \tag{5.110}$$

or

$$S_b = S_{b0}e^{-Kt} \tag{5.111}$$

where S_{b0} is the initial biodegradable VS concentration in the digester (kg m^{-3}) and t is the digestion time (d).

From Eqn (5.110) it is clear that a plot of ln S_{b0}/S_b versus t will give a straight line passing through the origin, with the slope giving the value of K. This indicates that in equal intervals of time, a certain definite fraction of biodegradable VS undergoes degradation. McCarty (1964) has shown that 0.35 m^3 of methane gas is produced per kg of COD destroyed, at STP (standard temperature and pressure). If the COD/VS ratio for a given waste is known, the amount of VS digested can be calculated from the quantity of methane gas produced.

During the batch digestion of MSW, it is difficult to measure accurately the VS content in the digester, at different intervals of time, to evaluate the batch digestion kinetics with the help of Eqn (5.110). Instead, if a relation is established between volatile fatty acids (VFA) (kg) (equivalent to acetic acid) produced per kg of COD or VS destroyed, it will be easy to evaluate the kinetic parameter K.

Theoretically 1 mol of CH_3COOH can produce 1 mol of CH_4 when the yield coefficient Y_{ma} is 0.37. Y_{ma} is defined as the quantity of methane, in m^3 (at STP), produced per kg of acetic acid consumed. In practice, however, Y_{ma} is generally less than 0.37. Therefore, theoretically, for the production of 0.35 m^3 of CH_4, 0.95 kg of acetic acid is required. In other words, 0.95 kg of acetic acid will be produced per 1 kg of COD destroyed. If Y_{ma} is different from 0.37, then per kg COD destroyed, the quantity of acetic acid produced will be

$$\frac{(0.95)\,(0.37)}{Y_{ma}} \quad \text{or} \quad \frac{0.35}{Y_{ma}}\,\text{kg}$$

If F_{TO} denotes the total acetic acid (in kg) which can be potentially produced through the complete digestion of the initial biodegradable VS in the digester (S_{b0}) and F_T is the cumulative acetic acid (in kg) produced in the first-stage batch digester at any time t, then the following relations hold true:

$$F_{TO} = \frac{0.35\beta S_{b0}}{Y_{ma}} \tag{5.112}$$

$$F_T = \frac{0.35\,(S_{b0} - S_b)\beta}{Y_{ma}} \tag{5.113}$$

where $\beta = $ COD/VS ratio (kg kg^{-1}). Then,

$$F_{TO} - F_T = 0.35 S_b \left(\frac{\beta}{Y_{ma}} \right) \tag{5.114}$$

Using Eqns (5.112) and (5.113), Eqn (5.110) can be expressed as

$$\ln \frac{F_{TO}}{F_{TO} - F_T} = Kt \tag{5.115}$$

$$\text{or} \quad F_T = F_{TO}(1 - e^{-Kt}) \tag{5.116}$$

Experimentally, K can be evaluated using Eqn (5.115). Substituting K in Eqn (5.116), the volatile acid production at any time t can be estimated. COD or VS destruction can be calculated from Eqn (5.113). In the absence of much more reliable experimental data, F_{TO} can be taken as $0.4/Y_{ma}$ kg kg^{-1} of biodegradable VS.

The volatile solids in MSW are not totally biodegradable, as they contain refractory fractions in varying degrees. Therefore, total volatile solid (TVS) content is defined as the sum of the biodegradable volatile solid (BVS) and refractory volatile solid (RVS) quantities. According to Jewell et al. (1980), the refractory fraction of the TVSs is defined as 'that portion of the TVS that is resistant to the microbial attack over long periods of time and remains after the rate of degradation of the TVS decreases to a very low value'. Chandler et al. (1980) showed that lignin controls the extent of substrate biodegradation during methane fermentation for a large number of substrates and developed a quick method for the determination of substrate biodegradability based on the estimation of VS lignin content.

Jewell et al. (1980) advanced the following conceptual models for biodegradable refractory fractions of the TVS in the digestion system:

$$S_0 = S_{b0} + S_{r0} \tag{5.117}$$

where S_0 is the initial TVS concentration (kg m^{-3}), S_{b0} is the initial BVS concentration (kg m^{-3}), and S_{r0} is the initial RVS concentration (kg m^{-3}). The refractory solids in both the initial and the digested biomass will be the same, as RVS is resistant to biodegradation. Therefore, we have

$$S_{r0} = S_{re} = S_r \tag{5.118}$$

where S_{re} is the RVS concentration in residues (kg m^{-3}) and S_r is the substrate RVS concentration (kg m^{-3}). Then, the character of the residues of the digestion system can be expressed as

$$S_T = S_b + S_{re} = S_b + S_{r0} \tag{5.119}$$

where S_T is the TVS concentration in residues (kg m^{-3}) and S_b is the BVS concentration in residues (kg m^{-3}).

As the estimation of S_{re} is difficult and mostly not possible, the value of S_b, which is very important in the digestion process, can be calculated from the values of S_T and S_{r0}, which can be more readily determined.

The refractory fraction or coefficient (R) of TVS in the feed material is expressed as

$$R = \frac{S_r}{S_0} \tag{5.120}$$

The anaerobic biodegradability of substrates is generally estimated through the long-term batch digestion process until no product (acetic acid or methane) is produced. The VS which remains undigested at the end of the experiment is considered the refractory fraction of the TVS.

In another empirical approach, it is assumed that as the solid retention time (SRT) approaches infinity, the biodegradable fraction of the substrate is completely destroyed, leaving only the refractory fraction. In batch digestion, SRT will be equivalent to the residence time t.

Direct estimation of S_T is difficult due to sampling problems. It can be indirectly calculated by subtracting the content of VS digested to produce the total acetic acid or biogas during time t from S_0.

Reliable kinetic data are essential for the determination of digester size, digestion capacity (reduction in COD or VS), product formation, etc. However, MSW is such a complex heterogeneous waste that its composition cannot be predicted accurately. As a result, approximation cannot be avoided during the designing of MSW treatment plants. Hence, to be on the safe side, a sufficient factor of safety should be provided during design calculations.

In the first stage of the two-stage biomethanation process, where hydrolysis and acid formation occur, the total solid concentration is substantially high and not less than 30%. Hydrolysis is the rate controlling process and the VS decay coefficient K is very much influenced by the TS concentration in the digester. At a low TS concentration, K may be taken as 0.09 d^{-1}, while at a high TS concentration, the value of K may become as low as 0.02 d^{-1}. The refractory coefficient R can be calculated experimentally from the following correlation:

$$\frac{S_T}{S_0} = a(S_0 t)^{-1} + R \tag{5.121}$$

where S_T is the TVS concentration at any time t and S_0 is the initial TVS concentration.

A plot of S_T/S_0 versus $(S_0 t)^{-1}$ will give a straight line relation, the slope of which will give the value of a and the interception at $(S_0 t)^{-1} = 0$ will give R. Depending on the nature of the MSW, R may vary from 0.3 to 0.5.

Depending on the TS concentration in the digester, the temperature, and the SRT, 30%–50% of the TVS gets degraded.

Second-stage Kinetics

In the first stage of the two-stage biomethanation system, MSW is hydrolysed and volatile acid is formed. This is leached by recirculating the liquid. VFA concentration reaches around 2000 g/m^3 in a short period. This VFA-rich leachate is the substrate for methane production in the second stage. In the previous section kinetic equations for the first stage have been presented based on the assumption that it is batch process. In this section the kinetic equation for biogas production will be presented considering VFA as the substrate. The second stage is operated semi-continuously, that is, the leachate is fed daily more than once into the reactor and the exhaust effluent is pumped back into the first-stage reactor to save water.

Cell growth on assimilable substrates is assumed to follow Monod kinetics and is expressed as

$$\mu = \frac{\mu_m S_\Lambda}{K_s + S_\Lambda} \tag{5.122}$$

where S_Λ is the equivalent acetic acid concentration (kg/m^3), K_s is the half-saturation constant with respect to the assimilable substrate (kg/m^3), μ is the specific growth rate (d^{-1}), and μ_m is the maximum specific growth rate (d^{-1}).

Under steady-state conditions of continuous digestion, without recycling, the following relationships hold true:

$$\mu = \frac{1}{\theta} \tag{5.123}$$

$$F = \frac{S_{\Lambda 0} - S_\Lambda}{\theta} \tag{5.124}$$

$$X = \frac{FY}{\mu} = (S_{\Lambda 0} - S_\Lambda)Y \tag{5.125}$$

where θ is the hydraulic retention time (d^{-1}), F is the volumetric substrate removal rate ($kg\ m^{-3}d^{-1}$), $S_{\Lambda 0}$ is the influent substrate concentration ($kg\ m^{-3}$), S_Λ is the substrate concentration in the effluent or in the digester ($kg\ m^{-3}$), X is the concentration of active cell biomass ($kg\ m^{-3}$), and Y is the biomass yield (kg cell mass/kg substrate). When acetic acid is the substrate, the refractory coefficient is negligible.

As per the derivation of Chen and Hashimoto (1978), if Y_m denotes the specific methane yield (in m^3/kg COD) or acetic acid at infinite retention time, the biodegradable COD or acetic acid in the digester will be directly proportional to $(Y_{m0} - Y_m)$ and Y_{m0} will be directly proportional to the biodegradable COD or acetic acid loading. Then the following relationship will hold:

$$\frac{(S_{\Lambda 0} - S_\Lambda)}{S_\Lambda} = \frac{Y_m}{Y_{m0} - Y_m} \tag{5.126}$$

Using Eqns (5.122) and (5.126), the following expression can be obtained:

$$\frac{Y_m}{Y_{m0}} = 1 - \frac{\dfrac{K_s}{S_{\Lambda 0}}}{\mu_m \theta - 1} \tag{5.127}$$

This equation shows that as θ approaches infinity, at constant influent substrate concentration, Y_m/Y_{m0} approaches unity.

The total volume of methane produced per unit volume of leachate, V_m, and the volumetric rate of methane production, M_v, can readily be calculated using the following expressions:

$$V_m = (S_{A0} - S_A)Y_m \tag{5.128}$$

$$M_v = \frac{(S_{A0} - S_A)Y_m}{\theta} \tag{5.129}$$

where V_m is the amount of methane produced per unit volume of leachate added ($m^3 \, m^{-3}$) and M_v is the methane production rate [m^3 (STP) m^{-3} (reactor vol.) d^{-1}]. Substituting Y_m from Eqn (5.127) in Eqn (5.129), we get

$$M_v = \left[\frac{Y_{m0}(S_{A0} - S_A)}{\theta}\right]\left[1 - \frac{\dfrac{K_s}{S_{A0}}}{\mu_m \theta - 1}\right] \tag{5.130}$$

The following are some useful kinetic data for the design of two-stage solid waste digesters: $K_s = 0.30$ kg m^{-3}, $Y_{m0} = 0.33$ m^3 (STP) kg^{-1}, $\mu_m = 0.44$ day^{-1}.

Though these values have been determined at 35°C and the influent concentration of acetic acid is 3.14 kg m^{-3}, for design purposes the same values can be used at other substrate concentrations also in the absence of more reliable data.

Summary

In this chapter different aspects of mass balance and reaction kinetics have been discussed. This provides the requisite quantitative approach without which no design calculation is possible. For example, for a definite quantity of industrial waste that needs to be treated every day, the time required for the desired biodegradation can be calculated from data on the mass balance input and output and from reaction kinetics. Thus, from mass balance and degradation time, the capacity of the reactor vessel can be found. Therefore, we shall now study as to how the information provided in Chapter 5 can be utilized in Chapters 6 and 7, which deal with the design of bioreactors for liquid effluents and solid wastes, respectively.

Review Questions

1. Without the knowledge of stoichiometry and reaction kinetics, will it be possible to design any reactor system? Justify your answer.
2. Equation (5.2), given in the text, involves three terms. Under what condition/s is this equation valid and when does accumulation become equal to zero?
3. Using the basic concept of mass balance, prove that Eqn (5.3) of the text is a balanced equation.
4. Derive Eqn (5.4) from Eqn (5.3) of the text.

5. A microbial cell mass is isolated from some spoiled food stuffs and is given to you. Explain the experimental procedure you will follow to determine the empirical formula of the isolated cell mass.

6. You have been provided with 10 g of cell mass whose empirical formula is $C_5H_7O_2N$ and, when fully digested, releases CO_2, H_2O, and NH_3. Determine the COD value of the cell mass.

7. Why are oxidation–reduction reactions so important in fermentation processes?

8. Why is the microbial growth rate higher in aerobic process than in anaerobic process?

9. Justify the presence of CO_2 as an essential requirement for methanogenesis.

10. Define the degree of reduction of a compound. What is its significance and what information do you get from its magnitude? What is the degree of reduction for alanine and casein?

11. Wastewater from a food products industrial complex is treated aerobically by the activated sludge process resulting in the production of a large amount of biomass, the disposal of which becomes a headache for the management. On consultation, a reputed biotechnologist suggested anaerobic degradation of the biomass to reduce the volume. By the experimental method, the empirical formula of the volatile content of biomass was found to be $C_5H_7O_2N$. When this biomass is experimentally digested under anaerobic conditions, the biochemical reaction gives the following generalized correlation (not balanced):

$$C_5H_7O_2N + H_2O \longrightarrow CH_4 + CO_2 + NH_4^+ + HCO_3^-$$

   (a) Through elemental mass balance, establish the balanced equation.
   (b) If the daily production of raw biomass is 1.0 t, which contains 12% volatile substance, what will be the daily production of methane gas in m^3 at STP?
   (c) What is the degree of reduction of biomass?

12. Prove that the theoretical yield coefficient is 0.42 when glucose is the substrate for biomass production under aerobic conditions. Also give the unit of this value.

13. A 10-L fermenter is operated continuously on the chemostat principle to produce a desired micro-organism. The cell density in the fermenter is maintained at 1 gL^{-1}. Glucose is used as the carbon source. The kinetic data available are yield coefficient = 0.5 g cells/g substrate, decay rate of organism = 0.15 d^{-1}, and specific rate of substrate utilization = 1.2 g glucose/g cells/day.
   (a) Calculate the specific growth rate.
   (b) What percentage of the true yield is the net yield? Explain the difference in these values.

14. Domestic wastewater contains biodegradable organic matter. By analysing its content of organic elements, the average empirical formula proposed is $C_{10}H_{19}O_3N$, which is the electron donor for biological conversion. Formulate the balanced equation for this electron donor.

15. If domestic wastewater is treated aerobically, where O_2 is the electron acceptor, determine the grams of O_2 required per electron equivalent of the organic compound (see Question 14) in wastewater, when 40% of the available electrons are used for cell synthesis, assuming sufficient ammonium is present for this purpose.

Hint:

(i) $\left(\dfrac{1}{4}\right)O_2 + H^+ + e^- = \left(\dfrac{1}{2}\right)H_2O$

(ii) $\left(\dfrac{1}{5}\right)CO_2 + \left(\dfrac{1}{20}\right)HCO_3^- + \left(\dfrac{21}{20}\right)NH_4^+ + e^- = \left(\dfrac{1}{20}\right)C_5H_7O_2N + \left(\dfrac{9}{20}\right)H_2O$

16. It is decided that the organic compound (question 14) in wastewater will be degraded anaerobically for the production of CH_4. Determine the percentage of methane that will be produced as biogass if 20% of the available electrons are used for cell synthesis. Assume that nitrate is available as the nitrogen source for cell synthesis and the bioconversion efficiency is 80%.

Hint:

(i) $\left(\dfrac{1}{28}\right)NO_3^- + \left(\dfrac{5}{28}\right)CO_2 + \left(\dfrac{29}{28}\right)H^+ + e^- = \left(\dfrac{1}{28}\right)C_5H_7O_2N + \left(\dfrac{9}{20}\right)H_2O$

(ii) $\left(\dfrac{1}{8}\right)CO_2 + H^+ + e^- = \left(\dfrac{1}{8}\right)CH_4 + \left(\dfrac{1}{4}\right)H_2O$

17. A sample of groundwater is found to be contaminated with nitrate and its concentration is found to be 84 mg/L. It is decided that by addition of methanol to groundwater, denitrification is to be achieved by anaerobic biological treatment. What should be the minimum concentration of methanol that needs to added to achieve complete reduction of nitrate to nitrogen gas? Assume that 30% of the available electrons are used for cell synthesis and no ammonium is present in the groundwater sample.

Hint:

(i) $\left(\dfrac{1}{6}\right)CO_2 + H^+ + e^- = \left(\dfrac{1}{6}\right)CH_3OH + \left(\dfrac{1}{6}\right)H_2O$

(ii) $\left(\dfrac{1}{28}\right)NO_3^- + \left(\dfrac{5}{28}\right)CO_2 + \left(\dfrac{29}{28}\right)H^+ + e^- = \left(\dfrac{1}{28}\right)C_5H_7O_2N + \left(\dfrac{11}{28}\right)H_2O$

(iii) $\left(\dfrac{1}{5}\right)NO_3^- + \left(\dfrac{6}{5}\right)H^+ + e^- = \left(\dfrac{1}{10}\right)N_2 + \left(\dfrac{3}{5}\right)H_2O$

18. Per equivalent of electron donors oxidizesd, would you expect to obtain more cell products from anaerobic conversion of glycerol to methane or from oxidation of lactate through reduction of sulphate to sulphite? Why?

Hint:

(i) $\left(\dfrac{3}{14}\right)CO_2 + H^+ + e^- = \left(\dfrac{1}{14}\right)CH_2OHCHOHCH_2OH + \left(\dfrac{3}{14}\right)H_2O$

(ii) $\left(\dfrac{1}{6}\right)CO_2 + \left(\dfrac{1}{12}\right)HCO_3 + H^- + e^- = \left(\dfrac{1}{12}\right)CH_3CHOHCOO^- + \left(\dfrac{1}{3}\right)H_2O$

(iii) $\left(\dfrac{1}{8}\right)CO_2 + H^+ + e^- = \left(\dfrac{1}{8}\right)CH_4 + \left(\dfrac{1}{4}\right)H_2O$

(iv) $\left(\dfrac{1}{5}\right)CO_2 + \left(\dfrac{1}{20}\right)HCO_3 + \left(\dfrac{1}{20}\right)NH_4^+ + H^+ + e^- = \left(\dfrac{1}{20}\right)C_5H_7O_2N + \left(\dfrac{9}{20}\right)H_2O$

(v) $\left(\dfrac{1}{8}\right)SO_4^- + \left(\dfrac{19}{16}\right)H^+ + e^- = \left(\dfrac{1}{16}\right)H_2S + \left(\dfrac{1}{16}\right)HS^- + \left(\dfrac{1}{2}\right)H_2O$

19. In oxidation–reduction reactions, which chemical compounds or elements are commonly used as electron acceptors for aerobic and anaerobic systems? Which is the best electron acceptor for the aerobic process? Why? Under what conditions do organic substrates act both as electron donors and as electron acceptors?

20. Define the types and rates of reactions. How are the different rates of reactions expressed with respect to their types?

21. (a) What is the difference between reversible and irreversible reactions? Which one is faster and why?
    (b) Define reaction rate constant.
    (c) How will you explain the order of a reaction? Develop kinetic expressions for first-and second-order reactions.

22. What is Monod's expression for bacterial growth? State its limitations.

23. Prove that for batch culture and the condition of substrate limitation, the rate of bacterial growth (r_g) is expressed by the following correlation:

$$r_g = \frac{\mu_m S}{K_s + S}$$

where μ_m is the maximum specific microbial growth rate, X is the concentration of micro-organisms, S is the concentration of growth-limiting substrates in the media, and K_s is the half-velocity constant (the substrate concentration at one-half the maximum growth rate).

24. In a CSTR system the hydraulic retention time θ can be expressed as

$$\theta = \left[\frac{\mu_m S}{K_s + S} - k_d \right]^{-1}$$

Derive this expression.

25. Prove that the cell mass concentration in CSTR under steady-state conditions is

$$X = \frac{\mu_m (S_0 - S)}{k(1 + \theta k_d)}$$

26. Derive an expression for substrate concentration in a CSTR under steady-state conditions.

27. If per kg COD destruction yields 0.35 m³ of methane gas at STP and the COD/VS ratio of the given solid waste is known, show how it is possible to determine the amount of VS digested from the volume of methane gas produced?

28. What are the two major constituents of TVS and how are they different from each other?

29. What is RVS?

30. It is decided that a solid food waste will be treated in a two-stage biomethanation system. To design the system, kinetic data are needed. If the waste sample is given to you, explain in detail, step-by-step, how you will obtain the required kinetic data necessary for the design of the system experimentally?

References

Chandler, J.A., W.J. Jewell, J.M. Gosett, P.J. Van Soost, and J.B. Robertson 1980, 'Predicting methane fermentation biodegradability', *Biotechnol. Bioeng. Symp.*, no. 10, pp. 93–107.

Chen, Y.R. and A.G. Hashimoto 1978, 'Kinetics of methane fermentation', *Biotechnol. Bioeng. Symp.*, no. 8, pp. 269–282.

Contois, D. 1959, 'Kinetics of bacterial growth relationship between population density and specific growth rate of continuous cultures', *J. Gen. Microb.*, vol. 21, pp. 40–50.

Jewell, W.J, S. Dell'Orto, K.J. Fanfoni, T.D. Hayes, A.P. Leuschner, and D.F. Sherman 1980, *Anaerobic Fermentation of Agricultural Residue: Potential for Improvement and Implementation*, Final Report NTIS, vol. 3, US Dept of Commerce, Springfield, VA.

McCarty, P.L. 1964, 'Anaerobic waste treatment fundamentals', Part I, *Chem. Microbiol.*, vol. 95, no. 9, pp. 107–12.

Metcaffe and Eddy 1979, *Wastewater Engineering Treatment Disposal Reuse*, 2nd edn, Tata McGraw-Hill, New Delhi.

Monod, J. 1950, '*La technique de culture continue; theorie et applications*', *Ann. Inst. Pasteur*, vol. 79, p. 390.

Rittmann, B.E. and P.L. McCarty 2001, *Environmental Biotechnology: Principles and Applications*, McGraw-Hill, New York.

Shuler, M.L. and F. Kargi 2002, *Bioprocess Engineering: Basic Concepts*, 2nd edn, Prentice-Hall Englewood Cliffs, NJ.

Tchobanoglous, G., H. Theisen, and S. Vigil 1993, *Integrated Solid Waste Management*, McGraw-Hill, New York.

6

Design of Bioreactors for Liquid Waste Treatment

Introduction

As already described in Chapter 2 the nature of a liquid waste may vary depending on its source. The major sources from which liquid effluents are discharged are municipality sewerage, dairy and other food processing industries, chemical industries, tanneries, and similar other industrial complexes. Though these effluents may differ from each other in terms of their composition, from the point of pollution abatement, they have one common characteristic on the basis of which they can be identified, which is the BOD (biological oxygen demand)/COD (chemical oxygen demand). This parameter also serves as a measure of the pollution level. Hence, while designing pollution control devices, the focus is on the reduction of the BOD/COD values of wastes.

Chapter 1 mentions that the bioreactor systems used for treatment of liquid wastes can be either aerobic or anaerobic in nature. Among the various aerobic treatment processes, the *activated sludge process* is most popular. In this chapter we will study this system in detail. We will also study the design of anaerobic bioreactor systems. Chapter 5 has already discussed material balance and kinetic expressions for biological systems, which are useful for the design of the bioreactor systems being presented in this chapter.

Canteen waste bioreactor

Activated Sludge Process

The activated sludge process (ASP) can be defined as a continuous or semi-continuous aerobic process used for biological wastewater treatment. The process primarily includes aeration of wastewater in the reactor for biological growth followed by the separation of this biomass from the treated wastewater in the settling tank. In this system a part of the biomass is removed as waste and the remainder is recycled.

The activated sludge process is currently one of the most widely used technologies for wastewater purification. This method involves a variable and mixed community of micro-organisms functioning in an aerobic aquatic environment. As discussed in Chapter 5 these micro-organisms derive energy from the carbonaceous organic matter present in aerated wastewater to produce new cells by a process known as *synthesis*. Simultaneously energy is

Activated sludge process

also released through the conversion of this organic matter into low-energy compounds such as carbon dioxide and water in the process called *respiration*. The system consists of different types of micro-organisms which obtain energy by converting ammonia nitrogen into nitrate nitrogen in a process termed *nitrification*. The biological components of these processes are collectively known as the *activated sludge* as it is actually a aggregation of micro-organisms. (We have already discussed the scientific aspects of these microbes in Chapter 3.)

Apart form these biological components, there are also a set of physical components that influence the efficiency of the process. The ASP comprises five essential interrelated equipment systems. These are as follows.

1. An *aeration tank or tanks* through which air or oxygen is pumped into the system to create an aerobic environment in order to meet the needs of the microbial community and ensure proper mixing of the activated sludge.

2. In order to ensure that there is adequate supply of oxygen into the tank(s) and that mixing takes place appropriately, an *aeration source* is required. This may be pure oxygen, compressed air, or provided by the help of mechanical agitation.

3. In *secondary clarifiers*, activated sludge solids are separated from the surrounding water by the process of flocculation and gravity sedimentation. In this process, flocs settle at the bottom of the clarifier in a quiescent environment. This separation leads, ideally, to the formation of a secondary effluent having a lower level of activated sludge solids, in suspension, in the upper portion of the clarifier and a thickened sludge comprising of flocs, termed *return activated sludge* or RAS, in the bottom portion of the clarifier.

4. RAS must be collected from the secondary clarifiers and pumped back into the aeration tanks before the quantity of DO (dissolved oxygen) finally gets reduced, hence *recycling the biomass*. In this way, the biological community needed to metabolize the influent organic or inorganic matter in the wastewater stream is replenished.

5. In the final step, the activated sludge containing an overabundance of micro-organisms, called *waste activated sludge* or WAS, must be removed or wasted from the system. This is accomplished with the use of pumps and is instrumental in controlling the food to micro-organism ratio in the aeration tank (see Chapter 5).

Figure 6.1 shows the flow diagram of a conventional activated sludge treatment system. The numbers in the boxes in Fig. 6.1 represent the various components of the treatment plant, which are explained next.

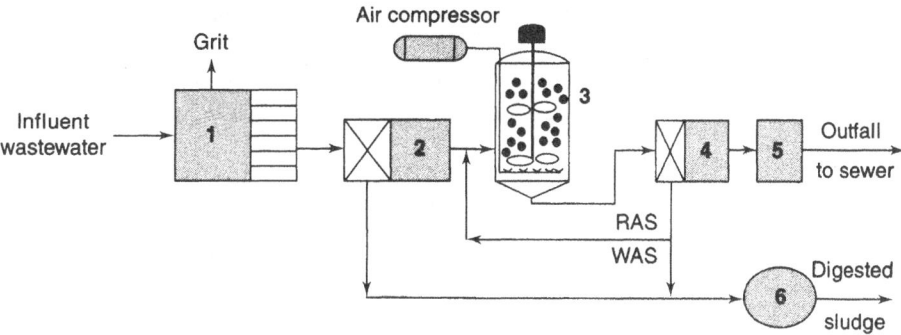

Flow diagram of a conventional activated sludge treatment system

Screening and grit units   The purpose of these units (1) is to remove large objects such as logs, branches, rags, and small fish, which can damage the pumps and clog pipes and channels. At this stage waste can also be ground to reduce particle size.

Primary settling tank(s)   Here (2), suspended particles are removed from wastewater based on the principle of gravity sedimentation. Particulates, found suspended in surface water, usually range in size from 10^{-1} to 10^{-7} mm in diameter, about the size of fine sand and small clay particles, respectively. Turbidity or cloudiness of water is caused by particles of size larger than 10^{-4} mm, while particles smaller than 10^{-4} mm contribute to the colour and taste of water. Such very small particles need to be considered for treatment purposes, to be dissolved rather than particulate. Water containing particulate matter is allowed to flow slowly through a sedimentation tank and is detained long enough for the clarified water to leave the tank over a weir at the outlet end. Particles that have settled to the bottom of the tank are removed manually or by mechanical scrapers. The detention time in a tank, which is usually 3 to 5 m deep, is typically 3 hours.

Aeration tanks   An aeration tank (3) is usually constructed with steel, poly, fiberglass, or concrete. The wastewater that flows into the aeration chamber is normally retained for 6 to 24 hours. The contents of the aeration tank are referred to as *mixed liquor*, and the solids are called *mixed liquor suspended solids* (MLSS). The latter includes inert material as well as living and dead microbial cells. In the aeration tank, micro-organisms are kept in suspension for 4 to 8 hours by mechanical mixers and/or diffused air. Their concentration in the tank is maintained by the continuous return of the settled biological floc (RAS) from the secondary settling tank (4) to the aeration tank and also by wasting a part of the activated sludge (WAS) through the biomass stabilization tank (6). Treated and settled wastewater is discharged through a tank (5) to

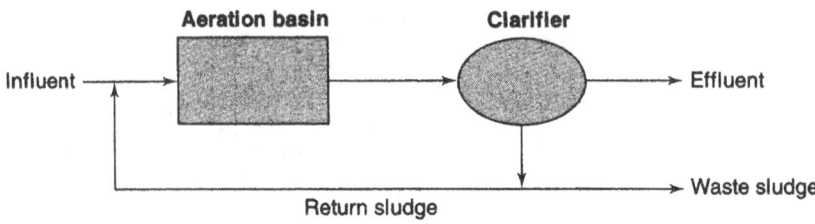

Flow diagram for complete mix activated sludge process

ensure the quality of effluent. Figure 6.2 shows the design of a commonly used aeration tank.

Complete mix   In this method wastewater may be immediately mixed throughout the entire contents of the aeration tank. It is the most common method being used today. Since the wastewater is completely mixed with bacteria and oxygen, the volatile suspended solids concentration and the oxygen demand are the same throughout the tank.

Contact Stabilization

Contact stabilization is a modified version of the activated sludge process as shown in Fig. 6.3. In this method micro-organisms consume the organic matter available in the contact tank.

To begin with, raw wastewater flows into the contact tank, where it is aerated and mixed with bacteria. Here, soluble materials pass through bacterial cell walls, while insoluble materials stick to the outside of the cell walls. Solids that settle down are wasted from the system or returned to the stabilization tank. Microbes digest the organic matter in the stabilization tank, and are then recycled back to the contact tank, because they need more food. The detention time is minimized, so that the size of the contact tank can be reduced. Accordingly, the capacity of the stabilization tank is also small because the basin receives only the concentrated return sludge and there is no incoming raw wastewater. In this type of plant, often, there is no primary

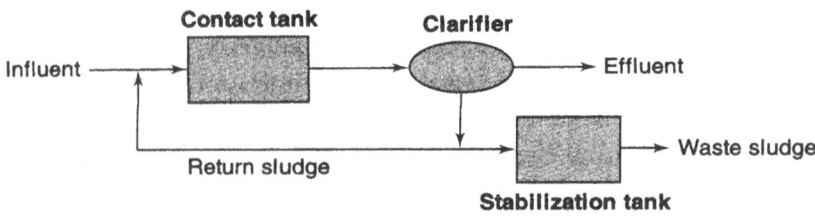

Fig. .3   Flow diagram for contact stabilization process

clarifier before the contact tank due to the rapid uptake of soluble and insoluble food.

Advantages

1. The method is highly adaptable—it can be used for a single household or even on an industrial scale as a huge plant.
2. Removes organics.
3. Effects oxidation and nitrification in the system.
4. Results in biological nitrification without requiring the addition of chemicals.
5. Removes biological phosphorus.
6. Causes the separation of solids and liquids.
7. Stabilizes sludge.
8. Is capable of removing up to 97% of suspended solids.
9. Is the most widely used wastewater treatment process.

Disadvantages

1. Does not remove colour from industrial wastes but may add colour, as it effects the formation of highly coloured intermediates during the process of oxidation.
2. Does not remove nutrients and hence tertiary treatment is necessary.
3. Does not provide well-settled sludge.
4. The recycled biomass maintains a high biomass concentration in the aeration tanks and thus necessitates higher power consumption for agitation.

The microbiological aspects of this process have been discussed in Chapter 3. Bacteria, fungi, protozoa, and rotifers constitute the biological component, or biological mass, of activated sludge. In addition, some metazoa, such as nematode worms, may be present. However, the constant agitation in the aeration tanks and sludge recirculation are deterrents to the growth of higher organisms.

Design of Activated Sludge Process

Designing any system including the activated sludge process is a creative activity that integrates a series of engineering judgements with mathematical equations that are based on the concepts already presented in Chapters 3 and 5. Most of the equations used in the design procedure have been derived in Chapter 5. In a few cases, however, a new relationship may be required, which we will now study along with the procedure. The two major equipment involved in an ASP are the *reactor* and *settler*. Let us first study the design approach in the case of the reactor.

Influent

First, the term influent must be understood clearly. The influent in this case is wastewater; its characteristics must be known before it can be treated. Before proceeding further, let us assume certain figures, which will help in design calculations.

(i) Flow rate of the inflow of wastewater = Q = 600 m^3/d
(ii) Substrate concentration in inflow of wastewater = S_0 = 300 mg BOD$_L$/L
(iii) Active biomass concentration in inflow of wastewater = X_0 = 0 mg VSS/L

S_0 can be expressed as BOD$_5$ also. BOD$_L$ helps in complete mass balance on the electron equivalents. Here S_0 is assumed to be the soluble or particulate matter that is hydrolysed to soluble BOD$_L$ within the activated sludge system. BOD means biological oxygen demand, with 'L' indicating ultimate, and VSS refers to volatile suspended solids.

Kinetic data and stoichiometric coefficients

These data and coefficients are given here along with numerical values as examples for easier understanding.

(i) Maximum rate of substrate consumption per unit mass of micro-organisms = k = 8 mg BOD$_L$ (mg active biomass)$^{-1}$ d^{-1}
(ii) Half-velocity constant (Monod's equation) or substrate concentration at one-half the maximum growth rate = K_s = 8 mg BOD$_L$/L
(iii) Yield coefficient for active biomass = Y = 0.3 (mg active VSS) (mg BOD$_L$)$^{-1}$
(iv) Decay rate of organisms = b = 0.1 (d)$^{-1}$
(v) Biodegradable fraction of the active biomass = f_d = 0.85

It is to be noted carefully that units must be consistent for all the parameters, say, mg for mass, litre for volume, day for time, BOD$_L$/BOD$_5$/COD for electron donor substrate, and VSS for biomass.

Design criteria

The purpose behind the design of the ASP is to control the pollution level of the effluent water to achieve the regulated effluent standards, such as BOD$_5$ less than 20 mg/L. For this, it is necessary to know the relation between BOD$_L$ and BOD$_5$ and also between VSS and SS (suspended solids).

Feasibility of the process

There are two parameters that determine the feasibility of a treatment process. These are $(\theta_c)_{min}$ (minimum mean cell residence time) and S_{min}. The first parameter determines the safety factor of the design and the second factor helps to decide whether or not the inflow substrate has been sufficiently digested in the reactor to a concentration well below the permissible effluent standard. The values of these two parameters are calculated as follows:

$$(\theta_c)_{min} = (kY - b)^{-1} = (0.3 \times 8 - 0.1)^{-1} = 0.435 \text{ d}$$

$$S_{min} = K_s b (\theta_c)_{min} = 8 \times 0.1 \times 0.435 = 0.35 \text{ mg BOD}_L/L$$

From these results it is clear that the design would be absolutely feasible if S_{min} << 20 mg BOD$_L$/L and $(\theta_c)_{min}$ is only a fraction of a day.

Cell retention time

Cell retention time or CRT also plays an important role in design calculations. In the literature, the following data are available for a conventional ASP. These are the normal ranges for the various factors.

Reactor loading = 0.6 kg BOD$_5$ m^{-3} d^{-1}

MLSS = 1000–3000 mg/L

$$\frac{F}{M} = 0.2\text{–}0.5 \text{ kg BOD}_5 \text{ (kg } X)^{-1} \text{ d}^{-1}$$

Typical BOD$_5$ removal efficiency = 95%

Typical θ_c = 4–14 d

Safety factor (SF) = 20–70

The design should not only be economical but also simultaneously result in the production of a high-quality effluent. Therefore, the safety factor (SF) should be selected accordingly from the previously given data towards the lower range, i.e., SF = 25. Now let us compute the design CRT.

$$\theta_c = SF(\theta_c)_{min} = 25 \times 0.435 = 10.875 \quad 10\text{–}1 \ 1 \ d$$

This value lies within the desired range and should therefore make it possible to produce a high-quality effluent.

BOD of effluent

With the help of the value of the design CRT (θ_c), the following relationship can be computed, which will help to determine the value of the BOD of the effluent.

$$S = \frac{K_s (1 + b\theta_c)}{Yk\theta_c - (1 + b\theta_c)} = \frac{8 \times (1 + 0.1 \times 10)}{[0.3 \times 8 \times 10 - (1 + 0.1 \times 10)]}$$

$$= 0.73 \text{ mg BOD}_L L^{-1}$$

As we can see, the value of S is well below the standard value of BOD$_5$ of effluent, i.e., 20 mg/L.

Hydraulic retention time

We will now determine the hydraulic retention time (HRT, θ). It has already been mentioned that in the activated sludge process θ and θ_c are independent of each other. At the same time the following correlation exists between the two factors:

$$X = \left(\frac{\theta_c}{\theta}\right)\frac{\{[1 + (1 - f_d) b\theta_c] Y (S_0 - S)\}}{(1 + b\theta_c)} \tag{6.1a}$$

$$\text{or} \qquad \theta = \frac{\theta_c \{[1 + (1 - f_d) b\theta_c] Y (S_0 - S)\}}{X(1 + b\theta_c)} \tag{6.1b}$$

Here X is the MLVSS (mixed liquor volatile suspended solid) in the reactor. Normally the ratio of MLVSS to MLSS varies between 0.8 and 0.9. In the section on CRT, we have seen that MLSS ranges from 1000 to 3000 mg/L. For a conventional ASP, higher loads are also permissible. Let us consider the average value for MLVSS within the range 2400 to 2700 mgVSS/L, i.e., $X = 2550$ mg/L. Substituting this in Eqn (6.1b), we get

$$\theta = \frac{(10) \{[1 + (1 - 0.85)(0.1 \times 10)] \times 0.3(300 - 0.73)\}}{2550(1 + 0.1 \times 10)}$$

$$= 0.2 \text{ d}$$

$$= 4.85 \text{ hours}$$

This value is acceptable for all practical purposes.

System volume

Next, the system volume V is computed as follows:

$$V = Q\theta = 600 \times 0.2 = 120 \text{ m}^3$$

Here, V refers to the volume of the reactor only, because the biomass is not considered to be active in the settler.

Concentration of active biomass

The concentration of active biomass (X_a) can be calculated from the following two correlations:

$$X_a = \frac{(\theta_c/\theta)Y(S_0 - S)}{(1 + b\theta_c)} \tag{6.2}$$

or

$$X_a = \frac{X}{[1 + (1 - f_d) b\theta_c]} \tag{6.3}$$

Substituting the values, we get

$$X_a = 2244.5 \text{ mg/L} \qquad \text{[from Eqn (6.2)]}$$

or

$$X_a = 2217.4 \text{ mg/L} \hspace{4cm} \text{[from Eqn (6.3)]}$$

Taking the average of the two values, we have

$$X_a = 2231 \text{ mg (VSS)}_a \text{ L}^{-1}$$

Here

$$\frac{X_a}{X} = \frac{2231}{2550} = 0.875 \text{ mg (VSS)}_a / \text{mg MLVSS}$$

MLSS

MLVSS is assumed to be 85% of MLSS. So,

$$\text{MLSS} = \frac{2550}{0.85} = 3000 \text{ mg SS/L}$$

For conventional loading this may be considered as the typical value, which gives X_a/MLSS equal to 74.4%.

Sludge recycle and SS in effluents

Two parameters that need to be estimated are the sludge recycle and the concentration of solids in the effluent. Without going into the details regarding the design of the settler at this stage, some judicious assumptions can be made, which are as follows. It is assumed that

$$X_e = 15 \text{ mgVSS L}^{-1}$$

$$X_r = 10,000 \text{ mgVSS L}^{-1}$$

$$\frac{X_a}{X} = 0.875$$

$$(X_a)_e = 13 \text{ mg (VSS)}_a \text{ L}^{-1} \ (= X_e \times 0.875)$$

$$(X_a)_r = 8750 \text{ mg (VSS)}_a \text{ L}^{-1} \ (= X_r \times 0.875)$$

$$(\text{SS})_e = 17.6 \text{ mg SS L}^{-1} \ (= X_e/0.85)$$

$$(\text{SS})_r = 11,765 \text{ mg SS L}^{-1} \ (= X_r /0.85)$$

Here we see that $(\text{SS})_e$ meets the effluent standard of 20 mg L^{-1}.

Sludge wasting rate

Now let us estimate the sludge wasting rate, by doing a mass balance on solids. Taking VSS as the basis, we have

$$\frac{VX}{\theta_c} = Q_w X_r + (Q - Q_w) X_e$$

Substituting the values, we get

$$120 \times 10^3 \times \frac{2550}{10} = Q_w \times 10{,}000 + (6 \times 10^5 - Q_w) \times 15$$

Solving this, we find

$$Q_w = 3005 \text{ L d}^{-1}$$

or 0.5% of the inflow of wastewater. This is an acceptable low quantity of waste, which is then treated for sludge dewatering and disposal. $Q_w X_r$ determines the mass rate of wasted sludge. This value is different for different types of solids—3.005×10^7 mg VSS d^{-1}, 2.63×10^7 mg (VSS)$_a$ d^{-1}, and 3.535×10^7 mg SS d^{-1} ('a' denotes active).

The rates of loss from effluents can be determined from $(Q - Q_w)X_e$. They are 8.955×10^6 mg VSS d^{-1}, 1.05×10^7 mg SS d^{-1}, and 7.836×10^6 mg (VSS)$_a$ d^{-1}.

Quality of effluent

If the quality of effluent is to be determined in terms of different types of oxygen demand (ODs), the values of S_e, X_e, and soluble molecules (SM) in the effluent, in mg L^{-1}, are required.

$$\frac{COD}{X} = 1.45 \quad \text{[refer to Eqn (5.4) and Illustration 5.1]}$$

$$\frac{BOD_L}{X_a} = 1.45 f_d$$

$$BOD_5 = BOD_L (1 - e^{-5k}) \tag{6.4}$$

where k is the first-order rate constant

$$= 0.23 \text{ d}^{-1} \text{ (for S)}$$

$$= 0.1 \text{ d}^{-1} \text{ (for } X_a)$$

$$= 0.03 \text{ d}^{-1} \text{ (for SM)}$$

$$\frac{BOD_5}{SM} = 0.14$$

So,

$$COD = [S_e + 1.45 \, (X_e) + (SM)_e] \text{ (mg/L)}$$

$$BOD_L = [S_e + 1.45 \, f_d \, (X_a)_e + (SM)_e] \text{ (mg/L)}$$

$$BOD_5 = [S_e \, (1 - e^{-5k}) + 1.45 \, f_d \, (1 - e^{-5k}) \, (X_a)_e + 0.14 \, (SM)_e] \text{ (mg/L)} \tag{6.5}$$

Sludge recycle rate

Another parameter that needs to be considered during design calculations is the sludge recycle rate. In Chapter 5, the mass balance around the settling tank has been considered. This yields the following simplified correlations:

$$R \text{ (recycle ratio)} = \frac{Q_R}{Q}$$

$$X = \frac{X_r R}{1 + R}$$

or

$$R = \frac{X}{X_r - X} \tag{6.6}$$

Substituting for $X = 2550$ mg/L (obtained during the calculation of HRT) and $X_r = 10,000$ mg/L (refer to the section on sludge recycle and SS in effluents), we get

$$R = \frac{2550}{(10,000 - 2550)} = 0.34$$

As we can see, Q_r is approximately one-third of Q.

The following information is useful for comprehensive design calculations. In general, for a well-settling activated sludge, the maximum value of X_r should lie between 10,000 and 14,000 mg/L. If the nature of the sludge is such that it compacts extremely well, the maximum value of X_r may be taken as high as 20,000 mg/L. However, for the types of sludge that do not settle very well, i.e., bulking sludge, the maximum value for X_r may be as low as 3000 to 6000 mg/L. From these correlations, it is clear that X_r influences not only R but also X.

Oxygen uptake rate

In the activated sludge process oxygen supply is a very important parameter because it is not only essential for operating the system but also the largest operating expense. The oxygen uptake rate (OUR) and the efficiency of the operating system together determine the energy requirement.

In Chapter 5 we have studied that for one electron equivalent of any substrate, the oxygen demand is 8 g. Therefore, the simplest way to calculate OUR is by formulating a mass balance equation on e^- equivalents, expressed as oxygen demand (OD). All the magnitudes are computed in mg as O_2 per day. The OUR is the difference between the equivalents entering and exiting the system. That is,

$$\text{OUR} = Q\,[S - (S_e + 1.45X_e)] \tag{6.7}$$

As S, S_e, and $1.45X_e$ have been expressed as mg COD or BOD per litre, we have

$$\text{OUR} = 6 \times 10^5\,[300 - (0.73 + 1.45 \times 15)] \text{ mg/d}$$

$$= 1.665 \times 10^8 \text{ mg/d}$$

Power requirement for aeration

Now we will estimate the power required for effecting aeration. Aeration serves the following purposes.

1. Supplies oxygen for (i) bacterial metabolism and (ii) contaminant oxidation.
2. Helps in mixing reactor contents, which in turn maintains the MLSS suspended and well distributed in the reactor. That is, it serves the dual purpose of agitation and oxidation.

BOD loading is limited by the ability of the aeration system. Oxygen is a sparingly soluble gas and therefore its transfer to water is an expensive proposition. The power required for its transition from the gas phase to the liquid phase is one of the major costs incurred during the operation of an ASP plant.

The rate of O_2 transition from the gas phase to the liquid phase is governed by the mass transfer from the bulk gas to the gas–liquid interface, and then from the interface to the bulk liquid. For sparingly soluble gases such as O_2, transfer to the gas–liquid interface is faster than that from the interface to the bulk. Thus, the liquid side transfer is rate limiting and the rate of the flux of O_2 from the gas to the liquid phase is given by the following expression:

$$R_{O_2} = K_L a(C_L^* - C_L) \qquad (6.8)$$

where R_{O_2} is the oxygen transfer rate per unit volume of the reactor (mg L^{-1} d^{-1}), $K_L a$ is the volumetric mass transfer rate coefficient (d^{-1}), C_L^* is the liquid phase oxygen concentration in equilibrium with the bulk gas phase (mg L^{-1}), and C_L is the liquid phase bulk oxygen concentration (mg L^{-1}). C_L^* is proportional to the partial pressure of oxygen in the air (C_G, atm) and Henry's constant (H_{O_2}, atm L/mg) for solubility of oxygen in water. Therefore,

$$C_L^* = \frac{C_G}{H_{O_2}}$$

Henry's constant is temperature dependent and in the case of pure water the following relationship is applicable:

$$\log H_{O_2} = 0.914 - \frac{750}{T} \qquad (6.9)$$

Here, T is the temperature measured in K.

However, the solubility of O_2 in wastewater may deviate from that in pure water due to the presence of salts and organic matter. A correlation factor β can be introduced here, which is expressed as

$$\beta = \frac{C_L^* \text{ (wastewater)}}{C_L^* \text{ (pure water)}} \qquad (6.10)$$

According to a literature report, β may vary from 0.7 to 0.98 and a value of 0.95 is considered to be ideal for municipal wastewater.

Now the following question arises: During the design of a treatment system, when the BOD is rate limiting, but C_L is not, what should be the concentration of dissolved oxygen (C_L) in the aeration tank? For efficient running of the system, C_L should be sufficiently in excess of K_s for oxygen. Normally K_s for oxygen may be less than 1 mg/L; to satisfy this criterion, if C_L is maintained around 2 mg/L, satisfactory results will be obtained.

Volumetric mass transfer coefficient, SOTE, and FOTE

Another very important parameter is the volumetric mass transfer coefficient ($K_L a$). This depends on various parameters such as temperature, design of aerator, and nature of wastewater. As oxygen solubility is affected by the nature of wastewater, so also is the value of $K_L a$, which is usually higher for pure water. To express this difference, a factor α is introduced, which is given as

$$\alpha = \frac{K_L a \text{ (wastewater)}}{K_L a \text{ (pure water)}} \tag{6.11}$$

Depending on the nature of aeration, the value of α varies. In the case of diffused aeration α may vary from 0.35 to 0.80, but when the aeration is done by mechanical means α may vary from 0.3 to 1.1. Irrespective of the aeration type, $K_L a$ is affected by the power input per unit volume. In tune with this, oxygen transfer is generally expressed as the mass of oxygen transferred under standard conditions per unit of power input to the aerator. Standard oxygen transfer efficiencies (SOTEs) are specified for standard conditions of 20°C ($C_L^* = 9.2$ mg/L), zero dissolved oxygen in the liquid phase ($C_L = 0$ mg/L), and pure water (α and $\beta = 1$). The values of SOTE generally lie between 1.2 and 2.7 kg O_2 per kWh.

For a design, the energy efficiency must be converted into field conditions. SOTE is converted to FOTE (field oxygen transfer efficiency) by using the following relationship:

$$\text{FOTE} = \frac{\text{SOTE} \times (1.035)^{20-t} \alpha (\beta C_L^* - C_L)}{9.2} \tag{6.12}$$

Here, temperature is measured in °C. 9.2 (mg/L) is the equilibrium concentration of O_2 at 20°C, $C_G = 0.21$ atm (partial pressure of O_2 in air), and $(1.035)^{20-t}$ is the temperature correction factor in mass transfer kinetics. FOTE is considerably less than SOTE.

It is to be noted that in the design calculation for power requirement, FOTEs should be used, not SOTEs, which are normally indicated by the manufacturers of aeration machines. When no data are available, as a thumb rule, the FOTE can be taken as 1 kg O_2 transferred per kWh power consumed for initial calculations.

Another point to be noted is that the purpose of aeration, as mentioned earlier, is also to keep the microbial floc in suspension, for which it is suggested that mixed liquor (ML) velocity in the aeration tank should not be less than 0.3 m/s, when the input air is 20–30 m^3 per min per 1000 m^3 of effective tank volume in the case of diffused aeration. However, for mechanical mixing, a power input of 15–30 kW per 1000 m^3 is required. With this information, the power requirement for the given example can be calculated as follows. FOTE is assumed to be 1 kg O_2 per kWh.

$$\text{Power required} = \frac{\text{OUR kg/d}}{\text{FOTE kg/kWd}}$$

$$= \frac{1.665 \times 10^8 \times 10^{-6}}{1 \times 24}$$

$$= 6.94 \text{ kW}$$

The installed aeration power should not be less than 7 kW.

Trickling Filter

A trickling filter (TF) consists of a bed of a highly permeable medium to which micro-organisms are attached. The wastewater is distributed through the bed medium and is allowed to trickle down. The concept of TF came from the use of contact filters, which were actually watertight basins filled with broken stones. The first TF unit was operated in England in 1893.

The filter media normally constitutes of materials such as rock or slag, varying in size from 25 mm to 100 mm. The diameter of the bed can be up to 60 m and the depth of the rock bed also varies, usually from 0.9 to 2.5 m and on an average is 1.8 m deep. Modern TFs also use plastic media, built in square or other shapes. These are narrower but deeper, more like towers. The depth usually varies from 4 to 12 m.

The standard method of wastewater distribution system is through rotary but a fixed distribution system can also be used. The plant also has a drain system at the bottom for collecting the filtrate as well as any biological solids that might become detached from the media during the process of treatment. This drain system also serves as a source of circulation of air through porous media for the growth and maintenance of micro-organisms on the filter bed. The treated wastewater and solids are passed to a settling tank where the solids are separated. Usually a part of the liquid from the settling tank is recirculated to dilute the strength of the wastewater, keep the filter bed moist, or often to optimize the process. It is essential that sufficient air is made available for the successful operation of the system. It has been found that natural draft and wind forces are normally sufficient if large enough ventilation ports are provided at the bottom of the filter and the medium has enough void spaces.

The organic material present in the wastewater is absorbed or degraded by a population of micro-organisms (aerobic, anaerobic, and facultative bacteria; fungi; algae; protozoa) attached to the filter medium as a biological film to the slime layer (~ 0.1 to 0.2 mm thick). This film is formed, as the wastewater flows over the media, by micro-organisms already present in the liquid, which gradually attach themselves to the rock, slag, or plastic surface. On the outer side of the biological slime layer, the organic material is degraded by aerobic micro-organisms. As the micro-organisms grow, the thickness of the slime layer increases and the diffused oxygen gets consumed before it can penetrate the full depth of the slime layer. Thus an anaerobic environment is established near the surface of the filter media. As the biological film continues to grow, the micro-organisms, next to the surface, lose their ability to cling to the media. Thus a portion of the slime layer falls off the filter. This is known as *sloughing* and is the main source of solids picked up by the drain system at the bottom. Figure 6.4 shows a multistage TF with plastic media.

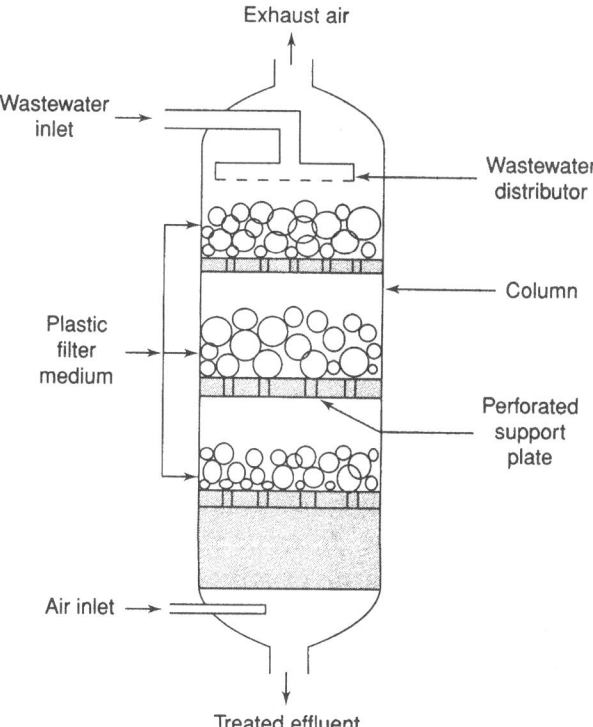

Fig. 6.4   Multistage trickling filter

Some of the main advantages and disadvantages of using trickling filter plants are given here.

Advantages

1. Process is simple and reliable; does not require large land area for wastewater treatment.
2. Effective in treating wastewater with high concentration of organics.
3. Appropriate for small- and medium-sized communities.
4. High degree of performance reliability and may qualify for secondary discharge standards.
5. Process elements are generally reliable.
6. Relatively low power requirement.
7. Level of skill and technical expertise needed to manage and operate the system are moderate.

Disadvantages

1. Additional treatment may be needed to meet more stringent discharge standards.
2. The sludge that is generated needs to treated and then disposed of.
3. Regular operator attention is needed.
4. Incidence of clogging is relatively high.
5. Relatively lower loading is required depending on the filter medium.
6. Has limited flexibility and control compared to activated sludge process.

Operation and Maintenance

Trickling filter is generally a reliable process. Nevertheless, the system may encounter some potential operational problems. Common among them are growth of biofilm, changes in wastewater characteristics, improper design, and equipment failure. Some problems commonly encountered are listed here.

1. Excessive organic load may cause anaerobic decomposition, and inadequate ventilation or aeration might produce disagreeable odours from the filters.
2. Excessive biological growth causes ponding on filter media.
3. Rotating distributors slow down or stop due to insufficient flow to turn distributor, clogging arms or orifices, or distributor arms not being levelled.

Rotary distributors are very reliable and easy to maintain. A gap of 0.2 to 0.3 m between the distributor arms and the top of the filter bed is required to allow the wastewater from the nozzles to spread out and cover the bed uniformly. The channelling must be checked.

It is essential that a TF system be pilot tested prior to installation to ensure that it satisfies effluent discharge standards.

Classification

Depending on the hydraulic and organic loading rates, TFs are classified into different types as shown here.

Low-rate filter   The design considerations are as follows:

>Hydraulic loading: 1–4 $m^3/m^2(d)$
>
>Organic loading: 0.08–0.32 $kg/m^3(d)$
>
>Depth: 1.5–3.0 m
>
>Recirculation ratio: 0
>
>Filter media: Rock, slag, etc.
>
>Power requirement: 2–4 $kW/10^3\ m^3$
>
>Dosing interval: Not more than 5 min

Intermediate-rate filters   The design considerations are as follows:

>Hydraulic loading: 4–10 $m^3/m^2(d)$
>
>Organic loading: 0.24–0.48 $kg/m^3(d)$
>
>Depth: 1.25–2.5 m
>
>Recirculation ratio: 0–1
>
>Filter media: Rock, slag, etc.
>
>Power requirement: 2–8 $kW/10^3\ m^3$
>
>Dosing interval: 15–60 s or continuous

High-rate filters   The design considerations are as follows:

>Hydraulic loading: 10–40 $m^3/m^2(d)$
>
>Organic loading: 0.32–1.0 $kg/m^3(d)$
>
>Depth: 1.0–2.0 m
>
>Recirculation ratio: 1–3
>
>Filter media: Rock, slag, synthetic materials, etc.
>
>Power requirement: 6–10 $kW/10^3\ m^3$
>
>Dosing interval: Not more than 15 s or continuous

Super-rate filters   This type of trickling filter has been developed as a result of the availability of various types of synthetic and wood packing media. The design considerations are as follows:

>Hydraulic loading: 40–200 $m^3/m^2(d)$

Organic loading: 0.8–6.0 kg/m^3(d)

Depth: 4.5–12 m

Recirculation ratio: 1–4

Power requirements: 10–20 kW/10^3 m^3

Dosing interval: Continuous

Physical Properties of Trickling Filter Media

The material of a filter media is considered most suitable if it has the following properties.

(i) High surface area to volume ratio
(ii) Cheap
(iii) High strength and reliability
(iv) Does not clog easily

ASCE manual 13 (Filtering Materials for Sewage Treatment Plants) gives detailed specifications for the different types of filter media. Information available in the literature is presented here in Table 6.1.

Process Design

It is difficult to propose a universal correlation for the design of trickling filters. However, there are some correlations available in the literature, which may

Table 6.1   Specifications for filter media

Type of medium	Size (mm)	Mass/unit volume (kg/m³)	Specific surface area (m²/m³)	Void space (%)
Conventional plastic	600×600×1200	30–100	80–100	94–97
High specific surface plastic	600×600×1200	30–100	100–200	94–97
Red wood	1200×1200×500	150–175	40–50	70–80
River rock (small)	25–65	1250–1450	55–70	40–50
River rock (large)	100–120	800–1000	40–50	50–60
Blast furnace slag (small)	50–80	900–1200	55–70	40–50
Blast furnace slag (large)	75–125	800–1000	45–60	50–60

be considered adequate to estimate the reduction of BOD in trickling filters. The equations involved in the design are presented here with the help of the following example.

An environmental biotechnologist has been assigned with the task of designing a modern trickling filter plant to treat wastewater from few small-scale fruit processing industries and a newly developed locality. For the design purpose, data were collected from the records of the local body and field tests. The engineer has decided to design a tower-type trickling filter with a plastic medium.

Nature and quality of wastewater

The first step is to determine the nature and quantity of the inflow of wastewater. As mentioned, the influent consists of domestic as well as fruit-processing wastewater. The data available are as follows.

(a) Domestic wastewater flow rate = 12,000 m^3/d
(b) Peak seasonal flow rate of fruit-processing (summer to pre-winter) wastewater = 6000 m^3/d
(c) Average BOD_5 (domestic) = 250 ppm
(d) Combined BOD_5 (domestic + industrial) = 600 ppm
(e) Temperature data (for design)
    (i) Summer to pre-winter = 30°C
    (ii) Winter = 15°C
(f) Effluent BOD_5 requirement = 25 ppm or less

The following data have been experimentally determined.

(g) BOD_5 removal rate constant for tower filter at 25°C = 0.1 m/d
(h) Temperature correction coefficient (from literature) = 1.08
(i) From Table 6.1, specific area of conventional plastic filter packing material (assumed) = 100 m^2/m^3
(j) Filter height to be restricted to 12 m

Equations for design of trickling filter

In the literature several empirical equations are available for the design of trickling filters. If their empirical constants are considered to be unity, the nature of all the proposed equations becomes similar. However, the following equation may be considered to be suitable for the design of a trickling filter:

$$\frac{S_e}{S_i} = \exp\left[-KZS_a^m\left(\frac{A}{Q}\right)^n\right]$$

(6.13)

where S_e is the BOD_5 of the treated and settled effluent from the filter (ppm), S_i is the BOD_5 of influent wastewater to the filter (ppm), K is the observed BOD_5 removal rate constant (temperature dependent) (m/d), Z is the total depth of the trickling filter bed (m), $S_a = A_s/V$ is the specific area of the filter medium per unit volume of filter bed (m^2/m^3), A_s is the specific area (m^2), V is the volume (m^3), A is the cross-sectional area of the filter bed (m^2), Q is the volumetric flow rate of wastewater (m^3/d), and m, n are the empirical constants (can be taken equal to unity if not available).

Temperature correction

The temperature-corrected value of K is determined from the following equation (see also Chapter 5):

$$\frac{K_T}{K_{15}} = (\theta)^{T-15} \tag{6.14}$$

In this equation K_{15} is the value of K at 15°C, K_T is that at T°C, and θ is the temperature coefficient.

$$\theta = 1.08$$

$$K_{15} = \frac{K_{25}}{(\theta)^{25-15}} = \frac{0.1}{2.16} = 0.046$$

Bulk volume of filter bed

First of all the bulk volume of the filter bed needs to be determined by using Eqn (6.13) and the data available:

$$\frac{S_e}{S_i} = e^{-x}$$

where

$$x = \frac{KS_a AZ}{Q}$$

Here AZ is the bulk volume of the filter bed, which is to be determined under two seasonal conditions, and the larger value is considered to be the controlling factor in the design of the trickling filter.

(a) Let us first consider the summer to pre-winter period.

$$K_{30} = K_{25}(1.08)^{30-25}$$

$$= 0.1 \times 1.46933 \text{ m/d}$$

$$= 0.147 \text{ m/d}$$

During this period $S_i = 600$ mg/L. This high value of BOD_5 may suffer from oxygen transfer limitations. To tackle this problem the inflow BOD_5 can be diluted by mixing a part of the treated effluent, which can be done by recirculating the treated effluent.

Assume that, after dilution, the S_i value is reduced from 600 ppm to 400 ppm. In that case the recirculation ratio is determined as follows:

$$600Q + 25Q_r = 400(Q + Q_r)$$

$$\frac{Q_r}{Q} = \frac{200}{375}$$

$$\alpha = 0.53$$

where $\alpha / \dfrac{Q_r}{Q}$. With this

$$Q = (1 + 0.53)(12{,}000 + 6000) \text{ m}^3/\text{d}$$

$$= 27{,}540 \text{ m}^3/\text{d}$$

$$\frac{S_e}{S_i} = \frac{25}{400} = 0.0625 = e^{-x}$$

Therefore,

$$x = \ln\left(\frac{400}{25}\right) = 2.77$$

$$KS_a\left(\frac{AZ}{Q}\right) = 2.77$$

or

$$AZ = \frac{2.77Q}{KS_a} = \frac{2.77 \times 27{,}540}{0.147 \times 100} = 5189.5 \text{ m}^3$$

[If no dilution is done, then $Q = 18{,}000$, $x = \ln(600/25) = 3.18$, $K = 0.147$, $S_a = 100$, and thus $AZ = (3.18 \times 18{,}000)/14.7 = 3894 \text{ m}^3$.] It can be seen that if no dilution is done the volume of the bed required is much less. Still, it needs to be checked whether the oxygen transfer is sufficient for handling such a high value of BOD_5.

(b) Now let us consider the bulk volume requirement during winter months.

The various design parameters are as follows:

$$K_{15} = K_{25}(1.08)^{T - 25} = 0.1(1.08)^{15 - 25} = 0.046 \text{ m/d}$$

$S_a = 100 \text{ m}^2/\text{m}^3$, $S_e = 25 \text{ mg/L}$, $S_i = 250 \text{ mg/L}$, $Q = 12{,}000 \text{ m}^3/\text{d}$, $x = \ln(250/25) = 2.3$. Substituting these values in design Eqn (6.13), we get.

$$AZ = \frac{xQ}{KS_a} = \frac{2.3 \times 12{,}000}{0.046 \times 100} = 6000 \text{ m}^3$$

Therefore the bulk volume of the filter bed required is 6000 m³, as it is the maximum among all the volumes calculated.

Area and depth of filter bed

Another parameter that influences the design calculation is the area and depth of the filter bed, which will be calculated for the desired bulk volume, which is 6000 m³. There are various methods by which this can be done. As such, multiple areas and the corresponding diameters give the same volume. However, all the combinations may not give satisfactory results, as performance depends on different parameters including cost. In such cases one can use the reported data available in the literature.

In an earlier section we have seen that trickling filters are classified on the basis of hydraulic loading rates. If the proposed column filter functions are considered, such as a high-rate filter, then the hydraulic loading rate can be set within 10–40 m³/m²(d). With this value, the area and depth required are calculated as follows.

(a) Area of a filter using a hydraulic loading rate of 35 m³/m²(d) for maximum inflow of wastewater is calculated as

$$A = \frac{27,540}{35} = 787 \text{ m}^2$$

So

$$\text{Depth } Z = \frac{6000}{787} = 7.6 \text{ m}$$

The depth is within the limit of 12 m. If the calculated depth is found to be more than 12 m, then the area needs to be increased or more than one filter has to be used.

(b) Let us now check the hydraulic loading rate during winter months.

$$\text{Hydraulic loading} = \frac{Q}{A} = \frac{12,000}{787}$$

$$= 15 \text{ m}^3/\text{m}^2 \text{ (d)}$$

This value is within the prescribed limit, i.e., 10–40 m³/m²(d).

(c) The recommended depth of the high-rate filter is around 1–2 m. If a multistage filter, as shown in Fig. 6.4, is considered and the number of stages is taken to be four, then the height of each stage bed should be (7.6)/4 m or 1.9 m.

(d) The diameter of the column is calculated from the area, which has already been determined:

$$\frac{\pi D^2}{4} = 787 \text{ m}^2$$

gives

$$D = 31.66 \text{ m}$$

Let us select 30 m as the diameter of the column; then the new area becomes

$$A = \pi (15)^2 = 706.5 \text{ m}^2$$

(e) With this new area, hydraulic loading becomes

$$= \frac{27{,}540}{706.5} \text{ m}^3/\text{m}^2\text{(d)}$$

$$= 39 \text{ m}^3/\text{m}^2\text{(d)}$$

Again, this value is also within the limit of 10–40 m^3/m^2(d) as prescribed for the high-rate filter.

(f) Let us now select the new depth of this filter bed.

$$Z = \frac{6000 \text{ m}^3}{706.5 \text{ m}^2}$$

$$= 8.5 \text{ m}$$

For a four-stage column filter, the depth of each stage bed selected is 8.5/4 m = 2.125 m.

(g) Design dimensions of the column filters are selected as follows:

Diameter = 30 m

Height of the column = depth of filter bed + bottom clearance of 1.0 m
$$+ 3 \times (\text{clearances of 0.5 m each between 4 beds})$$
$$+ \text{top clearance of 1.0 m}$$

$$= 2.125 \times 4 + 1.0 + 0.5 \times 3 + 1.0$$

$$= 12 \text{ m}$$

This is also within the limit of 12 m.

Rotating Biological Contactors

The rotating biological contactor (RBC) is a remediation technology used in the secondary treatment of wastewater. In this technology wastewater is allowed to come in contact with a biological medium in order to facilitate the removal of contaminants. Though several different designs are available, in its simplest form, an RBC consists of a series of discs mounted on a shaft, which is driven such that the discs rotate at right angles to the flow of the settled sewage. The discs are usually made of plastic (polythene, PVC, expanded polystyrene)

Rotating biological contactor

and are contained in a trough so that about 40 per cent of the disc area is immersed. The discs are arranged in groups or packs with baffles between each group to minimize surging or short-circuiting. Troughs are usually covered with small units, whereas larger units are often housed within buildings. This is done to reduce the effect of weather on the active biofilm that gets attached to the disc surfaces. The schematic diagram of an RBC is shown in Fig. 6.5.

RBC units are usually installed in a concrete tank so that the surface of the wastewater, while it passes through the tank, almost reaches the shaft. This means that about 40% of the total surface area of the discs is always submerged. The shaft continually rotates at 1 to 10 rpm, and a layer of biological growth,

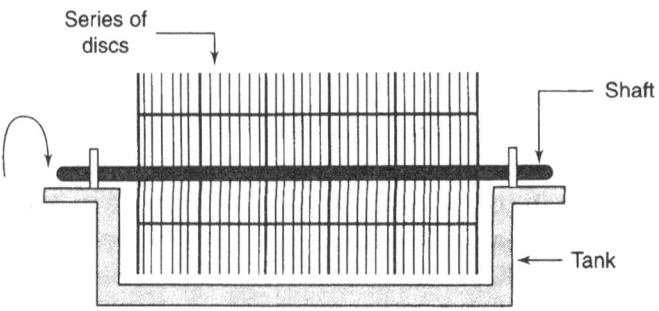

Fig. 6.5   Schematic diagram of a rotating biological contactor

2 to 4 mm thick, is soon established on the already wet surface of each disc. The biological growth that gets attached to the discs assimilates the organic material in the wastewater. Aeration is provided by the rotating action, which exposes the discs to the air after contacting them with the wastewater. Excess biomass is sheared off in the tank, where the rotating action of the discs maintains the solids in suspension. Eventually, the flow of the wastewater carries these solids out of the system and into a clarifier, where they are separated. By arranging several sets of discs in series, it is possible to achieve a high degree of organic removal and nitrification.

The primary advantages and disadvantages of using rotating biological contactors are listed next.

Advantages

1. Contact periods of relatively short duration are required because of the large active surface area.
2. Adaptable to handle a wide range of flows.
3. Sloughed biomass generally has good settling characteristics and can easily be separated from the waste stream.
4. Operating costs are low because plant operation demands little skill and expertise.
5. Shorter retention time.
6. Low power requirements.
7. Elimination of channelling, to which conventional percolators are susceptible.
8. Low sludge production and excellent process control.

Disadvantages

Practically, there are no major disadvantages of using the RBC system. An only exception is that the shaft bearings and mechanical drive units need regular maintenance.

Performance of Biological Contactors

The performance of RBC systems depends on temperature, concentration of the pollutants, and the rate at which the treatment is expected to proceed. Studies have shown that, in terms of BOD removal, the system performance peaks with a critical hydraulic retention time of 3 hours and any further increase in the retention duration results in little or no improvement in performance. It has also been shown that as the applied organic loading rate increases above, about, 5 g BOD/m^2/d, the deviation from 100 per cent efficiency becomes more pronounced. The performance characteristics of RBCs are frequently expressed as applied loading/removal rate curves.

The oxidation of ammonia is also an important feature that determines the performance of any biological reactor. The degree of nitrification that is achieved by RBCs depends on several factors. Studies have shown that the removal of ammoniacal nitrogen is related to the hydraulic loading rate, and it has been reported that full nitrification can only be achieved when the organic loading rate is less than 5 g $BOD/m^2/d$. Research has highlighted two aspects of nitrification by RBCs. One aspect is that the process is influenced and limited by the availability of oxygen during the months of summer and that proper BOD removal is necessary prior to nitrification. Failure to fulfil these two conditions results in the amount of disc surface available for colonization by the nitrifying species being significantly restricted.

Design of Rotating Biological Contactors

Though RBCs can be customized, for economic reasons, discs of standard dimensions and ready-made RBCs are preferred. The following parameters need to be borne in mind while designing an RBC unit or when selecting a ready-made unit.

1. Organic loading rate (g BOD per m^2 disc area per day)
2. Hydraulic loading rate (m^3/d)
3. Hydraulic retention time (d)
4. Diameter of discs (m)
5. Number of discs
6. Depth of submergence of rotating disc (m)
7. Rate of rotation (rpm)

In practice, the loading rates and liquid residence time or HRT may vary widely and, therefore, the designing of an RBC is not only a science but also an art. On the other hand, unlike the design of the activated sludge system, the design of RBCs is simple and straightforward. This is because the major design parameter considered during the design of an RBC system is the amount of soluble BOD removed per unit surface area. Most industrial wastewaters may contain soluble BOD of the order of 150 ppm or more. To reduce this large amount of soluble BOD to 50 ppm or less in the final discharge effluent, sufficient surface area must be provided in the design. Thus, a single-stage unit may not be practical. Depending on the desired quality of the discharged effluent, the required number of treatment stages is decided. The following design information available in the literature can be used as guidelines.

1. The influent wastewater should have a surface BOD loading rate of around 12–14 $g/m^2/d$ in the first stage of the treatment process. If BOD loading in the influent is higher, then it needs to be diluted by recycling a certain amount of low-BOD effluent through the influent wastewater.

2. A second stage is required to reduce the soluble BOD of the first stage effluent from 50 ppm or more to the desired effluent level.
3. Lower the desired BOD value of the final discharged effluent, higher the number of stages. For example,
   (a) a single-stage unit is sufficient if the desired effluent is greater than 25 ppm,
   (b) for 10–25 ppm, two stages are required, and
   (c) if the final effluent should have less than 10 ppm, the number of stages required will be three or four.
4. However, for a ready-made RBC unit, the process design curves supplied by the manufacturers can be used to determine the number of stages.
5. Thickness of the microbial growth normally needs to be maintained at around 2 to 4 mm and thus the gap between two adjacent discs should be between 5 to 10 mm.
6. Biomass concentration attached to the surface needs to be as high as 50,000 mg/L, and on being sloughed from the surface it should provide a suspended microbial cell concentration of 10,000–20,000 mg/L. At such a high microbial cell concentration, the degree of BOD removal is very high even with a short liquid retention time. This high concentration of biomass enables the RBC system to withstand shock loads caused due to high hydraulic or organic loading rates.

   It is interesting to note that, compared to trickling filters, microbial cell density is much higher in RBC systems and thus less surface area is required for cell support.

7. The Reynolds number of the water mass moving past the disc surface varies from around 8000 at the outer perimeter of the disc to around 5000 at the inner edge of the microbial growth on the surface. This exerts more shearing force than in trickling filters. This, in turn, results in reduced mass transfer resistance for oxygen and substrate, and thus increases the BOD removal rate.
8. On an average the rate of BOD removal in an RBC unit is two to three times greater than that in a standard trickling filter.
9. Another important parameter is the selection of disc media for assembling the RBC unit, which should have a sufficiently large surface area to accommodate microbial growth. The medium should be packed up to a length of 8 to 9 m on a single shaft and have up to 9000–10,000 m^2 of surface area. High-density media, having 50% more surface area than conventional media, are not suitable for high-strength wastewater, because excessive microbial growth will lead to clogging. However, they are useful for low-strength wastewater.

Illustration 6.1

An RBC system is to be designed to treat a sample of wastewater based on the following information.

1. Hydraulic loading rate (Q) = 6000 m³/d
2. Influent organic concentration (S_i) = 300 mg BOD_L/L
   (assume 50% BOD_L is soluble and BOD_5/BOD_L ratio is 0.68)
3. Number of stages = 3
   (use standard density media for stages 1 and 2 and high-density media for stage 3)
4. Organic loading rate in the first stage is either
   60 kg BOD_L/1000 m² (disc area) d (maximum)

   or

   10 kg BOD_L/1000 m² (disc area) d (overall)
5. Tip velocity (commonly assumed for the full-scale unit) = 20 m/min
   [Increasing the tip velocity increases the oxygen transfer rate (the proportionality is roughly linear), but the energy requirement also increases.]
6. The outflow rate in the settler for average flow = 20 m/d (assumed) (RBC systems are almost always operated in a series mode with three to five stages. After this the effluent is passed through a settler to reduce suspended solids. The settler is normally designed with an overflow rate of 16 to 32 m/d.)
7. This is a full-scale RBC, hence the diameter of the medium is approximately 3.6 m.
8. The specific surface area (m²/m³) of the medium ranges between 110 m⁻¹ (standard density) and 170 m⁻¹ (high density). (High-density media, which have smaller channels for liquid and gas penetration, are more susceptible to clogging and should be used only with low loading rates.)
9. The medium is attached to a steel shaft (or axle) that is supported on bearings and rotated by a direct mechanical drive in most cases. In a few cases, the trough is bubble aerated, and the bubbles are trapped in 'cups' to create air-driven rotation. Modules of the medium are attached to the shaft to give a total medium length of approximately 8 m. [RBCs are generally sold in units of 'shafts' of approximately 8 m length].
10. A shaft of standard density medium has approximately 9300 m² of surface area, while a high-density shaft has 14,000 m² of surface area. Generally, high-density media are used only in the later stages of an RBC system, when the BOD concentration has been reduced enough, so that the smaller pour openings are not clogged.
11. Typical per cent submergence is 25 to 40. An increased submergence increases the oxygen transfer rate, but also requires greater energy to

maintain a given rotating speed. The rotating speed for a full-scale unit is normally around 2 rpm, though the most widely used criterion for setting the rotating speed is the *tip velocity*.

12. Tip velocity = rpm × (D_m), where D_m is the diameter of the medium.
13. Another approach of RBC design can be based on biofilm kinetics, for which each stage should be treated as a completely mixed biofilm reactor. The biofilm-kinetic approach is supplemented with empirically derived design criteria as mentioned earlier. Based on experience, an overall surface loading in the range of 3–15 kg soluble BOD_5/1000 m^2 d, with a typical range of 5–8 (in the same units) is desirable.
14. A first stage loading limit of 45 kg total BOD_5/1000 m^2 d can also be employed as additional design criteria, in order to avoid excessive growth, oxygen depletion, and odour.
15. A hydraulic loading (Q/A, where A is the medium surface area) was considered earlier as the key design loading. Typical values were 0.04 to 0.16 m/d. This design criterion alone, however, is not sound enough.

Solution

This numerical illustration should help to determine how many shafts are needed for each stage, the rpm of the shaft, and the settlers' surface area.

(a) Total BOD_L mass loading

$$= (300 \text{ mg } BOD_L/L)(10^{-6} \text{ kg/mg})(10^3 \text{ L/m}^3)(6000 \text{ m}^3/d)$$
$$= 1800 \text{ kg/d}$$

(b) The soluble BOD_L loading (50% of total BOD_L) = 900 kg BOD_L/d
(c) To satisfy the first stage maximum surface load, the first stage should have an area of at least

$$A_{min} = \frac{900 \text{ kg } BOD_L/d}{60 \text{ kg } BOD_L / 1000 \text{ m}^2 \text{ d}} = 15,000 \text{ m}^2$$

(d) To satisfy the overall surface loading, the total area must be

$$A_{total} = \frac{900 \text{ kg } BOD_L/d}{10 \text{ kg } BOD_L/1000 \text{ m}^2 \text{ d}} = 90,000 \text{ m}^2$$

(e) Dividing the total area by 3 gives an average of 30,000 m^2 area per segment. In this case, the overall loading determines the area for the first stage.
(f) If we consider a maximum of 10,000 m^2 per shaft for standard media, then the first and second stages require (30,000/10,000) or three shafts each.
(g) The third stage uses high-density media with a maximum of 15,000 m^2/shaft. This requires two shafts.

(It should be noted that fractional number of shafts are rounded off to integral higher values.)

(h) To maintain a tip speed of 20 m/min, the 3.6-m-diameter shafts require a rotating speed of

$$rpm = \frac{(20 \text{ m/min})}{\pi \times 3.6 \text{ m}} = 1.8$$

(i) The settler requires an overflow rate of 20 m/d, which gives a surface area of

$$A_{settler} = \frac{6000 \text{ m}^3/d}{20 \text{ m/d}} = 300 \text{ m}^2$$

If the settler is circular, a radius of about 10 m is required.

Anaerobic Treatment of Wastewater

We have studied in Chapter 1 that anaerobic treatment processes refer to those biological treatment processes that occur in the absence of oxygen. Bacteria that can survive in the absence of dissolved oxygen take part in anaerobic treatment processes and are known as obligate anaerobes. Anoxic denitrification is also an anaerobic process by which nitrate nitrogen is converted biologically to nitrogen gas in the absence of oxygen.

In Chapter 1 it has also been mentioned that the anaerobic treatment process produces less sludge (0.1 to 0.2 kg biomass or sludge per kg BOD removed) compared to the aerobic treatment process (0.5 to 1.5 kg biomass per kg of BOD removed). It has also been mentioned that the anaerobic method is preferred for treatment of wastes with high BOD values. It is usually recommended as a pre-treatment process to reduce the BOD from 600 ppm or more to 200–300 ppm. Anaerobic treatment units are more energy efficient.

Methane gas produced in an anaerobic process can be used as an energy source. Again, the energy needed for mixing in the anaerobic processes is much less than the energy required for the aeration of aerobic processes. However, the slower rate of reaction in anaerobic processes makes it necessary to use treatment plants of larger sizes.

Design of Anaerobic Digestion System

Anaerobic digestion process is an age-old process used for the stabilization of sludge. It involves the decomposition of organic and inorganic matter in the absence of molecular oxygen. Its main application has been the stabilization of the concentrated sludge produced from the treatment of wastewater as well as the treatment of some industrial wastes.

In this process the organic matter present in the wastes is biologically converted to methane (CH_4) and carbon dioxide (CO_2). The process needs to be carried out in an airtight reactor. Slurry is introduced continuously or intermittently and retained in the reactor for varying residence time. The treated slurry, which is withdrawn from the digester continuously or intermittently, is non-putrescible in nature and its pathogen contents are also largely reduced. The digester can be designed as a *standard-rate* or *high-rate* digester. In the former case, the contents of the digester are not externally heated or mechanically mixed. The average hydraulic residence time varies from 30 to 60 days. On the other hand, the contents of the high-rate digester are heated and thoroughly mixed. The required retention time is less than or equal to 15 days.

In the two-stage biomethanogenesis process, hydrolysis occurs in the first stage and methanation takes place in the second stage (refer to Chapter 3). The microbiological and biochemical aspects of the anaerobic digestion process have been elaborately presented in Chapter 3.

The typical kinetic coefficient values for anaerobic digestion of various common substrates are presented in Table 6.2.

The pros and cons of using the anaerobic treatment process, as compared to the aerobic process, lie in the fact that the rate of growth of methanogenic bacteria is slow. This signifies that only a small amount of organic matter is used during cell synthesis. As a result of low cellular growth rate and the conversion of organic matter to biogas, the resulting solid matter is well stabilized. This can be disposed of without further treatment. However, the sludge obtained from aerobic digestion must either be digested anaerobically or dewatered and incinerated because of the presence of a higher percentage of cellular organic matter. A small amount can be sold as fertilizer after heat drying.

Anaerobic digester

Table 6.2 Kinetic coefficients for anaerobic digestion

Substrates	Coefficients	Range	Typical value
Domestic sludge	Y	0.04–0.1	0.06
	k_d	0.02–0.04	0.03
Fatty acid	Y	0.04–0.07	0.05
	k_d	0.03–0.05	0.04
Carbohydrates	Y	0.02–0.04	0.024
	k_d	0.025–0.035	0.03
Protein	Y	0.05–0.09	0.075
	k_d	0.01–0.02	0.014

* Y is in mg VSS/mg BOD and k_d is in d^{-1}.

The anaerobic digester tank size can be designed on the basis of any of the following methods.

(a) Mean cell residence time
(b) Volumetric loading factors
(c) Volume reduction

All the three are empirical methods. Ideally, of course, the design should be based on the utilization of the concepts of biochemistry and microbiology as discussed in Chapters 3 and 5. Let us now study each of these methods.

Design based on mean cell residence time

The design principle behind mean cell residence time has been discussed in Chapter 5. In anaerobic digestion, the respiratory and oxidative end product is biogas, which contains mainly CH_4 and CO_2. The theoretical conversion factor for the amount of methane produced from the digestion of 1 kg of BOD_L can be estimated by taking glucose ($C_6H_{12}O_6$) as the base material for BOD_L. The following equations will give the solution.

$$C_6H_{12}O_6 + 6O_2 = 6CO_2 + 6H_2O$$
$$(180) \qquad (192)$$

$$C_6H_{12}O_6 = 3CO_2 + 3CH_4$$
$$(180) \qquad (132) \quad (48)$$

Therefore,

$$\frac{kg\ CH_4}{kg\ BOD_L} = \frac{\dfrac{48}{180}}{\dfrac{192}{180}} = 0.25$$

This shows that per kg BOD_L digested, 0.25 kg methane is obtained. Therefore,

$$V_{CH_4} \text{ (at NTP)} = \left[\frac{(0.25 \times 10^3)}{16} \right] \times (22.4 \text{ L}) \times 10^{-3} \text{ m}^3$$

$$= 0.35 \text{ m}^3 \text{ CH}_4 \text{ per kg BOD}_L$$

The empirical correlation for the quantity of methane gas produced at NTP from anaerobic digestion of wastewater or sludge is the following:

$$V_{CH_4} = (0.35)(EQS_i - 1.42P_x) \tag{6.15}$$

where E is the efficiency of waste utilization (normally varies from 0.6 to 0.9), Q is the volumetric flow rate of wastewater (m^3/d), S_i is the influent BOD_L kg/m^3, P_x is the net mass of cell tissue produced per day (kg/d), and 1.42 is the conversion factor for the cell tissue into BOD_L.

In Chapter 5 it has been stated that the average formula for cell mass can be taken as $C_5H_7O_2N$. The BOD_L value of the cell mass can be estimated as follows:

$$C_5H_7O_2N + 5O_2 = 5CO_2 + 2H_2O + NH_3$$
$$(113) \qquad (160)$$

Therefore,

$$\frac{\text{kg BOD}_L}{\text{kg cell mass}} = \frac{160}{113} = 1.42$$

Considering that the digester is a continuous flow stirred tank reactor (CSTR) without recycle, the cell mass synthesized daily (P_x) can be estimated using the following correlation:

$$P_x = \frac{YQ \, (ES_i)}{1 + k_d \theta_c} \tag{6.16}$$

where Y is the yield coefficient [kg cell mass produced/kg BOD_L consumed (see Table 6.2)], k_d is the endogenous coefficient [d^{-1} (see Table 6.2)], and θ_c is the mean cell residence time [d (see Table 6.3)].

Table 6.3 gives the suggested magnitude of mean cell residence time (θ_c) applicable for the design of CSTRs. It may be noted that for CSTRs, θ_c should be the same as the hydraulic retention time θ.

Design based on volumetric loading factors

This method of deciding the size of the digester is not only very simple but also one of the most common methods. The loading factors are decided from the experience gained in the laboratory and also from the various field trials conducted under various conditions. Among the various loading

Table 6.3 Mean cell residence time for design purpose

Operating temperature (°C)	θ_c (d, as suggested for design)
18	28
24	20
30	14
35	10
40	10

factors that have been proposed, two of them are most widely accepted. These are based on

1. kg volatile solids fed per day per m^3 of effective digester volume and
2. kg of VS charged per day per kg of volatile solids in the digester.

This is similar to the F/M (food to micro-organism) ratio discussed in Chapter 5. While applying the loading factors in determining the digester capacity, hydraulic retention time should also be checked for its relationship with organism growth and wash-out effect. Also, the type of digester being used cannot be neglected. For example, it has been observed that for the standard rate, single-stage digester, only 50% or less of the volume is effective. Stratification occurs in this type of digester. The supernatant remains at the top of the layer, the active digestion zone is in the middle, and the bottom layer contains thickened sludge. For this reason, the volumetric loading rate in the conventional standard rate digester is low, and the retention time based on this may vary from 30 days to more than 90 days. The recommended solid loading for this type of digester is 0.5 to 1.6 kgVS/m^3 d. On the other hand, for high-rate digesters, the loading rate lies between 1.6 and 6.4 kgVS/m^3 d and the retention time varies from 10 to 20 days.

Design based on volume reduction

This method is not as popular as the other two. However, there are certain empirical correlations that can be derived to determine the digester capacity.

$$V = \left[V_f - \frac{2}{3}(V_f - V_d) \right] t \tag{6.17}$$

where V is the digester volume (m^3), V_f is the fresh sludge feed rate (m^3/d), V_d is the digester sludge removal rate (m^3/d), and t is the digestion time (d).

Illustration 6.2

This example will facilitate better understanding of the procedure of estimating digester volume and its performance. The following data have been made available.

1. 10,000 m^3/d of low BOD_L wastewater is first treated aerobically (preliminary treatment).
2. The sludge obtained after step 1 from the treatment plant is to be treated anaerobically in a continuous flow stirred tank digester.
3. Dry solids forming the sludge in the treated wastewater = 0.20 kg/m^3.
4. Moisture content of the sludge = 90% and specific gravity of sludge = 1.02.
5. BOD_L removed from the treated wastewater = 0.15 kg/m^3.
6. Reactor will be operated as a continuous flow stirred tank and the temperature is maintained at 35°C (room temperature).
7. From Table 6.3, θ_c = 10 d (in this case same as hydraulic retention time θ).
8. As mentioned earlier, for a properly managed aerobic digester, BOD_L removal efficiency E ranges from 0.6 to 0.9. In the present example, it can be assumed that $E = 0.75$.
9. Another assumption that can be made here is with respect to the availability of nitrogen and phosphorus in the waste for microbial growth. Let it be assumed that these two essential elements necessary for the synthesis of new cells exist adequately in the waste.
10. From Table 6.2, we take Y = 0.06 kg cells/kg BOD_L consumed and k_d = 0.03 d^{-1}.
11. Room temperature in this case is 35°C.

Solution

1. Let us first estimate the volume of sludge produced per day after the preliminary treatment of wastewater (10,000 m^3/d).

   (a) Volume of sludge formation

   $$V_s = \frac{0.2 \times 10,000}{(0.1)(1.02)(1000)}$$

   $$= 19.6 \ m^3/d$$

   (b) BOD_L loading from the wastewater

   $$QS_i = 0.15 \times 10,000$$

   $$= 1500 \ kg/d$$

2. From the volumetric flow rate of the sludge ($V_s = 19.6 \ m^3/d$) and the mean cell residence time (10 d), the volume of the digester can be computed.

   Volume of the digester $(V) = V_s \theta$

In this problem,

$$V/V_s = \theta = \theta_c = 10 \text{ d}$$

Therefore,

$$V = V_s\theta_c$$
$$= 19.6 \times 10 = 196 \text{ m}^3$$

3. Now let us determine the volumetric loading of the digester.

$$\text{kg BOD}_L/\text{m}^3\text{d} = \frac{1500}{196} = 7.65 \text{ kg/m}^3\text{d}$$

This value is 120% of the recommended highest value (6.4 kg/m³ d) for high-rate digesters. In such situations it is advisable to increase the hydraulic retention time. For high-rate digesters, the hydraulic retention time varies from 10 to 20 d. It is proposed, therefore, to take care of volumetric loading of VS as 6 kg/m³d for the purpose of design calculations.

4. With this new condition of volumetric loading of BOD$_L$ (6.0 kg/m³ d), the effective volume of the digester is calculated as

$$V = \frac{1500}{6.0} = 250 \text{ m}^3$$

With this digester volume and the volume of sludge to be treated per day (i.e., 19.6 m³/d), the hydraulic retention time θ and mean cell retention time θ_c can be computed as follows:

$$\theta = \theta_c = \frac{250}{19.6} = 12.76 \text{ d} \quad (\text{say, 13 d})$$

(For a continuous flow stirred tank reactor, θ and θ_c are practically the same.)

5. Now let us calculate the quantity of volatile solids (VS) produced per day from Eqn (6.16).

$$P_x = \frac{YQ\,(ES_i)}{1 + k_d\theta_c}$$

where $Y = (0.06 \text{ kg cells produced})/(\text{kg BOD}_L \text{ utilized})$, $Q = 10,000 \text{ m}^3/\text{d}$ (vol. flow rate of wastewater), $E = 0.75$ (waste utilization efficiency), $S_i = 0.15 \text{ kg/m}^3$ (influent BOD$_L$ concentration), $k_d = 0.03 \text{ d}^{-1}$ (endogenous coefficient), and $\theta_c = 13 \text{ d}$ (mean cell residence time, in this case the same as hydraulic retention time). Substituting these values in Eqn (6.16), we get

$$P_x = \frac{0.06 \times 10,000 \times 0.75 \times 0.15}{1 + 0.03 \times 13} = 48.6 \text{ kg/d}$$

6. Let us now estimate the per cent stabilization. In the anaerobic digestion process, per cent stabilization depends on the conversion of BOD_L to biogas:

Total BOD_L consumed$/d = QES_i$ (kg/d)

$$= 10,000 \times 0.75 \times 0.15 \text{ kg/d}$$

$$= 1125 \text{ kg/d}$$

BOD_L utilized for new cell mass production $= 1.42 P_x$ kg/d

$$= 1.42 \times 48.6 \text{ kg/d}$$

$$= 69 \text{ kg/d}$$

BOD_L utilized for biogas production $= (1125 - 69)$ kg/d $= 1056$ kg/d

Total BOD_L loading to the digester $= QS_i = 10,000 \times 0.15$ kg/d

$$= 1500 \text{ kg/d}$$

$$\text{Per cent stabilization} = \left(\frac{1056}{1500}\right) \times 100 = 70.4\%$$

7. Let us now determine the volume of methane gas produced per day. For this we can use Eqn (6.15). However, in step 6, we found that 1056 kg BOD_L/d is utilized for biomethanation. In the section on design based on mean cell residence time, we have studied that per kg of BOD_L consumed, 0.35 m^3 of methane (CH_4) gas is produced at NTP. So,

$$\frac{\text{Methane produced at NTP}}{d} = 1056 \times 0.35 \text{ m}^3/d$$

$$= 370 \text{ m}^3/d$$

8. Now we will estimate the volume of biogas produced. We know that biogas normally contains two-thirds methane. Therefore, the total volume of biogas will be

$$\text{Total biogas volume} = \frac{370}{0.67} \text{ m}^3 = 552 \text{ m}^3$$

9. Let us now see how much volume of methane gas can be produced per kg of cell mass. 1 kg of cell mass is equivalent to 1.42 kg of BOD_L. Therefore, 1 kg of cell mass, if anaerobically digested, yields

$$1.42 \times 0.35 = 0.5 \text{ m}^3 \text{ methane gas}$$

Anaerobic Filter

It has been mentioned earlier that the wastes having a high BOD can be stabilized economically and most efficiently by the anaerobic treatment process. The anaerobic

Anaerobic filter

filter process used for the treatment of both domestic and industrial wastes is the most common anaerobic treatment process anaerobic treatment process in which active microbes are attached or immobilized on the packing medium.

In the area of wastewater treatment, the anaerobic filter is a new development. The system consists of a column filled with various types of solid media used for the treatment of carbonaceous organic matter present in wastewater. The wastewater flows upward through a column, which contains the medium on which anaerobic bacteria grow and are retained. As the bacteria are retained on the medium and practically not washed off, the mean cell residence time is generally in the order of 100 days. In such systems large values of θ_c can be achieved with short hydraulic retention times. Thus anaerobic filters can be used for the treatment of low-strength wastes at room temperature. A simplified schematic diagram of the system is shown in Fig. 6.6.

An anaerobic filter, as shown in Fig. 6.6, is similar to the packed-bed digester system, which is a high-rate anaerobic digestion process. In spite of its high efficiency the packed-bed filter has not been able to enjoy popularity in the field of industrial waste treatment. The upward flow limits the diameter of the digester, to usually a maximum of 6 m, to ensure uniform flow distribution. Therefore the process is uneconomical and not technically lucrative for the treatment of large volumes of waste, which is normally the case in industrial waste treatment. Also, for a large digester, clogging is a major problem, which is difficult to solve. Since there is no mixing in the digester, the difficulty in keeping a uniform temperature through the digester seriously limits the process application in temperate or semi-temperate regions. However, the system will be suitable for small or medium capacity treatment plants.

As mentioned earlier, the packing media for the digester are normally similar to those used in trickling filters. These can be small pieces of stone or various forms and sizes of plastic rings specially designed to achieve high void space and

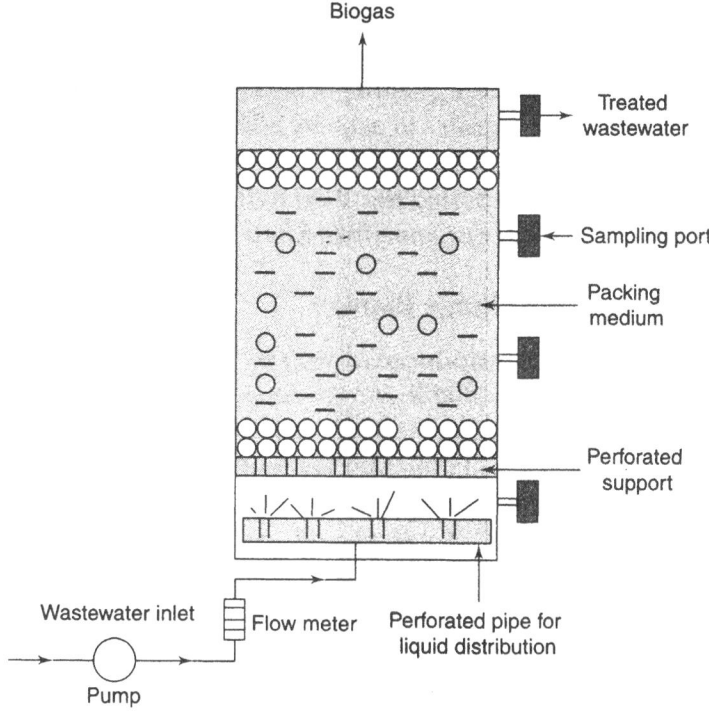

Fig. 6.6 Anaerobic filter

high surface area for bacterial growth. Stone media are much cheaper than plastic media, but are less efficient and require a digester of very strong structure to withstand their high load and lateral pressure. In tropical countries, such as India, small bamboo rings can be selected as packing media, since they are easy and inexpensive to procure. Table 6.4 gives the physical characteristics of some common packing media.

Table 6.4 Physical characteristics of common packing media

Type	Size (mm)	Arrangement	Specific surface area (m^2/m^3)	Void space (%)
Raschig rings	31–100	Random	52–127	71–77
Blast furnace slag	80–130	Random	40	50
Smooth rock basalt	80–130	Random	40	53
Vertically subdivided PVC tubes	—	—	220	94
Bamboo rings	ID 16 OD 27 H 27.5	Random	106.4	67.6

Source Charadej (1980).

As mentioned in the case of the trickling filter system, the bacterial mass gets formed as slimy films around the medium and as flocs trapped in the voids or gaps. Thus, a large quantity of bacterial mass is retained in the digester and this enables the digester to achieve a high rate of digestion within a very short retention period compared to that required by conventional digesters. Table 6.5 presents the reported results of industrial wastewater treatment using the packed-bed digester or anaerobic filter.

Upflow Anaerobic Sludge Blanket

This system was first introduced in 1979 by Lettinga et al. in Netherlands and has been found to be suitable for the treatment of industrial and municipal wastewaters. The schematic diagram of a typical UASB is shown in Fig. 6.7.

In a UASB reactor, a sludge blanket is allowed to form in the bottom one-third to half-portion of the reactor and a three-phase separator is provided at the top for separation of gas, liquid, and flocs. The wastewater inlet is provided at the bottom and outlet near the top of the reactor. From top to bottom, several sampling ports are provided to check the performance of the reactor. The most important point to be noted, during the design of the UASB system, is the upflow velocity of the wastewater, which should be such that

Table 6.5 Treatment conditions in anaerobic filter and results (ε is the porosity and μ is the specific surface area)

Types of waste and condition	Packing media	Loading rate (kg COD/m³/d)	HRT (h)	% COD removal	Gas yield (m³/kg COD fed)
Brewery press liquor (35°C); COD: 6000–27,000 mg/L	Crushed limestone $\varepsilon = 0.45$	0.8–6.4	15–33	30–97	
Pharmaceutical waste (95% methanol, 37°C); COD: 1250–16,000 mg/L	25–38 mm stone $\varepsilon = 0.47$	0.22–3.52	12–48	94–98	
Potato processing waste (19–22°C); COD: 3000 mg/L	38 mm stone	0.53–2.3	13–59	41–79	
Soluble fraction of pig waste (20°C); COD: 470–22,200 mg/L	Random pack $\varepsilon = 0.98$, $\mu = 200$ m²/m³	0.168–8	66.7	55–78	
Screened and diluted pig waste (20°C); COD: 15,000 mg/L	Random pack $\varepsilon = 0.98$ $\mu = 200$ m²/m³	5.47–63.89	5.6–68	63.9–84	0.06–0.2

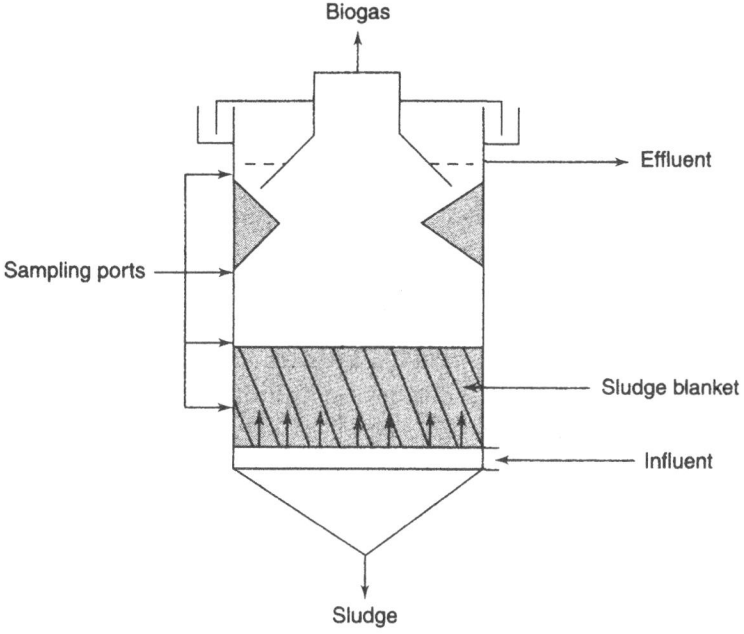

Fig. . Schematic diagram of upflow anaerobic sludge blanket reactor

it does not disturb the sludge blanket. To achieve this condition the upward velocity of the wastewater should be less than or equal to the free settling velocity of the biomass particles or flocs. From this analysis the diameter of the UASB column is determined. The interesting point is that the flexibility of operation of the UASB is limited by the free settling velocity and the upward liquid velocity. Depending on the column design, the sludge blanket can be established either at the lower part of the column or at the upper end. When the blanket is found in the upper part, the diameter of the column is accordingly increased in the upper part of the column, so that the velocity of the liquid drops down and the particles start settling and remain suspended in the upper region.

The UASB system is very efficient for the simple reason that in this method SRT is very high but HRT can be kept low. Many UASB systems are being used with a great deal of success for the treatment of a variety of food processing industry wastewater as well as for treating wastes from paper and chemical industries. Design loading lies, typically, in the range of 4 to 15 kg COD/m^3 d. At times, in the UASB system granules or biosolids are formed that do not settle well within the reactor, and hence a separate settler is provided as a safeguard against excessive loss of biosolids from the reactor.

Anaerobic Rotating Biological Contactor

In the section on design of rotating biological contactors, the use of RBCs for the treatment of wastewater under aerobic conditions has been discussed. In this section the same RBC system will be discussed with some modifications in the system, which makes it suitable for the treatment of wastewater under anaerobic conditions also. The RBC system is not normally covered under aerobic methods of treatment. In the case of an AnRBC the top must be closed with an airtight cover. The schematic diagram of an AnRBC has been presented in Fig. 6.8.

As in an RBC, in an AnRBC also discs are mounted on a shaft, which are rotated at low speed (revolutions per minute, rpm) with the help of an electric motor that is fitted inside the reactor. The discs remain immersed in the reactor content. Scrapers are provided in the reactor to maintain the uniform growth of a biofilm on the discs. A feed inlet, effluent outlet, and biogas outlet are provided suitably in the system.

Wastewater treatment plant

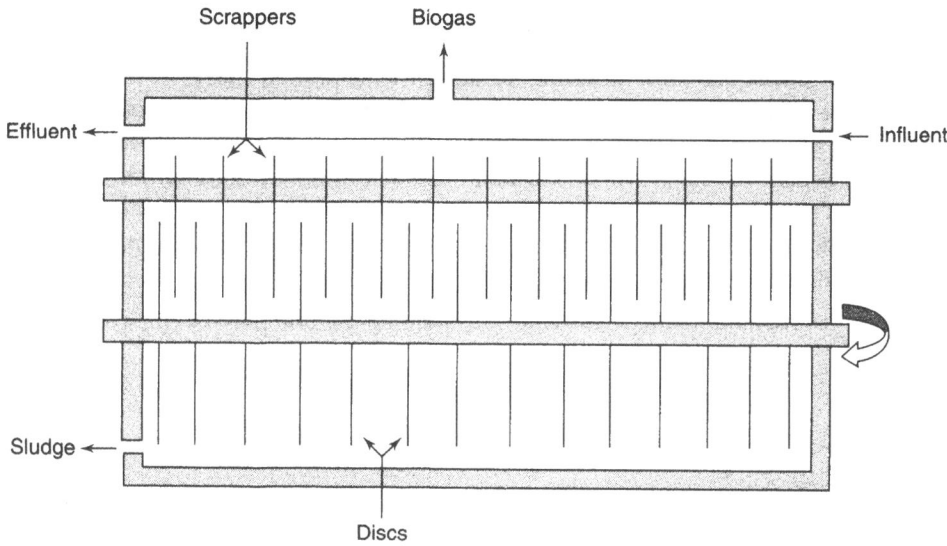

Schematic diagram of anaerobic rotating biological contactor

The design considerations for RBC and AnRBC are different. In an RBC, along with biofilm formation, the aeration of the reactor is equally important. However, in the case of an AnRBC, biofilm formation is the desired requirement along with the maintenance of anaerobic conditions. For this reason, the entire surface area of an AnRBC discs is allowed to be immersed in the reactor content. This is not possible in the case of an RBC because of oxygen requirement. So, in the latter case only 40% of the disc area is allowed to be immersed.

Summary

This chapter presents the treatment processes and design aspects of bioreactors for wastewaters, including their advantages and limitations. It discusses the most popular and practical bioreactors used for wastewater stabilization. Among these, the activated sludge bioreactor is most suitable for the aerobic process. The trickling filter bioreactor can be used for the aerobic process as well as the anaerobic treatment process.

Rotary biological contactors are suitable for both aerobic and anaerobic treatment of wastewater (40% of the disc's surface area is submerged in the case of an aerobic process reactor, whereas 100% area is kept under liquid in the anaerobic process). Another modern bioreactor suitable only for anaerobic treatment processes—the upflow anaerobic sludge blanket—has been described in this chapter.

The design procedures for most of these bioreactors have been explained step by step with a numerical example.

Review Questions

1. With a simple diagram, describe the steps involved in an ASP.
2. List the advantages and disadvantages of the activated sludge process.
3. If the regulated standards demand that the BOD_5 level of the wastewater after treatment should be less than 10 mg/L, which two parameters would help to determine the feasibility of the biological treatment process? Give the mathematical expressions for these two parameters.
4. From the kinetic data given here, calculate the minimum cell retention time and the minimum substrate concentration, as mg BOD_L/L, in the outgoing effluent of a biological treatment plant.

   Maximum rate of substrate consumption per unit mass of micro-organism $(k) = 10 \text{ d}^{-1}$.

   Yield coefficient $Y = 0.6$

   Half-velocity constant $(K_s) = 8$ mg BOD_L/L

   Decay rate of organisms $(b) = 0.2 \text{ d}^{-1}$

5. 1000 m^3 per day of wastewater is released from a phenol manufacturing unit. As a biotechnologist, your responsibility is to design an activated sludge plant to reduce the phenol content of this water from 150 mg/L to 0.05 mg/L or less, which is the requirement of the pollution control board. For design purposes, following biological coefficients are applicable:

   $X_o = 0$

   $Y = 0.65$ g $(VSS)_a$/g phenol

   $k = 8$ g phenol/ g $(VSS)_a$ (d)

   $b = 0.1/d$

   $K_s = 1$ mg phenol/L

   (a) Determine the cell retention time θ_c for the design of this plant and also justify any assumptions made.
   (b) Determine the reactor volume using the calculated θ_c. Justify assumptions made.

6. In the activated sludge process, waste sludge can be discharged from two places. Which are these two parts of the ASP? Which of these two is preferred and why? Will there be any change in the overall mass balance? Justify your answer.
7. If a sample of wastewater is required to be treated in the ASP system and you have been asked to design the plant, what information would you need to design the reactor system?
8. Design an ASP system based on the data given here.
   (a) Influent values:

   $Q = 5000 \text{ m}^3/d$

   $S_0 = 400$ mg/L

   $X_o = X_i = 50$ mg VSS/L

   (b) Kinetic parameters obtained from laboratory experiments and stoichiometry are as follows:

   $Y = 0.5$ g $(VSS)_a$ /g

   $k = 20$ g/g $(VSS)_a$ (d)

$K_s = 25$ mg/L

$b = 0.15$ /d

$f_d = 0.85$

(c) Design factors:

Safety factor $= 30$

MLVSS $= X_a + X_i = 4000$ mg/L

$X_r = (X_r)_a + (X_r)_i = 20,000$ mg/L

$X_e = (X_e)_a + (X_e)_i = 20$ mg VSS/L

Compute the following:
  (i) $(CRT)_{min}$
  (ii) CRT
  (iii) S_e
  (iv) kg of substrate consumed per day
  (v)   kg of active VS, inactive VS, and TVS produced per day in the aerator
  (vi) kg of active VSS, inactive VSS, and TVSS in the aerator assuming θ_c is the same for active and inactive mass.
  (vii) HRT
  (viii) $X_a/(X_a + X_i)$, assume θ_c is the same for X_a and X_i
  (ix) X_a in the aeration tank
  (x) Q_w, if sludge is wasted from the return line
  (xi) kg of active VSS, inactive VSS, and TVSS wasted per day (take into account solids in the influent and effluent)
  (xii) Required Q_r and R
  (xiii) Volumetric loading rate in kg/m^3(d)
  (xiv) BOD$_L$ of the solids in the effluent, assuming $(VSS)_a$ in the effluents exerts BOD
  (xv) Volume of the aerator

9. With a neat sketch, describe the different parts of a trickling filter and its operational principle.
10. What is a multistage trickling filter? State its advantages and possible disadvantages.
11. Classify trickling filters.
12. List the desired properties of a trickling filter medium. Name a few popular filter media which have a high specific surface area.
13. Design a tower-type trickling filter to treat wastewater released from a soft drink production unit, for which you may choose plastic medium. The data available are as follows.

(a) Wastewater flow rate $= 10,000$ m^3/d
(b) Average BOD$_5 = 300$ ppm
(c) Temperature $= 25°C$
(d) Effluent BOD$_5$ requirement $= 20$ ppm or less
(e) BOD$_5$ removal rate constant for tower filter at 25°C $= 0.1$ m/d
(f) The filter height to be restricted to 10 m.

Any other data may be assumed, if required; give reasons.

14. With a schematic diagram, describe the RBC system and explain why RBC should be preferably used for the secondary treatment of wastewater?
15. What are the advantages of an RBC? Are there any disadvantages of using this system?
16. Describe the functioning of an RBC unit.
17. Design an RBC system from the data provided here.
    - (i) $Q = 12,000$ m^3/d
    - (ii) $S_i = 250$ mg BOD$_L$/L (60% of BOD$_L$ is soluble and BOD$_5$/ BOD$_L$ = 0.7)
    - (iii) Number of stages = 3
    - (iv) Organic loading rate (average of all the three stages) = 20 kg BOD$_L$/1000 m^2 d
    - (v) Tip velocity = 50 m/min
    - (vi) Outflow rate in the settler for average flow = 30 m/d
    - (vii) Diameter of medium (i.e., disc) = 4.8 m
18. Explain and compare the three methods that can be used to design an anaerobic digester.
19. What are the advantages and disadvantages of the anaerobic system of treatment of wastewater over the aerobic method?
20. Discuss briefly how the following anaerobic treatment plants function.
    - (a) Anaerobic filter
    - (b) UASB
    - (c) Anaerobic RBC

References

Bhattacharyya, B.C., 'Biotechnology of anaerobic digestion, unpublished manuscript.

Bhattacharyya, B.C. 1979, 'Design and development of gobar (bio) gas plant for rural population' Monograph, Chemical Age of India.

Charadej (1980). 'Anaerobic filter for biogas production', *J. Energy Heat Mass Transfer*, vol. 2, no. 1, p. 34.

Chen, Y.R. and A.G. Hashimoto 1978, 'Kinetics of methane fermentation, *Biotechnol. Bioeng. Symp., no. 8* pp. 269–82.

Chin, E.S.K. and F.B. DeWalle 1977, 'Treatment of high strength acidic wastewater with a completely mixed anaerobic filter', *Water Res.*, vol. 11, no. 3, pp. 295–304.

Hartman, R.B. 1979, 'Sludge stabilization through aerobic digestion', *Journal of Waste Pollution Control Federation*, vol. 49, p. 2353.

Metcalf and Eddy, Inc. 1991, *Waste Water Engineering: Treatment, Disposal and Reuse*, 3rd edn, McGraw-Hill, New York.

Rittman, B.E. and P.L. McCarty 2001, *Environmental Biotechnology: Principles and Applications*, McGraw-Hill (International edition).

APWA–AWWA–WPCF 1976, *Standard Methods for the Examination of Water and Waste Water*, 14th edn, John D. Lucas, Baltimore.

CHAPTER 7

Solid Waste Management

Introduction

In Chapter 6 liquid waste treatment methods have been discussed. In the present chapter treatment procedures for another form of waste, i.e., solid waste will be presented.

Solid Waste Management

Before proceeding to the study of treatment of solid wastes, it is necessary to have a clear picture of the nature of solid wastes and what is meant by solid waste management. Solid waste refers to all kinds of wastes that are normally solid in nature and are produced as a result of various activities of humans and animals, and are discarded as useless or unwanted. The term 'solid waste' as used here encompasses the heterogeneous mass refused by the urban populace and also the more homogeneous accumulation of agricultural, industrial, and mining wastes.

Solid waste management encompasses all the activities related to the control of production, storage, collection, transportation, processing, and disposal of solid wastes (with special reference to the urban setting, where the accumulation of solid wastes is a direct consequence of lifestyle) in a manner that is in accord with the best principles of public health, economics, technology, and environmental factors. Within the scope of solid waste management fall All administrative, planning, financial, and engineering functions that are involved in providing solutions to problems related to the disposal of solid wastes fall within the scope of solid waste management.

Some basic functional elements of the waste management system are listed here.

1. *Waste generation* This involves activities aimed at identifying material that is not useful and needs to be collected for disposal.
2. *Waste handling and separation* These activities are directed towards the management of wastes until they can be placed in storage containers for further collection. Handling also implies the movement of loaded containers to the point of collection. Separation of waste material is also an important step of storage and further processing. For example, it is necessary to separate organic and inorganic components.
3. *Collection* This important function includes the gathering of solid wastes and recyclable materials. These are transported to the location where all the collected material is emptied from the vehicles. Collection may account for as much as 50% of the total annual cost of urban solid waste management.
4. *Processing of solid waste* This often includes the separation of bulky items, separation of waste components on the basis of size by using screens or by manual separation, size reduction by shredding, separation of ferrous metals by magnets, and volume reduction by compaction and combustion.
5. *Disposal* This is the final step of the solid waste management system. Nowadays all types of solid wastes are ultimately disposed by the methods of landfilling or land spreading. A modern sanitary landfill is not merely a dump yard but rather an engineered facility used for disposing of solid wastes on land without posing a threat to public health and the contamination of groundwater.

Treatment Processes for Solid Wastes

There are two methods by which solid wastes can be disposed of. These are

1. Thermal conversion process
2. Biological conversion process

The thermal conversion system of solid waste is used for both volume reduction and energy recovery. This can be achieved by three methods, namely, *incineration*, *pyrolysis*, and *gasification*. In the thermal conversion process, the inclusion of an energy recovery system is essential.

Among the biological conversion processes, the important systems are *aerobic composting, low-solid anaerobic digestion, high-solid anaerobic digestion, anaerobic treatment of MSW (municipal solid waste)*. While discussing the biological conversion processes energy production will also be considered.

Thermal Conversion Process

Let us first understand the meaning of thermal processing of solid waste. It is defined as the conversion of solid waste into gaseous, liquid, and solid products, with the concurrent or subsequent release of heat energy. Depending on the availability of the supply of air (oxygen), thermal processes can be categorized as *incineration* (air supply available is more than the stoichiometric amount), *gasification* (air supply is less than that required for complete combustion), or *pyrolysis* (no air supply, indirect heating). Stoichiometric quantity is defined as the theoretically required quantity of oxygen (air) for effecting complete oxidation of the combustible materials present in the solid waste.

Incineration

Incineration is the general term used for the process of complete combustion of solid waste. This may either be *mass-fired* or *processed solid waste fired*.

In a mass-fired combustion system, the solid waste requires very little processing; such wastes are known as unprocessed solid wastes (USW). While designing the mass-fired incinerator, it should be borne in mind that anything present in the solid waste may enter the system, including bulky or oversized non-combustible objects. For this reason the design of the system should be such that it takes care of these objectionable wastes without damaging the equipment or system operating personnel. Therefore the system is required to be over-designed. Moreover, due to the non-uniformity of the contents of mass-fired wastes, the energy content is also extremely variable. In spite of the disadvantages of the mass-fired incineration system, this technology, because of its simplicity, is most popular. It may be noted that the most important component in the mass-fired incineration system is the design of the grate. It takes care of several functions including the movement of wastes through the system, mixing of wastes, and injection of combustion air.

In the case of processed solid wastes fired (PSWF) incinerators, the waste is typically burned on a travelling grate stoker. The grate provides a platform on which the PSW can be burnt and also facilitates the introduction of underfire air for the promotion of turbulence and uniform combustion.

Compared to the uncontrolled nature of unprocessed solid wastes, PSW can be produced with fair consistency to meet specifications in terms of energy, moisture, and ash content. PSW can be produced in shredded or fluff form, or as compressed pellets or cubes. The later type, though costlier to produce, are easier to transport and store. USW and PSW can be compared in the following manner.

1. PSW has higher energy content compared to USW.
2. Due to the reasons already mentioned PSW incinerators are smaller in size than USW incinerators, for treatment of the same quantity of combustible solid waste.
3. A PSW system can be controlled more efficiently than a USW system, as the nature of PSW is more homogeneous, which allows for better combustion control.
4. Air pollution control devices perform better in the case of PSW systems as compared to USW systems.
5. System design is simpler for USW compared to PSW.

Gasification

The term gasification is generally used to define the process of incomplete or partial combustion of a solid combustible material, which is deliberately done by using less than the required stoichiometric quantity of air or oxygen. Although the process has been known for over hundred years, it has only recently been applied to the processing of solid wastes.

The gasification process is described as an energy efficient technique for reducing the volume of solid waste accompanied with the recovery of energy. A carbonaceous fuel, in this process, is partially burnt to generate fuel gas, rich in CO, H_2, and to some extent CH_4. This low calorie fuel gas can be used in IC engines, gas turbines, or boilers. During the gasification process the following reactions take place:

$$C + O_2 \longrightarrow CO_2 \quad \text{(exothermic)}$$
$$C + CO_2 \longrightarrow 2CO \quad \text{(endothermic)}$$
$$CO + H_2O \longrightarrow CO_2 + H_2 \quad \text{(exothermic)}$$
$$C + H_2O \longrightarrow CO + H_2 \quad \text{(endothermic)}$$
$$C + 2H_2 \longrightarrow CH_4 \quad \text{(exothermic)}$$

To sustain the endothermic reactions, energy is made available from the exothermic reactions. The product obtained from a gasifier is a low calorie fuel gas typically containing 10% CO_2, 20% CO, 15% H_2, and 2% CH_4 by volume; the rest is N_2. Due to the diluting effect of N_2 in the input air, the calorific value of the low calorie gas is about 5600 kJ/m^3. When pure oxygen is used instead of air, gas with a higher calorific value can be produced with an energy content of about 11,000 kJ/m^3.

Different types of gasifiers are used for solid waste treatment. The most common among them are vertical fixed bed (VFB), horizontal fixed bed (HFB), and fluidized bed (FB) gasifiers.

Vertical fixed bed gasifier

The VFB gasifier is very simple in design. It is normally cylindrical or rectangular in structure and usually made of fire-clay brick. The top end is closed for gas collection, whereas the bottom manhole is used for ash removal. Air is supplied from the bottom sidewall from different points. This gasifier requires low capital investment. It produces low calorie fuel. The operating temperature varies from 650°C–800°C. The schematic diagram of a VFB gasifier is shown in Fig. 7.1.

VFB gasifiers can also be operated with pure oxygen instead of air. This will produce a high calorie fuel (10,000–12,000 kJ/m^3) and the average gas composition would be 50% CO, 30% H_2, 14% CO_2, and 4% CH_4; the rest is nitrogen and hydrocarbons. In this case, the temperature of the fuel bed is relatively high, about 1400°C–1600°C.

Horizontal fixed bed gasifier

HFB gasifiers are most popular commercially. They consist of two major components, namely, *primary* and *secondary combustion chambers*. In the primary chamber, solid waste is gasified under limited oxygen (air) supply. Thus a low calorie fuel gas is produced, which then flows into the secondary combustion chamber, where complete combustion occurs in the presence of excess air. As a result, in the secondary chamber a high-temperature flue gas is produced, which can be used in the waste heat boiler for process steam or hot water. The flue gas temperature ranges from 650°C to 850°C.

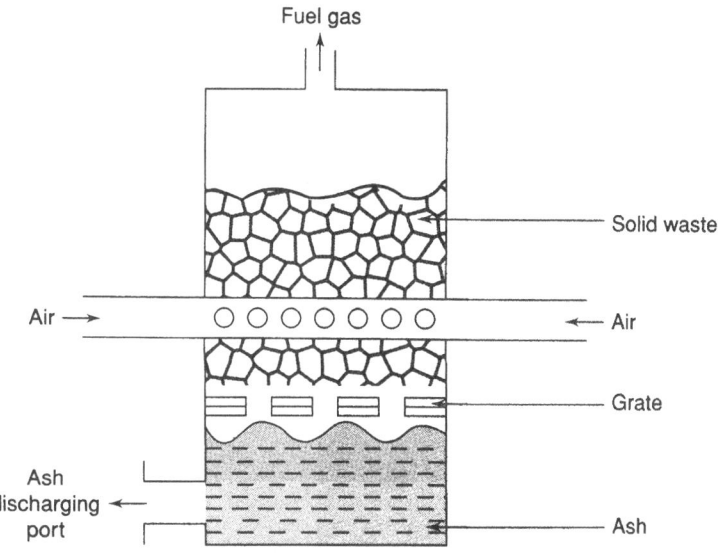

Fig. 7.1   Schematic diagram of VFB gasifier

Standard sizes of HFB gasifiers can handle solid wastes ranging from 50 kg to 4000 kg per hour. The smaller size units are manually operated in the batch mode, while larger HFB units are operated using the continuous feeding technique and ash removal. The schematic diagram of an HFB system is presented in Fig. 7.2.

Fluidized bed gasifier

The fluidized bed system is a modern development used for the combustion of MSW. If excess air is used for fluidization, then the system becomes an incinerator. If the air supplied is less than the stoichiometrically desired quantity, the system can be used as a gasifier.

Pyrolysis

Pyrolysis refers to the strictly anaerobic thermal processing of solid waste. Though both gasification and pyrolysis systems are used to convert solid wastes into gaseous, liquid, and solid fuels, they differ from each other in their functioning. While the pyrolysis system uses an external source of heat to drive the endothermic pyrolysis reaction under strict anaerobic conditions, the gasifier is actually an aerobic self-sustaining system generating the heat required for the endothermic reaction from the partial combustion of solid wastes.

Pyrolysis is based on the conception of thermal cracking and condensation reactions of thermally unstable organic substances in an oxygen-free atmosphere into gaseous, liquid, and solid fractions. Most organic substances are known to be thermally unstable. Pyrolysis is highly endothermic in nature, while

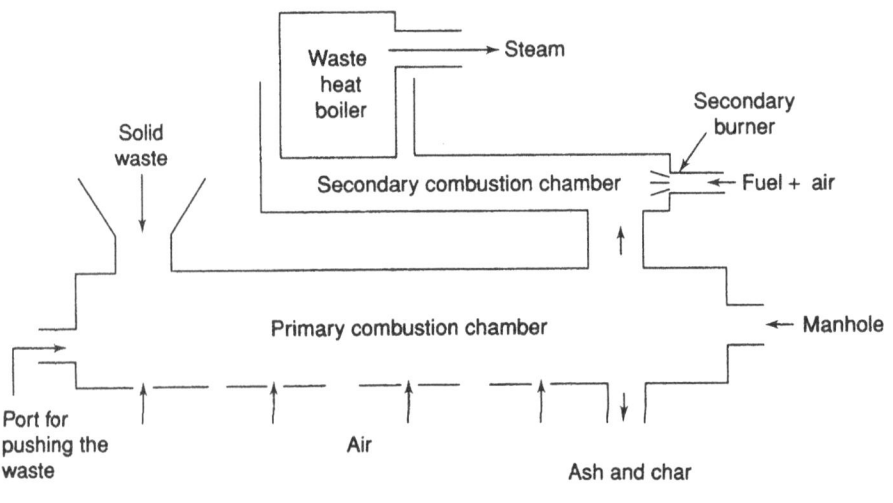

Fig. .   Horizontal fixed bed gasifier

combustion or gasification is highly exothermic. Another name for pyrolysis is *destructive distillation*.

The following products are obtained as a result of pyrolysis or destructive distillation of organic wastes.

1. Hydrogen, methane, carbon monoxide, carbon dioxide, to name a few, in gaseous state.
2. Acetic acid, acetone, methanol, and complex hydrocarbons in the liquid stream. These can be used as synthetic fuel oils.
3. A part of the solid waste is char, which mainly contains pure carbon and inorganic substances.

The chemical reaction that occurs during pyrolysis is the following:

$$3C_6H_{10}O_8 \longrightarrow (8H_2O + 2CO + 2CO_2 + CH_4 + H_2) + C_6H_8O + 7C$$

(Cellulose   (Gaseous products)   (Liquid) (Solid)
representing
solid waste)

The product fractions obtained vary drastically with the temperature at which pyrolysis is done. The energy content of pyrolytic oils is usually 21,000 kJ/kg and under conditions of maximum gasification, the energy content of the gas may be around 26,000 kJ/m^3.

Heat Recovery Systems

Every type of waste treatment process involves expenditure. Therefore, presently efforts are being made to cull out any benefits that can be obtained from the treatment process, so as to reduce the ultimate expenses on the waste treatment process. For example, with some additional arrangements, heat can be recovered from the hot flue gas generated as a result of the combustion of MSW or PSW.

The two most common methods of heat recovery from flue gas are either by arranging water-wall combustion chambers or by using waste heat boilers. As per the requirement, a plant can be designed to produce hot water or steam. Hot water can be used for low-temperature heating applications in various process industries. Steam has also got versatile applications such as heating and generation of electricity. These measures will help to reduce the fuel cost for process industries.

Further, heat recovery will also help to reduce the capital and operational expenses of the air pollution control equipment. For example, in the absence of any heat recovery mechanism, an MSW combustion system needs 100 to 200 per cent excess air to ensure complete combustion and the desired turbulence to avoid the accumulation of any material on the walls. This will produce a

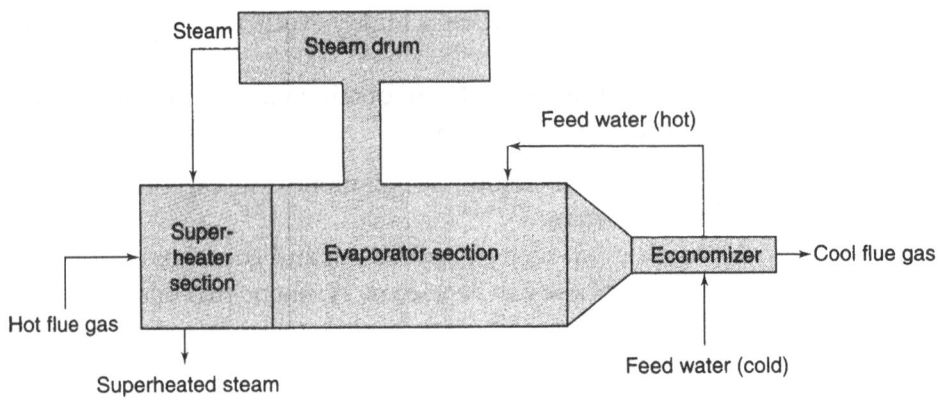

Fig.   .3   Waste heat boiler

greater volume of flue gas, which in turn will require higher capacity air pollution control equipment. If the heat recovery system is used, then only 50 to 100 per cent excess air is sufficient. Moreover, due to heat reduction, the flue gas volume will also be further reduced, resulting in the requirement of smaller capacity air pollution control equipment.

In the water-wall combustion chamber, the walls of the combustion chamber are lined with boiler tubes that are arranged vertically. As water is circulated through the tubes, heat is absorbed and steam is produced. The waste heat boiler is designed as a separate chamber located on the external side of the combustion chamber. Hot flue gas produced in the furnace of the combustion chamber is introduced into the superheater section of the waste heat boiler. The design outline of a waste heat boiler is shown in Fig. 7.3.

Biological Conversion Process

Currently, biological conversion of solid wastes in general and MSW in particular is the most preferred and economical method of waste treatment. Using this method the volume and weight of the solid wastes can be reduced; a humus-like material, compost, an ideal soil-conditioner is produced; and depending on the process condition methane gas can also be obtained.

The organisms involved in the biological transformation of organic wastes are mainly bacteria, fungi, yeasts, and *actinomycetes*. Depending on whether the operation is performed in the presence or absence of oxygen, the process can either be aerobic (known as aerobic composting) or anaerobic (anaerobic digestion). Accordingly, the nature of the end products also differs, as the conversion reactions of aerobic and anaerobic processes are differently accomplished. Sufficient amount of oxygen is a prerequisite for the aerobic conversion process.

Aerobic Composting of Solid Wastes

Aerobic composting of solid wastes is normally an uncontrolled process of biological conversion of organic matter such as MSW. The extent and the time period over which decomposition occurs is controlled by various factors such as the availability of oxygen, nutrients, moisture content, and the nature of waste. Under controlled conditions, however, within a reasonably short period of 4 to 6 weeks time, the yard waste and the organic fraction of SW (solid waste) can be converted into a stable organic residue known as *compost*.

The transformation reaction that occurs during aerobic composting of SW can be represented as follows:

$$\text{Oganic matter} + \text{nutrients} + O_2 \xrightarrow{\text{micro-organisms}} \text{new cell mass}$$
$$+ \text{residual organic matter}$$
$$+ CO_2 + H_2O + NH_3$$
$$+ SO_4^{--} + \text{heat} \qquad (7.1)$$

The residual organic matter is the compost consisting mainly of cellulose, lignin, and inorganic matter. A better theoretical analysis can be presented if we assume the decomposable organic matter of SW on a molar basis to be $C_aH_bO_cN_d$ and the hard-to-decompose residual organic matter to be $C_wH_xO_yN_z$. With this assumption it would be possible to estimate the amount of oxygen required for the stabilization of the biodegradable organic fraction of SW by using the following correlation. The production of new cells and sulphate during the process needs to be ignored.

$$C_aH_bO_cN_d + 1/2\ (ny + 2s + r - c)O_2 = nC_wH_xO_yN_z + sCO_2 + rH_2O$$
$$+ (d - nx)NH_3 \qquad (7.2)$$

Composting

where $s = (a - nw)$, $r = (1/2)[b - nx - 3(d - nz)]$, and the number of moles of $O_2 = (1/2)(ny + 2s + r - c)$. If no residual organic matter exists after the aerobic biological conversion of SW, then the expression can be given as

$$C_aH_bO_cN_d + \left\{\frac{4a + b - 2c - 3d}{4}\right\}O_2 = aCO_2 + \left\{\frac{b - 3d}{2}\right\}H_2O + dNH_3 \quad (7.3)$$

Illustration 7.1 shows how the amount of oxygen required for the aerobic stabilization of solid waste producing compost can be estimated numerically.

Illustration 7.1

A fruit and vegetable processing unit generates 1 t of solid waste that needs to be stabilized aerobically. Estimate the amount of oxygen required to oxidize the waste. It may be assumed that the initial composition of the biodegradable organic material to be decomposed is $[C_6H_7O_2(OH)_3]_5$ and the final composition of the residual organic matter is $[C_6H_7O_2(OH)_3]_2$. After the oxidation process, 40% of the material is available as compost.

Solution

The oxygen requirement can be estimated with the help of Eqn (7.2). From the data given in the problem, the initial molar composition of SW according to Eqn (7.2) will be $C_{30}H_{50}O_{25}N_0$ and the molar composition of the residual matter is $C_{12}H_{20}O_{10}N_0$. According to Eqn (7.2), 1 mol of SW yields n mols of compost. Again from the data given in the problem, we find that 1 t of SW yields 400 kg of compost. On the other hand,

$$1 \text{ t of SW} = \left[\frac{1000}{30 \times 12 + 50 \times 1 + 16 \times 25 + 0 \times 14}\right]$$

$$= 1.23 \text{ mol}$$

$$400 \text{ kg of compost} = \left[\frac{400}{12 \times 12 + 20 \times 1 + 10 \times 16 + 0 \times 14}\right]$$

$$= 1.23 \text{ mol}$$

Thus we find $n = \dfrac{1.23}{1.23} = 1$

Further, as per Eqn (7.2), moles of oxygen required per mole of SW are given by

$$(1/2)(ny + 2s + r - c)$$

where $n = 1$, $y = 10$, $s = a - nw = 30 - 1 \times 12 = 18$, $c = 25$:

$$r = (1/2)[b - nx - 3(d - nz)]$$

$$= (1/2)[50 - 1 \times 20 - 3(0 - 1 \times 20)]$$

$$= 15$$

Substituting values in the equation used for calculating the number of moles of oxygen, we have

Moles of O_2 required $= (1/2) (1 \times 10 + 2 \times 18 + 15 - 25) = 18$

So the total amount of oxygen required

$= 1.23 \times 18$ mol

$= 22.14$ mol

$= 708.5$ kg

In the absence of nitrogen in SW, no ammonia will be formed. Thus, process input consists of SW and oxygen, and process output contains compost, carbon dioxide, and water. From material balance, we get

Input $= 1000$ kg (SW) $+ 708.5$ kg (O_2) $= 1708.5$ kg

Output $= 400$ kg (compost) $+ 1.23 \times 18 \times 44$ kg (CO_2)

$+ 1.23 \times 15 \times 18$ kg (H_2O)

$= 1706.3$ kg

Unaccounted mass $= 2.2$ kg

The modern technique used for composting of solid waste such as MSW for a large-scale operation has three basic steps. These are pre-processing; biological decomposition; and preparation and marketing of the compost.

Pre-processing of MSW

Pre-processing of MSW for composting includes receiving, removal of inorganic or non-biodegradable materials, size reduction, adjustment of carbon–nitrogen ratio, and addition of moisture (to adjust to 50–60%) and nutrients, if required.

Biological decomposition of MSW

Techniques such as windrow, static pile, and in-vessel composting constitute the decomposition step.

Windrow composting In the case of windrow composting, the pre-processed MSW or industrial solid waste is placed in a windrow in an open field. Windrows are formed by simply dumping the SW into small heaps 2.5 to 3 m high and 6 to 7.5 m wide at the base. A smaller heap size will ensure better control. Composting period is normally 4 to 5 weeks, during which the windrows are required to be turned once or twice per week for ensuring proper supply of oxygen and maintenance of temperature (around 55°C). The varied types of micro-organisms occurring naturally in the surrounding environment cause the decomposition of the organic matter. Completion of the decomposition process is indicated by a drop in the temperature. To ensure complete stabilization of the solid waste, the composted material should be cured for another 2 to 4 weeks in open windrows.

Static pile system   The aerated static pile system consists of a grid of exhaust perforated piping (to facilitate aeration) over which the pre-processed organic fraction of MSW is dumped. Typically, the pile heights are about 2 to 2.5 m. To control the odours emitted at the initial stage and also to provide insulation, a layer of screened compost may be placed on top of each of the newly formed piles. For effective control of aeration, each pile should be provided with an individual blower. Air is introduced usually through disposable plastic drainage pipes, which provide the oxygen needed for biological conversion and help to control the temperature within the pile. About 3 to 4 weeks are required for composting the organic material, which is further cured for another 4 weeks.

In-vessel composting   As the name indicates this process is accomplished inside a vessel or enclosed container of any size, shape, or design including vertical towers, and horizontal circular and rectangular tanks. The system can either be the plug flow type or the agitated bed type. The plug flow system operates in a first-in, first-out basis, while in the case of the agitated bed system, the material is mixed mechanically during processing.

Preparation and marketing of compost

Preparation and marketing of compost, the third stage in the composting process, starts after the compost has been cured and stabilized. This includes fine grinding, screening, air classification, blending with various additives to improve the quality and other properties, granulation, bagging, storage, shipping, or direct marketing of the compost.

Design aspects of aerobic composting

To design an efficient composting system, following aspects need to be borne in mind.

Particle size   A reduced particle size increases the biochemical reaction rate during the composting process. Though the most desirable particle size for composting is less than 5 cm, larger particles can also be composted. Small particle size increases bed density and also the cost of size reduction.

Carbon to nitrogen ratio   For any biological transformation process, including aerobic composting, C/N ratio is very critical. The ideal C/N ratio for any biological process lies between 25 and 30. As discussed in Chapter 2, C/N ratios for different solid wastes are different. In some cases the magnitude is very low and in some other cases the ratio is very high. In such situations, it is a prerequisite to blend carbon and nitrogen in proper proportions to bring the ratio to around 30. This is exemplified in Illustration 7.2. Blending is not only necessary for optimizing the C/N ratio, but also to appropriate the moisture content. Some wastes may be too dry or some other material may contain too much water. None of these conditions is favourable for microbial composting. For this reason dry material should be blended with wet material to achieve the desired moisture content (50–60%). Moisture less than 40% is undesirable.

Seeding or inoculation   Though the wastes contain micro-organisms, their population density may not be sufficient to effect rapid composting of the solid waste. For this reason it is essential that a desired volume of microbial culture be added to effect the decomposition of the organic fraction of the solid waste at a faster rate.

Mixing or turning   The purpose of mixing or turning is to achieve the following:

(a) uniform distribution of micro-organisms and nutrient in the compost bed,
(b) increase or decrease the moisture content,
(c) supply oxygen for the biological process to continue, and
(d) temperature control.

In the case of organic wastes with 55–60% moisture content and a composting period of about 15 days, the first turning can be done on the third day and thereafter on every alternate day.

Temperature   As aerobic composting is a microbial process, the type of organisms involved are either mesophilic or thermophilic. That means the composting bed needs to be maintained at either 30–38°C for mesophilic or 55–60°C for thermophilic microbes. The process is exothermic; therefore, the bed temperature also rises automatically. In aerated static piles and the in-vessel composting system, temperature is regulated by monitoring the temperature and controlling the air flow. In windrow composting, temperature can only be controlled indirectly—by varying the frequency of turning based on temperature measurement. In general, the pile temperature drops down to 5–10°C after every turning, but returns to its previous level within some hours. With the eventual oxidation of readily available biodegradable organic material, the temperature in the windrow composting system starts to decrease after 10–15 days.

Air requirement   In the case of the aerated static pile and in-vessel systems, the essential design parameters are total air requirement and air flow rate. Illustration 7.3 shows the computation of total air requirement and air flow rate for an in-vessel composting system. A similar procedure can be adopted for the computation of air flow rate and total air requirement for an aerated static pile system also.

*p*H **control**   Another important parameter for microbial waste stabilization is *p*H, as the composting of solid waste depends on the functioning of micro-organisms. Like temperature, the *p*H value of the composting bed also varies with time as the process continues. The initial *p*H of the organic fraction of solid waste or MSW is normally between 5 and 7. At the beginning of the composting process, the *p*H drops to 5 or less. At this stage the mesophilic microbes begin to multiply and as a result the temperature rises rapidly from the ambient condition. Initially, organic acids such as butyric acid, propionic acid, caproic acid, valeric acid, and acetic acid are produced, which are responsible for the drop in *p*H. In about 3 days time, the temperature reaches the thermophilic range, and the *p*H begins to rise to about 8 or 8.5 and this continues for the remainder of the aerobic process. In the final condition, when the bed temperature again starts to decrease and the compost is matured, the *p*H value settles between 7 and 8. However, if there is insufficient aeration, then anaerobic conditions will prevail and the *p*H will drop to about 4–5. This condition retards the composting process.

Degree of decomposition   The various methods by which the degree of decomposition can be measured are the following.

(a) Final drop in temperature
(b) Amount of decomposable and resistant organic matter in the compost
(c) Rise in redox potential in the bed
(d) Oxygen uptake rate
(e) Starch–iodine test

However, analysis of COD and the lignin test can provide a quicker way of assessing the degree of decomposition. A low COD value and high lignin content (> 30%) indicate that the compost is stable.

Illustration 7.2

Plant leaves can be a good source for the production of compost. For the composting of any organic material, its C/N ratio should be around 25–30 and moisture content 50–60%. However, the leaves provided as a sample have a much higher C/N ratio (~50). In order to use these leaves for composting, it is necessary to blend the leaves with some other organic wastes whose C/N ratio is much lower. For this purpose waste activated sludge has been selected with C/N ratio ~10. With this information, determine the proportion in which the two components need to be blend, so that the desired C/N ratio and moisture content can be achieved in the blended mix. Some more data are provided as given here.

Leaves: (a) Moisture content = 50%
              (b) Nitrogen content = 0.8%
Sludge: (a) Moisture content = 70%
              (b) Nitrogen content = 5.0%

Solution

1. Let us assume that the required proportion of leaves is P. That means, for P kg of leaves, 1 kg sludge should be mixed.

   (a) P kg leaves contain $0.5P$ kg moisture. So,

   Weight of dry leaves $= P - 0.5P = 0.5P$ kg
   Nitrogen content $= 0.008 \times 0.5P$ kg $= 0.004P$ kg
   Carbon content $= 50 \times 0.004P$ kg $= 0.2P$ kg

   (b) 1 kg sludge contains 0.7 kg moisture. So,

   Dry weight of sludge $= (1 - 0.7)$ kg $= 0.3$ kg
   Nitrogen content $= 0.05 \times 0.3$ kg $= 0.015$ kg
   Carbon content $= 10 \times 0.015$ kg $= 0.15$ kg

2. Now let us determine the value of P, assuming $C/N = 30$:

   Carbon content of blended mix $= (0.2P + 0.15)$ kg
   Nitrogen content of the blended waste $= (0.004P + 0.015)$ kg

   So,

   $$30 = \frac{0.2P + 0.15}{0.004P + 0.015}$$

   Solving this equation, we get

   $P = 3.75$ kg leaves/kg sludge

3. Let us now check for the C/N ratio and moisture content of the blended waste.

   (a) For 3.75 kg leaves,

   Water $= 3.75 \times 0.5$ kg $= 1.875$ kg
   Dry matter $= (3.75 - 1.875)$ kg $= 1.875$ kg
   Nitrogen $= 1.875 \times 0.008$ kg $= 0.015$ kg
   Carbon $= 0.015 \times 50 = 0.75$ kg

   (b) For 1 kg sludge,

   Water $= 1 \times 0.7 = 0.7$ kg
   Dry matter $= 1 - 0.7 = 0.3$ kg
   Nitrogen $= 0.3 \times 0.05 = 0.015$ kg
   Carbon $= 0.015 \times 10 = 0.15$ kg

   (c) $C/N = \dfrac{0.75 + 0.15}{0.015 + 0.015} = \dfrac{0.9}{0.03} = 30$

(d) Moisture $= \dfrac{1.875 + 0.7}{3.75 + 1} = \dfrac{2.575}{4.75} = 0.542$ or 54.2%

This is within the range of 50–60%.

Illustration 7.3

0.6 t of solid waste from the market place is treated to produce compost in an in-vessel composting system. If forced aeration is provided, determine the quantity of air required for composting. Other information is as follows: composition of solid waste: $C_{60}H_{95}O_{40}N$, moisture content $= 30\%$, VS (volatile solids) $= 0.9 \times$ TS (total solids), BVS (biodegradable volatile solids) $= 0.6 \times$ VS, expected conversion efficiency of BVS $= 95\%$, composting period $= 5$ days. The oxygen demand can be assumed for the successive five days of composting as 20%, 30%, 30%, 15%, and 5%. The ammonium produced is released into the atmosphere. Air contains 23% O_2 by mass and its specific weight is 1.2 kg/m^3. A factor of 1.8 times the quantity of theoretical air required is needed to ensure that the oxygen content of the exhaust air does not fall below 45% of its original value.

Solution

First the quantity of BVS in 0.6 t of solid waste needs to be computed:

TS $= 600$ kg $\times (1 - 0.3) = 420$ kg

VS $=$ TS $\times 0.9 = 420 \times 0.9 = 378$ kg

BVS $=$ VS $\times 0.6 = 378 \times 0.6 = 226.8$ kg

Expected BVS converted $= 226.8 \times 0.95 = 215.5$ kg

Using Eqn (7.3), the amount of oxygen required per kg of converted BVS can be calculated.

$$C_a H_b O_c N_d + \left\{ \frac{4a + b - 2c - 3d}{4} \right\} O_2 = a CO_2 + \left\{ \frac{b - 3d}{2} \right\} H_2 O + d NH_3$$

From the given chemical composition, $a = 60$, $b = 95$, $c = 40$, $d = 1$. The balanced equation will be

$C_{60}H_{95}O_{40}N + 63O_2 = 60CO_2 + 46H_2O + NH_3$

(1469)   (2016) (2640)   (828)   (17)

So, oxygen required per kg BVS converted $= 2016/1469 = 1.37$ kg O_2/kg BVS converted. Theoretically, the total oxygen required for conversion of 215.5 kg BVS $= 1.37 \times 215.5 = 295$ kg. So, the theoretical amount of air required to treat 600 kg of solid wastes containing 226.8 kg BVS $= 295/0.23 = 1282.6$ kg.

Volume of air theoretically required $= 1282.6/1.2 = 1068.8$ m^3

Volume of air actually supplied $= 1068.8 \times 1.8 = 1923.84$ m^3

The required capacity of air blower, expressed in m^3/min, based on the maximum consumption of oxygen in a day = (total air actually supplied) $\times$ (maximum percentage of total oxygen demand in a day) $\div$ (total minutes in a day)

$$= \frac{(1923.84)\,(0.30)}{1440}$$

$$= 0.4\ m^3/min$$

Anaerobic Composting of Solid Waste

Anaerobic composting of solid waste is done by treating solid waste biologically in the absence of oxygen. This can also be termed as anaerobic fermentation or anaerobic digestion. In this method, along with the compost, methane gas is also produced.

Three groups of micro-organisms are involved in this process. One group of micro-organisms is responsible for hydrolysing organic polymers and lipids to fatty acids, monosaccharides, amino acids, and other compounds. A second group of anaerobic bacteria ferments the chemicals produced by the first group to simple organic acids such as acetic acid. The second group of organisms consists of facultative or obligate anaerobes identified as *acetogens*. The third group of microbes, which are strictly anaerobes, convert the hydrogen and acetic acid formed by the acetogens into biogas (methane and CO_2). The organisms of the third group are known as methanogens.

For efficient performance of the anaerobic composting system, the methanogenic and non-methanogenic bacteria should be in a state of dynamic equilibrium and this can be achieved by keeping the reactor content free from dissolved oxygen and excess ammonia. The *p*H should be maintained between 6.5 and 7.5, for which the digestion mixture needs to be accordingly alkaline in nature. The required alkalinity ranges from 1000 to 5000 mg/L and the volatile fatty acid concentration should be below 250 mg/L. However, in the case of high-solids anaerobic digestion process, the values for alkalinity and VFA can be as high as 12,000 mg/L and 700 mg/L, respectively. For proper growth of the biological community, sufficient amount of nutrients such as nitrogen and phosphorus must also be present.

Details regarding the biochemical reactions and pathways have already been discussed in Chapter 3. It is known that methanogens use substrates such as $CO_2 + H_2O$, formate, acetate, methanol, methyl amines, and CO for growth and energy, i.e., oxidation – reduction reactions. The general anaerobic transformation of solid wastes can be described by using the following correlation:

Organic matter of SW + H_2O $\longrightarrow$ new cells
+ inoculum + nutrient   + residual organic matter + CO_2
    + CH_4 + NH_3 + H_2S + heat   (7.4)

For practical purposes, the equation can be represented in the following form:

$$C_aH_bO_cN_d \longrightarrow nC_wH_xO_yN_z + mCH_4 + sCO_2 + rH_2O + (d-nz)NH_3 \quad (7.5)$$

where $s = a - nw - m$ and $r = c - ny - 2s$. The terms $C_aH_bO_cN_d$ and $C_wH_xO_yN_z$ represent the organic matter of the solid wastes on their molar basis, before and after digestion, respectively. If the organic wastes are stabilized completely, the corresponding expression is given as

$$C_aH_bO_cN_d + \left\{ \frac{4a - b - 2c + 3d}{4} \right\} H_2O = \left\{ \frac{4a + b - 2c - 3d}{8} \right\} CH_4 \quad (7.6)$$

$$+ \left\{ \frac{4a - b + 2c + 3d}{8} \right\} CO_2 + dNH_3$$

Normally, the biogas generated in the digester contains 50–60% methane and the volume of the gas produced is about 0.6 to 1.0 m³ of gas/kg of volatile acid consumed. Illustration 7.4 shows the computation of the volume of gas produced by complete anaerobic stabilization of solid wastes.

Illustration 7.4
Using the data provided here, estimate theoretically the volume of biogas that can be produced by anaerobic treatment of 100 kg of solid waste.

(a) Organic material (VS) in SW = 80% (including moisture)
(b) Moisture content in VS = 30%
(c) Biodegradable VS = 95% (dry basis)
(d) Chemical formula of BVS = $C_{60}H_{95}O_{40}N$
(e) Specific weights of methane and carbon dioxide = 0.7112 and 1.9607, respectively.

Solution
Using Eqn (7.6), theoretically the amount of CH_4 and CO_2 that is produced can be determined as shown here.

$$C_aH_bO_cN_d + \left\{ \frac{4a - b - 2c + 3d}{4} \right\} H_2O = \left\{ \frac{4a + b - 2c - 3d}{8} \right\} CH_4 +$$

$$+ \left\{ \frac{4a - b + 2c + 3d}{8} \right\} CO_2 + dNH_3$$

From the data, we get $a = 60$, $b = 95$, $c = 40$, $d = 1$. With these values the given equation becomes

$$C_{60}H_{95}O_{40}N + 17H_2O \longrightarrow 31.5CH_4 + 28.5CO_2 + NH_3$$
$$(1469) \qquad (306) \qquad (504) \qquad (1254) \qquad (17)$$

From these numerical values the weights of CH_4 and CO_2 can be calculated as follows:

$$\text{Total BVS in SW} = 100 \times 0.80 \times (1 - 0.30) \times 0.95 = 53.2 \text{ kg}$$

So,

$$\text{Weight of } CH_4 \text{ produced} = \frac{504 \times 53.2}{1469} = 18.25 \text{ kg}$$

$$\text{Weight of } CO_2 \text{ produced} = \frac{1254 \times 53.2}{1469} = 45.41 \text{ kg}$$

$$\text{Volume of } CH_4 = \frac{18.25}{0.7112} = 25.66 \text{ m}^3$$

$$\text{Volume of } CO_2 = \frac{45.41}{1.9607} = 23.16 \text{ m}^3$$

$$\text{Total volume of gas produced} = (25.66 + 23.16) \text{ m}^3 = 48.82 \text{ m}^3$$

$$\text{Volume \% of } CH_4 = \left(\frac{25.66}{48.82}\right) \times 100 = 52.56\%$$

$$\text{Volume \% of } CO_2 = (100 - 52.56)\% = 47.44\%$$

Based on BVS, gas generated per unit weight, $48.82/53.2 = 0.92$ m^3/kg BVS consumed. Based on total solid wastes, gas production $= 48.82/100 = 0.4882$ m^3/kg SW treated anaerobically. It may be noted that in practice the theoretical amount of 0.92 m^3/kg BVS may not be achievable.

Types of anaerobic digestion

Anaerobic digestion can be either of low-solids or of high-solids type. These two differ in the concentration of solids in the digester. In the case of low-solids anaerobic fermentation, the solid concentration equals to or is less than 4–8%, whereas the solid concentration for the high-solids anaerobic digestion process is 22% or more.

The production of methane from the low-solids organic fraction of solid wastes consists of three basic steps. The first step involves the preparation of organic fractions of the solid waste such as receiving, sorting and separation, and size reduction. The second step is very important as it includes the addition of moisture and nutrients, blending, pH adjustment to about 6.8, and heating of the slurry if the digestion is to take place in the thermophilic range of 55–60°C. The third step comprises the separation of the gas and the dewatering and disposal of the digested sludge.

Similarly, the high-solids anaerobic digestion of solid wastes also has three steps. However, the main difference between the two is that for the high-solids process, less effort is required to dewater and dispose of the digested sludge. Further, in both the cases, to prevent ammonia toxicity, proper adjustment of the C/N ratio of the input feedstock is required. It is obvious that the high-solids anaerobic digestion system is capable of producing more gas and stabilizing more organic wastes per unit volume of the reactor than the low-solids process. Illustration 7.5 shows the amount of methane that can be produced by means of high-solids anaerobic digestion process.

Illustration 7.5

The data given here are for the anaerobic digestion of an industrial solid waste whose solid concentration is high: organic fraction of SW, i.e., TS (total solids) = 1.6 t, moisture content in TS = 25%, digestion period = 30 days, VS = 93% of TS, BVS = 75% of VS, expected BVS conversion efficiency = 95%, gas production = 0.5 m^3/kg of BVS destroyed, energy content of biogas = 18,630 kJ/m^3 = 4440 kcal/m^3, kerosene equivalent of biogas = 0.62 L/m^3, price of kerosene = Rs 10,000/kl. Determine the volume of gas produced from 1.6 t of organic solid waste, energy content of the gas, and its market value.

Solution

$$VS = 1.6 \times 1000 \times (1 - 0.25) \times 0.93 \text{ kg} = 1116 \text{ kg}$$
$$BVS = 1116 \times 0.75 \text{ kg} = 837 \text{ kg}$$
$$BVS \text{ converted} = 837 \times 0.95 \text{ kg} = 795.15 \text{ kg}$$
$$\text{Volume of biogas produced from 1.6 t of organic SW} = 795.15 \times 0.5 \text{ m}^3$$
$$= 397.5 \text{ m}^3$$
$$\text{Kerosene equivalent} = 397.5 \times 0.62 \text{ L} = 246.5 \text{ L}$$
$$\text{Market value of the gas produced in rupees} = 246.5 \times 10 = \text{Rs } 2465$$

It may be noted that anaerobically digested sludge can be sold as good organic manure.

Vermicomposting

During the green revolution era, the large-scale use of chemical fertilizers did solve the food problem of our country to a great extent. But at the same time the unscrupulous use of these chemical fertilizers left the agricultural lands acidic, thereby adversely affecting the production in the long run. The unrestrained use of chemical fertilizers has had a detrimental effect on the health of people and has led to the rejection of Indian agricultural produce in the international market. As a result, a viable alternative to the use of existing chemical fertilizers was highly sought after.

Traditional backyard manure-producing techniques did work as an alternative to the existing chemical system but lacked the ability to match the convenience and efficiency of agriculture that came with the use of chemical fertilizers. Agro-residues, yard wastes, and kitchen wastes, which make up a major portion of the solid waste menace, were utilized in the traditional backyard manure-producing methods. Although much of this organic waste was recycled to yield valuable manure, the traditional aerobic backyard composting techniques generally employed in small scale were not appropriate and were often inconvenient, unhygienic, and inefficient.

Vermicomposting, or composting using earthworms, was soon discovered to be an excellent alternative to the traditional backyard technique for recycling food waste, composting yard wastes in backyards as well as in households. Vermicomposting is an ecologically safe method that naturally converts several types of organic wastes into extremely environment-friendly products.

The term vermicomposting is an aggregate of two words. The prefix *vermi* is Latin for worm. Earthworms have been recognized for long as friends of farmers. As the name itself suggests, vermicomposting refers to the utilization of the digestive process of earthworms to make compost. Vermicomposting employs various species of earthworms (e.g., *Pontoscolex corethrurus*, *Megascolex konkanensis*, *Lampito mauritii*, *Drawida willsi*, *Lumbricus rebellus*, *Eisensia foetida*, *Octochaetona surensis*, *Amynthas corticis*, and *Metaphire houlleti*) that work in harmony with other compost organisms to decompose organic matter.

Advantages of vermicomposting

Vermicomposting has several advantages over traditional backyard composting techniques. Some of them are listed here.

1. It reduces household garbage as well as agro-residues and yard waste disposal costs.
2. Produces fewer odours and attracts fewer pests than those that can be attributed to food and yard wastes (which have high organic matter) deposited into open garbage containers or pits.
3. The method saves water and electricity that is consumed by kitchen sink garbage disposal units.
4. It produces a free, high-quality soil amendment (compost) which has a lower C/N ratio than traditional backyard compost.
5. The compost produced requires only little space, labour, and maintenance (as upturning is not needed).
6. The process is much faster.
7. It spawns worms for fish feed and poultry feed, free of cost.

Properties of vermicompost as an ideal fertilizer

1. It has a lower C/N ratio, which reflects in its ability to enhance fertility much better than other manures.

2. Vermicompost has been reported to have the ability to make a higher percentage of potassium, phosphorus, and nitrogen available to plants.
3. It contains immobilized enzymes such as amylase, protease, lipase, lichenase, and chitinase, which catalyse the biodegradation of macromolecules in the soil and thereby enable quicker uptake of nutrients from the soil.
4. Immobilized microflora in the vermicompost help in soil conditioning.
5. The compost is rich in vitamins, antibiotics, and growth hormones.
6. It is generally free from pathogenic micro-organisms.
7. Vermicasts have earthworm cocoons which promote earthworm population in the soil and thereby ensure a continuous production of vermicasts in the soil itself.
8. It maintains soil moisture content by absorbing moisture from air.
9. It gives structural stability to the soil.

Categories of earthworms

Earthworms are normally categorized depending upon their natural habitat. They are selected for vermicomposting on the basis of their digestive efficiency and the need of the entrepreneur or farmer. Depending upon the depth of the soil at which they are found, earthworms are divided into three categories.

1. Epigeic (surface dwellers)
2. Mesogeic (surface and sub-surface dwellers)
3. Endogeic (sub-surface dwellers)

Endogeic species invariably are very large in size and have very poor turnover ability. As they are sub-surface dwellers, they are characterized by strong burrowing muscles. *Mesogeic* species are moderately large and are characterized by moderate turnover. They too are characterized by well-developed burrowing muscles. Though not preferred for vermicomposting due to their not so efficient vermicast turnover ability, they can be employed for decompacting the soil.

Various greenhouse and field trials have shown significant increase in plant production when *Pontoscolex corethrurus* was inoculated, but its long-term use has also been shown to be detrimental to the system under specific soil and climatic conditions, and in the absence of soil-decompacting earthworm species.

Epigeic species (surface dwellers) are most favoured for vermicomposting. These species are selected as they have a very high turnover rate and a high rate of reproduction.

Biology of the earthworm

Earthworms play an important role in our environment by converting organics into valuable organic fertilizer—vermicompost. The earthworm has a long, thin, soft, and rounded body with a pointed head and a slightly flattened posterior (Fig. 7.4).

The body of an earthworm is made up of rings with grooves between them. Each ring is called a segment. Rings that surround the moist, soft body allow the earthworm to twist and turn, especially since it has no backbone. The earthworm has no true legs. Instead it has bristles (setae) on the body which move back and forth and allow the worm to crawl. The earthworm breathes through its skin. Worms need a dark, moist environment for the skin to function properly so that they can get sufficient oxygen from the air.

Earthworms are hermaphrodites, i.e., they have both the male and female sex organs, but they still require another earthworm to mate. The wide band (clitellum) that surrounds a mature breeding earthworm secretes mucus (albumin) after mating. Sperm from another worm is stored in sacs. As the mucus slides over the worm, it encases the sperm and eggs inside. After slipping free from the worm, both ends seal, forming a lemon-shaped cocoon, approximately 3 mm long. Two or more baby worms hatch from one end of the cocoon in about 3 weeks time. Baby worms are whitish to almost transparent and 12 to 25 mm long. Red worms take 4 to 6 weeks to become sexually mature. They have a long, thin

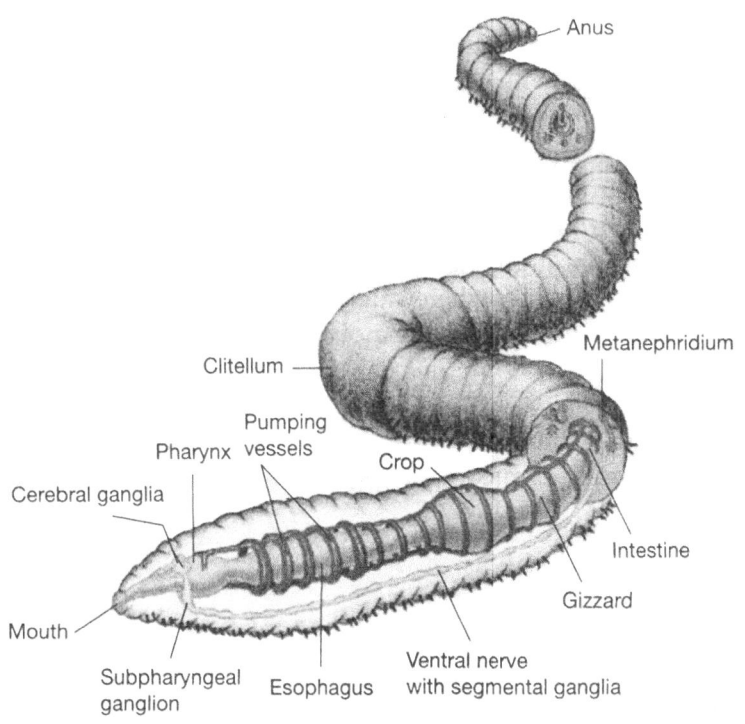

Fig. 7.4   Structure of an earthworm

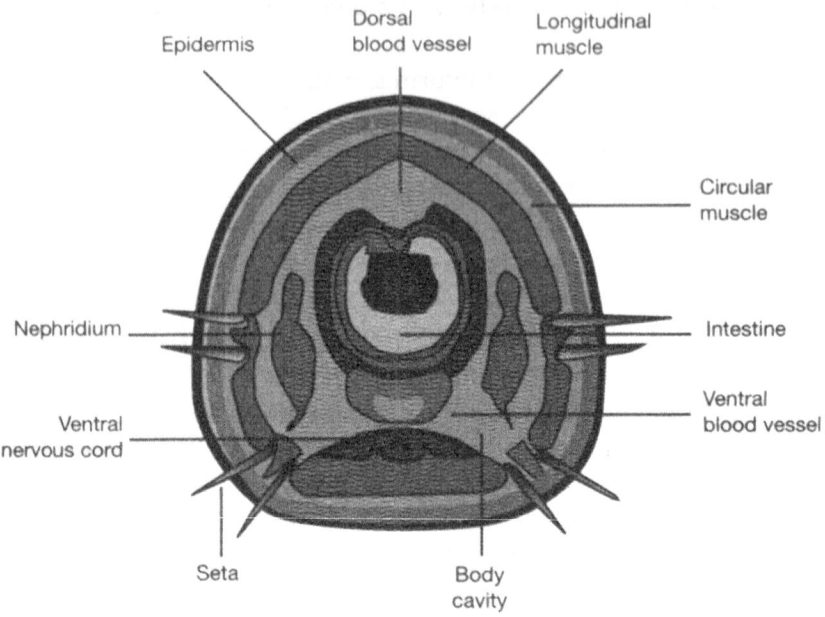

Epidermis   Dorsal blood vessel   Longitudinal muscle

Circular muscle

Nephridium

Intestine

Ventral nervous cord

Ventral blood vessel

Seta

Body cavity

Fig. 7.5   Internal anatomy of an earthworm (cross section)

shape and a soft body with no bones. Despite the fact that it has no eyes an earthworm can sense light. It is averse to exposure to light.

Worms generally like to move head first, hence a careful observation of the movement of a worm will help to determine which end is the head. Also, the clitellum, or the thick band around an adult worm, is closer to the head. Food is ingested through the mouth into the stomach (crop) (Fig. 7.5). Later the food passes through the gizzard, where it is ground up by ingested stones. After passing through the intestine for digestion, the undigested matter is eliminated. This is the rich vermicast of our interest. It lacks the foul odour of organic substrates. Instead it is moist, light, rounded, with the sweet smell of moist clay. Table 7.1 compares garden compost and vermicompost.

Technique of vermicomposting

The technique of vermicomposting consists of the following steps.

1. Preparation of worm pit
2. Bedding material
3. Addition of worms
4. Addition of organic waste
5. Controlling moisture and temperature
6. Maintenance of bin
7. Harvesting worms and vermicompost

Table 7.1 Contents of garden compost and vermicompost

Parameter*	Garden compost	Vermicompost
pH	7.80	6.80
EC (mmhos/cm)[†]	3.60	11.70
Total Kjeldahl nitrogen(%)[‡]	0.80	1.94
Nitrate nitrogen (ppm)[¶]	156.50	902.20
Phosphorous (%)	0.35	0.47
Potassium (%)	0.48	0.70
Calcium (%)	2.27	4.40
Sodium (%)	< 0.01	0.02
Magnesium (%)	0.57	0.46
Iron (ppm)	11,690.00	7563.00
Zinc (ppm)	128.00	278.00
Manganese (ppm)	414.00	475.00
Copper (ppm)	17.00	27.00
Boron (ppm)	25.00	34.00
Aluminium (ppm)	7380.00	7012.00

*Units: ppm = parts per million, mmhos/cm = millimhos per centimetre.
[†]EC = electrical conductivity is a measure (millimhos per centimetre) of the relative salinity of soil or the amount of soluble salts it contains.
[‡]Kjeldahl nitrogen is a measure of the total percentage of nitrogen in the sample including that in the organic matter.
[¶]Nitrate nitrogen refers to the quantity of nitrogen in the sample that is immediately available for plant uptake through the roots.

Preparation of worm pit Worm pits should be prepared in a shady place (or a shade can be built over the pit). The pit may be made from concrete rings or can be dug in the ground. Bins made of wood or plastic, or from recycled containers such as old bathtubs, barrels, or trunks may also be employed. They can be located inside or outside, depending on preferences and circumstances. For commercial vermicomposting, the earthworms generally employed are the epigeic species. Hence bins need not be more than 20 to 30 cm deep. Bedding and food wastes tend to pack down in deeper bins, forcing air out. The resulting anaerobic condition can cause emission of foul odours and death of the worms. The length and width of the bin depends on whether it is to be stationary or portable. It also depends on the amount of waste produced. A good thumb rule is to provide 2000 sq. cm of surface area per kg of waste in the bin or pit.

Most often wooden bins are preferred, as they are better absorbents and also provide better insulation. However, woods such as redwood or other highly aromatic woods should not be employed for bin construction as these may kill the worms. Plastic material tends to keep the compost too moist and hence extra effort is needed to maintain the moisture content at the right level. However, plastic has the advantage that it is less messy and easier to maintain. Care should be taken that the containers being used have never been previously used for storing pesticides or other such chemicals, which may have a detrimental effect on the earthworm population.

Drilling air/drainage holes (6 to 12 mm diameter) in the bottom and sides of the bin will ensure good water drainage and air circulation. The bin can be placed on bricks or wooden blocks in a tray to make it convenient to collect the excess water that drains from the bin. The resulting compost tea (vermiwash) can be used as a liquid fertilizer around the home landscape.

Each bin or pit should have a cover to conserve moisture and exclude light. Worms prefer darkness. Pits or bins can be covered with a straw mulch or moist burlap to ensure darkness while providing good air ventilation at the same time. Bins placed outside in the open may also require a lid to keep out scavengers and other unwanted pests.

Outdoor bins should be insulated from the cold to protect the worms. One option is to dig a rectangular pit 30 cm deep and line its sides with wooden planks. The bottomless box can then be filled with appropriate bedding material, food waste, and worms. Food waste can be continually added as and when they accumulate. The pile should be kept damp and dark to optimize worm activity. During winter, soil can be piled against the edges of the bin and straw can be placed on top to protect the worms from cold weather.

Bedding material Bedding for bins can be made from a diverse range of organic substances such as shredded newspapers (non-glossy), computer paper, or cardboard; shredded leaves, straw, hay, or dead plants; sawdust; peat moss; or compost or aged (or composted) manure. Depending upon the material used for bedding, different kinds of pretreatments are needed. For example, peat moss first needs to be soaked for 24 hours in water, and then lightly wrung out to ensure it is sufficiently moist. Grass clippings should be allowed to age before they can be used because otherwise they may decompose too quickly, causing the compost to heat up.

Bedding materials, high in cellulose content, are considered to be most ideal because they help to aerate the bin, thus enabling the worms to breathe. Varying the bedding material provides a richer source of nutrients. Some soil or sand can be added to help provide grit for the worms' digestive system. The bedding material should be allowed to set for several days to make sure it does not heat up (and allow it to cool sufficiently before adding worms). Also the bedding material should be thoroughly moistened (about to the consistency of a damp sponge) before adding worms to it.

Vermicomposting bedding material

Addition of worms Under optimum conditions, in one day epigeic earth-worms can eat food equal to their own weight from food scraps and bedding. On an average, however, it takes approximately 2 kg of earthworms to recycle 1 kg of food waste in 24 hours. The same quantity of worms require about 0.1 m^3 of bin space to process the food waste and bedding (0.03 m^3 of worm bin/500 worms).

Composting worms can be purchased from dealers, or the endemic species present in the soil can also be utilized instead of using a pure culture. The worms should be added to the top of the moist bedding as soon as they are brought. The worms can be seen disappearing into the bedding within a few minutes.

Addition of organic waste Earthworms feed on a wide range of organic wastes including forest litter, coffee grounds, tea bags, vegetable and fruit waste, pulverized egg shells, grass clippings, manure, and sewage sludge. Night soil, bones, dairy products, and meats should be avoided as these may attract pests and flies. Limited amount of citrus products can be added, but excess quantity can make the compost too acidic. If possible the compost should be maintained at a *p*H of 6.5, with upper and lower limits of 7.0 and 6.0, respectively. Overly acidic compost can be corrected by adding crushed egg shells.

Addition of chemicals (including insecticides), metals, plastics, glass, soaps, pit manures, and oleanders or other poisonous plants, or plants sprayed with insecticides should be consciously avoided as it is detrimental to the health of the worms.

Food waste should be added to the bin by pulling back the bedding material and burying it. It should be covered to avoid flies and pests. Also covering with some mat or leaves helps to retain the moisture in the pit or bin. For better and more efficient composting, shredding is done to reduce the particle size. This helps the worms to ingest the materials more easily.

Controlling temperature and moisture in bin   The optimum temperature range for earthworms is between 13°C and 25°C. The worms should never be allowed to freeze. Bins kept outside may have to be insulated with straw during winter to keep the worms from freezing. Portable bins can be kept by the side of a hot water heater in the garage during the cold season to keep them warm.

The bin contents should be kept moist but not soaked. Rain water should never be allowed to run-off the roof into the bin. This could cause the worms to drown. A straw covering may be needed to protect the exposed sites and keep the bin from drying out during hot summer weather. A small light can be kept over the pit so that the worms do not escape from the pit during night.

Maintenance of bin   Food scraps can be continually added to the bin for up to 2 to 3 months, or until the bedding material disappears. When the bedding material disappears, the worms are harvested and the finished compost is collected and removed. The bins can again be refilled with new bedding material. The bins should not be overloaded, which may cause anaerobic conditions, resulting in a foul odour. If such conditions arise, further addition of wastes should be temporarily stopped. Excessively moist food waste and bedding also produce odour. In such a case the bedding should be aerated by slowly turning it and the water holes should be checked. Soggy beddings attract flies and an array of pests.

Harvesting of compost and worms   Harvesting involves shifting the finished compost and worms to one side of the bin and adding new bedding material and food waste from the other side. Worms in the finished compost should be allowed to move over to the new bedding area which has fresh food waste. The finished compost can then be removed for further use.

Troubleshooting problems in vermicomposting

Despite the apparent ease of the vermicomposting process it can be disturbed by various agents (both biotic and abiotic). Though prevention is the best way to avoid these agents, it is often unavoidable and troubleshooting has to be done. Some of the common problems in vermicomposting and the ways to tackle them are discussed here.

Sudden death of worms and/or attempt by the worms to escape There may be three probable reasons for such an occurrence. The bed is too dry, too wet, or has been used up. If the bedding is too dry, it has to be moistened; if it is too wet, then more bedding material has to be added. The bin needs to be harvested if the bedding is used up.

Foul smell from the bin This implies that composting is taking place in the absence of oxygen, or an anaerobic condition has developed in the bin. This may be due to excess of food/substrate and water or lack of aeration. In the first case further addition of food has to be stopped and the exit holes have to be checked. Mild turning may also help.

Fruit flies This may be due to the presence of exposed food on the surface. The food added has to be buried to avoid this problem.

Apart from these problems a number of techniques are used for efficient management of vermicompost. The pH can be maintained by addition of calcium carbonate if the medium is too acidic and addition of citrus remains and other acidic food if the bin contents turn too basic (although this is quite unlikely).

With efficient management, vermicomposting can prove itself an ideal alternative to the existing chemical procedures.

Landfill Bioreactor for Solid Waste Treatment

Landfill bioreactors are most suited for large-scale treatment of solid wastes or garbage. In the early stages of its development, the process required large land area because the conventional method of dumping waste was adapted. New, with the availability of a better modified design, the reactor size can be reduced significantly.

Conventional landfill refers simply to the disposal of solid waste on the surface soils of the earth. The modified design of landfill system is the sanitary landfill plant, which is an engineered facility for the disposal of solid refuse. Landfilling is actually the process by which solid wastes (SW) are deposited to fill a piece of land.

To prepare a landfill system, the total available area is divided into various cells of smaller areas. These cells are then filled up with solid waste, one after another. After filling up a cell to a certain depth, the top is covered with native soil, compost, or cattle dung. It is necessary to cover it to maintain the anaerobic condition of the bed. After filling up a particular cell, the final layer of cover is put in place. This usually consists of either multiple layers of soil or some membrane-materials designed to enhance surface drainage, intercept percolating water, and support surface vegetation.

Landfill site

One important aspect to be noted is that due to the biochemical reactions taking place in the solid bed that cause the solubilization or liquefaction of organic solid wastes, leachate is formed, which contains a variety of chemicals and biochemical products. Besides chemicals, the leachate can also include the water initially contained in the SW. This leachate percolates through the solid bed and gets collected at the bottom of the bed. As is known, anaerobic digestion of organic matter generates biogas containing methane and CO_2. The biogas produced from a landfill system is also known as landfill gas. The bottom of the bed is treated and made leak-proof to prevent the diffusion of landfill leachate and landfill gas.

The first step in the landfilling process is the preparation of the site for the construction of a landfill. Site preparation includes proper drainage of the landfill area, roads, weighing facilities, and installation of fences.

The second step is the excavation and preparation of the landfill bottom and sub-surface sides. To minimize cost, cover material can be obtained from the stock pile of excavated material. The landfill reactor bottom should be properly shaped to facilitate drainage of leachate and its bottom surface should be lined with low-permeability material to prevent loss of leachate.

After the site has been prepared and excavated, the solid waste is dumped into the cells one after another. The total depth of the waste contained in a cell may vary from 2.5 m to 3.5 m, with multiple layers. The thickness of each layer may be 0.5 to 0.75 m after compaction; each such layer is called a *lift*. The width of a cell varies from 3 to 10 m and a suitable length may be chosen. After filling up the cells with solid waste and compacting, the top is covered with a 15–30-cm-thick layer of soil or same other suitable material.

The gas collection system is designed by excavating horizontal trenches filled with gravel and laying perforated plastic pipes in each of the trenches. These trenches are provided at different heights and finally all of them are connected with vertical gas extraction wells. All the pipes are connected with a manifold and the gas is routed to energy recovery facilities such as an electricity generator or for heating purposes.

The additional leachate that is collected at the bottom of the cell can be either spread back over the solid wastes to enhance gas production or used in the secondary reactor for the production of methane gas.

Illustration 7.6

At IIT Kharagpur, the total population is 10,000. The concerned authority has decided to design a landfilling bioreactor system to treat solid wastes. To facilitate the collection of organic solid wastes, two vats were placed at each collection point, one for biodegradable solid wastes and another for non-biodegradable waste. On an average it was found that the biodegradable solid waste generated per capita per day was, approximately, 1 kg. It was decided that the average depth of the compacted solid waste should be about 5 m, and experimentally it was found that the specific weight of the compacted solid waste is 400 kg/m^3. With this information, estimate the area of land required for the proposed landfilling operation.

Solution

Daily biodegradable solid waste generation

$$= 10{,}000 \times \frac{1}{1000} \, t/d$$

$$= 10 \, t/d$$

Volume required per day for landfilling

$$= 10 \times \frac{1000}{400} \, m^3/d$$

$$= 25 \, m^3/d$$

Area required per day

$$= \frac{25}{5} \, m^2/d$$

$$= 5 \, m^2/d$$

So, the minimum area required for landfilling for one year = $5 \times 365 \, m^2 = 1825 \, m^2$ or 0.45 acre/year. Some additional space is required for roads, partitioning, etc.

Landfill Gas

The system operates practically under anaerobic conditions. Except for the presence of the initially entrapped air in the SW, no air is supplied during the treatment of solid wastes in a landfill reactor. Due to this reason, the two major constituents of landfill gas are methane (CH_4, 50–60%) and carbon dioxide (CO_2, 40–50%). Other components that may be found in landfill gas are nitrogen, oxygen (in the upper layer), ammonia, sulphide, hydrogen, carbon monoxide, etc. The heating value of the gas may vary from 14,900 kJ (3560 kcal) to 20,490 kJ (4890 kcal) per standard m^3.

The principle of anaerobic digestion has been discussed in Chapters 3 and 5 and also in an earlier section of this chapter. First the biodegradable components of solid waste undergo microbial decomposition. In the initial stage, biological degradation occurs mainly under aerobic conditions due to the presence of entrapped oxygen (air) in the untreated SW. The principal sources of both aerobic and anaerobic micro-organisms are the soil material, compost, or cattle dung which have been used for the daily and final cover of solid waste after it has been dumped in the landfill reactor. Recycled leachate and waste sludge from the activated sludge process of a wastewater treatment plant are the other potential sources of micro-organisms.

Anaerobic conditions develop with the depletion of oxygen content in the landfill bed. As discussed in Chapter 5, under anaerobic conditions, nitrates and sulphates serve as electron acceptors in the bioconversion reactions and these are reduced to nitrogen gas and sulphide, respectively. By measuring the redox potential of the waste in the landfill, the time of onset of anaerobic condition can be determined. Table 7.2 lists the redox potential values that indicate the onset of anaerobic conditions and methane production, as methanogens are strictly obligate anaerobes.

After the enzyme-mediated hydrolysis or solubilization stage of SW comes the acid-forming phase. Hydrolysates are transformed by acid-formers to various organic acids, predominantly acetic acid (CH_3COOH). In this phase, the gas produced contains mainly CO_2. At this stage the micro-organisms responsible for acid formation are non-methanogenic facultative and obligate anaerobes. Here, the *p*H value of the leachate drops down to 5 or even less than that. This is due to the presence of organic acids and increase in the concentration of CO_2 in the landfill bed. Further, the presence of dissolved organic acids in the leachate significantly increases the BOD_5 or COD value along with the conductivity of the leachate. The low *p*H value of the leachate helps to solubilize inorganic constituents such as metals and nutrients of the landfill bed that are

Table 7.2 Redox potential values

Redox potential (mV)	Reducing conditions resulting
–50 to –100	Reduction of nitrate and sulphate
–150 to –300	Occurrence of methane production

washed out along with the leachate. If this leachate is not recycled back into the landfill, the bed will become lean in nutrient concentration, which will affect the rate of decomposition of solid waste, as the micro-organisms will not get sufficient food for their growth.

After the formation of acids, methane gas is formed in the next stage. In this phase, the methane-forming bacteria become more active. This group of micro-organisms converts acetic acid and hydrogen gas, formed by the acidogens in the previous stage, into CH_4 and CO_2. These methanogenic bacteria are strictly anaerobic. It is interesting to note that because of the consumption of the acid and hydrogen gas by methane-forming bacteria, the pH of the landfill bed in this stage approaches the neutral value in the range of 6.8 to 7.8. Under the influence of methanogens, BOD_5 or COD and the conductivity of the leachate get reduced. At neutral or high pH, the inorganic constituents such as metals precipitate out, making the leachate free from heavy or other metals.

In all the stages described so far, most of the biodegradable organic material is ultimately converted into biogas ($CH_4 + CO_2$). But the residual digested mass is not easily biodegradable. At this stage the leachate may contain humic and fulvic acids, which are also not easily biodegradable. Hence, it can be said that the residual biomass is biologically stabilized and can be disposed without the fear of it causing any pollution menace.

To increase the calorific value of the landfill gas, CH_4 can be separated from CO_2 by the absorption of CO_2 in water under pressure; by chemical reaction of CO_2 in some weak alkaline chemical; or by the use of a semi-permeable membrane. In all the cases, pure CO_2 is obtained, which can be made available for some other potential use.

Leachate in Landfill Bioreactor

The term leachate refers to the solvent or liquid, enriched with soluble and suspended materials of the solid waste, which percolates through solid particle bed. In the case of a landfill bioreactor the solid particle bed is nothing but processed solid waste and the solvent or liquid is mainly water, such as rain water, surface drainage, and the liquid produced from the biological decomposition of solid waste. The chemical composition of the leachate varies depending on the age of the landfill. For a new landfill system whose age is less than 2 years, the constituents of the leachate can be listed as follows:

BOD_5 = 2000–30,000 mg/L (average 10,000)

TOC = 1500–20,000 mg/L (average 6000)

COD = 3000–60,000 mg/L (average 18,000)

TSS = 200–2000 mg/L (average 500)

Organic nitrogen = 10–800 mg/L (average 200)

Ammoniacal nitrogen = 10–800 mg/L (average 200)

Total phosphate content = 5–100 mg/L (average 30)

Alkalinity as $CaCO_3$ = 1000–10,000 mg/L (average 3000)

pH = 4.5–7.5 (average 6)

For an old landfill these values are much less.

Bioremediation

Bioremediation is a newer approach, directed towards the treatment of contaminated solids such as groundwater aquifers, soils, and sediments. It has been generally found that contaminants are either adsorbed on a solid surface or dissolved in a liquid phase and in turn trapped in a solid matrix. Bioremediation primarily deals with the strategies that can be employed to clean up these contaminants biologically.

In other words, bioremediation is nothing but an augmented natural process of decontamination. Soils, sediments, and aquifers have friendly micro-organisms which can act on contaminants, but due to their low population density, the rate of decontamination is too slow. If a favourable environment is created by supplying deficient items such as nutrients, trace elements, and oxygen, the population density of micro-organisms will increase, which in turn will speed up the decontamination process. This is the underlying principle of a bioremediation strategy. A good example is the landfill bioreactor. In the case of conventional landfilling, the stabilization of solid wastes takes about 10–15 years, which is too long a period. However, a controlled landfill reactor stabilizes solid wastes in a couple of months. The reason is the same as mentioned earlier.

Bioremediation strategies are of three types, namely, *engineered in situ*, *intrinsic in situ*, and *engineered ex situ*. The main difference between in situ and ex situ bioremediation is that in the in situ process the contaminated solids are retained in their original place during the process. On the other hand, engineered bioremediation refers to the employment of appropriate techniques to supply stimulating materials for enhancing the efficiency of the system. Intrinsic bioremediation, however, relies greatly on the intrinsically generated supply of substrates and nutrients and depends on the intrinsic population density of active micro-organisms. It should be noted that the main objective of any kind of bioremediation process is to increase the rate of biotransformation, so that the clean-up operation time is reduced.

Engineered bioremediation uses engineered measures to supply the materials necessary to significantly increase the rate of biotransformation and thus reduce the time required to clean up the contaminants Therefore, engineered bioremediation is very cost-effective. Though intrinsic (i.e., naturally occurring) bioremediation does not accelerate the rate of clean-up, it prevents the spread of contaminants from their place of origin. This happens due to the naturally developing biological activity. To design an in situ bioremediation strategy, knowledge about the site in terms of hydrology, extent and type of contamination, intrinsic microbial activity, and intrinsic supply rate of key materials is a prerequisite.

Bioventing

Bioventing is used to treat contaminants trapped in the porous soil above the water table. This process is aerobic; hence oxygen is supplied by pulling air through the porous soil. Perforated pipes are inserted deep into the contaminated soil or solid waste to supply oxygen. Suction pressure is created through a vacuum pump, as a result of which air rushes into the soil from the surroundings. The design of the piping network depends, however, on the permeability factor of the soil and the capacity of the suction pump. A vacuum must be created all over to ensure the delivery of oxygen to all contaminated points.

Along with air supply, it is also necessary to sparge moisture and nutrients into the soil or waste layers to prevent the soil from drying due to air circulation. Moisture is very important for proper growth and efficient activity of microorganisms. When the permeable soil surface is sprayed with water or nutrient solution, water percolates into the inner layers due to vacuum. Air sparging can also be accomplished by supplying compressed air from the bottom of the contaminated soil layer through perforated pipes.

Ex Situ Bioremediation

In this process contaminated soil is removed from its place of origin and treated above the ground by various treatment methods. Ex situ bioremediation is most suited for small but highly contaminated sources as it leads to rapid site clean-up. The various methods of ex situ bioremediation are the slurry reactor system, composting, and land farming.

Slurry reactor system

In this system 5% of the contaminated solids are vigorously agitated to effect mixing and aeration. This system has the advantage that it maximizes the rate of biodegradation. However, a major disadvantage is that it is more expensive, as the operation and dewatering of decontaminated solids are costly procedures.

Slurry reactor

Composting

As discussed earlier, composting is an aerobic biodegradation process occurring in the solid state. However, it is not a dry solid condition. Moisture content must be 50–60 per cent and air is circulated through a porous solid bed to supply oxygen. During the peak period of biodegradation, when more air needs to be circulated to meet the increased oxygen demand, the temperature rises to 60–70°C.

Land farming

Land farming is a simple form of ex situ bioremediation. In this process contaminated solids are mixed with a top soil layer, where aerobes are predominant. Mixing results in moisturization and aeration of the contents. This system is found to be most effective for solids contaminated with organics.

Bioremediation of Xenobiotics

The word *xenobiotic* literally means *foreign to the biological world* (*xenos* is Greek for foreign), i.e., substances which are not, or negligibly, present in nature but are introduced into the environment due to anthropogenic (human) activities. Organisms are normally not exposed to these chemicals and as a result their biological mechanisms do not, as a rule, have a pathway for the metabolism of such compounds. Consequently, a large number of such chemicals are not easily degraded by the indigenous microflora and fauna.

Depending upon their convenience and ease of bioremediation, the chemicals used for various purposes such as pesticides, paints, and dyes, or in some other processes can be categorized into two types, namely, biodegradable and non-biodegradable (recalcitrant).

A *biodegradable chemical* can be converted by microbial action into a non-toxic compound within a few months and hence may not be considered too hazardous. However, a *recalcitrant chemical* may persist in the environment for several years in the toxic form. These chemicals are introduced into the environment mainly by untamed human activities although some natural processes such as erosion, oil seepage, and volcanic eruptions are also known to be contributors. *Many of these recalcitrant compounds are highly persistent in nature and lead to the removal of various elements from biogeochemical cycles.* With increase in environmental awareness, degradation of these xenobiotics has become an area of intense research.

Xenobiotic substances include a family of semi-volatile organic pollutants such as polycyclic aromatic hydrocarbons (PAHs). Some examples are anthracene, pyrene, benzo(a)pyrene, benzo[b]fluoranthene, chlorobenzenes, dioxins, 2,2-bis (4-hydroxyphenyl) propane (bisphenol A), and chlorophenols. There are typically two main sources of PAHs. One of them is spilled or released petroleum products and the other is the chemicals used for various materials such as pesticides, paints, dyes, and so on.

Xenobiotic compounds normally have unusual chemical or physical properties that make them resistant to biodegradation. For example, 2,4-dichlorophenoxy-acetic acid can be biodegraded within days, while 2,4,5-trichlorophenoxyacetic acid, differing only in the presence of one additional chlorine substitution, is non-biodegradable and persists for several months. The additional substitution interferes with the hydroxylation and cleavage of the aromatic ring.

Many xenobiotic compounds are highly toxic in nature. Adding to the toxicity is their *bioconcentration* in organisms and *biomagnification* along with the food chains, which finally affects human beings. The concentration of the xenobiotic chemicals in the environment, when diluted, may vary from ppm (parts per million) to ppb (parts per billion) levels, and at still lower levels they may no effect at all. However, the compound may become progressively more concentrated in the bodies of certain animals as it moves or is transferred up the food chain. This process is called *biomagnification*. Prominent examples are phthalate esters and PCBs (polychlorobiphenyls) generally used as pesticides. This phenomenon was first observed in California, where a lake had been treated with the pesticide DDD (related to DDT) to kill some insects. Later on the fish that ate the phytoplankton, containing DDD, as well as the birds that ate the fish started dying.

Many physical, chemical, and biological processes are being tried for the removal of xenobiotics. Biological approaches that involve using the extensive biodegrading capabilities of certain micro-organisms have been extremely successful in many cases, even for in situ remediation. However, for bioremediation to be effective, the pollutant must be amenable to metabolic transformation, the end products must be safe, the environmental conditions should be favourable for microbial activity, and the process should be cost-effective and eco-friendly. To achieve harmony among these varying factors is a difficult task, often as the environmental conditions are hardly conducive for microbial growth. However, after the optimization of the different environmental parameters and with the help of some suitable modifications, biodegrading microbes can be made to grow optimally. When suitable cultures of micro-organisms have to be added to the soil to enable efficient degradation, a process called *bio-augmentation* is employed.

Phytoremediation and *bio-augmentation* are two approaches of bioremediation by which the exudate derived from plants is used to stimulate the survival action of bacteria and subsequently lead to more efficient degradation of the pollutants. The natural pollutants present in the soil can be decomposed by different groups of micro-organisms depending upon the nature of the different pollutants and the degree of degradation. However, if the pollutants are xenobiotics, then the degradation is not very fast, but depends on the nature of bacteria and the bacterial load.

Bio-augmentation is a technique by which the rate of degradation is improved, where xenobiotic pollutants are present, by introducing bacteria that are efficient in degrading polyaromatic compounds, chlorinated aliphatics, and aromatics, particularly nitro-aromatic compounds. The strains used may be wild or genetically engineered. Some of the pollutants can be toxic to the organism and may hamper the growth of the organism. Thus, the concentration of the contaminant pollutant plays a very important role in bio-augmentation. Another important parameter influences the growth and degradation of pollutants is the concentration of micro-organisms. If the inoculum concentration is too low, then it will have some adverse effects on the rate of degradation of pollutants. Sometimes, some protozoa are present in the soil which feed upon the bacterial population and thus reduce the rate of degradation. Often the microbes present in the soil need an additional carbon source, which may also adversely affect the rate of degradation. Many microbes secrete certain metabolites such as biosurfactants. It has been seen that surfactants sometimes enhance the degradation efficiency of xenobiotic compounds. Surfactants are amphipathic molecules with both hydrophobic and hydrophilic parts. They accumulate at the interface and lower the surface tension of the interface and thereby form micelles. One of the potential strains producing biosurfactants is the *Pseudomonas* strain.

Bioremediation of Heavy Metals

Bioremediation of heavy metals is a major concern regarding which a lot of research is being carried out. Heavy metals originate mainly from municipal waste incinerators, car exhausts, as residues from metalliferous mining industry, smelting industry, urban compost, pesticides, fertilizers, sludge, and sewage. The rapid increase in industrisalization in residential areas is one of the major causes for contaminants being discharged into sewage, resulting in large-scale environmental pollution. Urban sewage containing industrial effluents was found to carry relatively high amounts of heavy metals such as Ni, Cr, Pb, Cd, and Co, and a salt load capable of causing salinity and alkalinity hazards, leading to a decrease in soil microbial activity, soil fertility, and yield losses.

Many metal ions are an essential component of nutrition as trace elements but higher concentrations result in metal toxicity. Heavy metals cannot be chemically or biologically degraded and are ultimately indestructible, and hence are difficult to remove from the environment. Today, many heavy metals constitute a global environmental hazard. For example, environmental pollution by cadmium, released mainly from mining and smelting, sewage sludge, and from the use of phosphate fertilizers, is on the rise.

The heavy metals present in soil can be removed either by the conventional chemical methods or by bioremediation. Conventional technology employs stringent physico-chemical agents, which can dramatically inhibit soil fertility and subsequently damage the ecosystem. Numerous methods have been proposed to remove heavy metals from sewage sludge, including chlorination, use of chelating agents, and acid treatments at high temperatures. However, those methods are difficult to apply practically due to high cost, operational difficulties, and low metal-leaching efficiency. In fact, the use of micro-organisms and plants for the decontamination of heavy metals has attracted growing attention because of several problems associated with pollutant removal using conventional methods.

An alternative to chemical methods is the bioremediation method involving microbial leaching. In this process, the hazardous wastes are removed using micro-organisms or plants and it is the safest method of removing pollutants from the soil. Bioremediation strategies have been proposed as an attractive alternative owing to their low cost and high efficiency.

Biomineralization can play a major role in several sectors ranging from bioremediation of contaminated groundwater to bioleaching of metals present in trace amounts. Remediation of metals through biosorption is the process by which metals are absorbed and/or complexed to either living or dead biomass. Many fungi can be quite effective for biomineralization, and an integrated approach to increase soil fertility along with bioremediation can lead to the decontamination of sites facing the problem of heavy metal toxicity. Micro-organisms can be used to clean up metal contamination by removing metals from contaminated water and waste streams, sequestering metals from soils and sediments, or solubilizing metals to facilitate their extraction. In general, many plant species have been effectively used to remove heavy metals from contaminated sites, for which the term phytoremediation is used.

Phytoremediation is an emerging technology based on the use of plants to clean up polluted sites. *Phytoextraction* is the removal of metals (or other pollutants) from contaminated soils whereby the metal is extracted from the soil and then translocated to and concentrated in the harvestable parts of the plant. Phytoremediation technologies are becoming recognized as cost-effective methods for remediating sites contaminated with toxic metals at a fraction of the cost of conventional technologies, such as soil replacement, solidification, and washing strategies.

Phytoremediation can be categorized under five major subgroups as follows.

Phyoextraction   Removal and concentration of metals into harvestable plant parts. Hydrophobic contaminants are more suitable for this kind of extraction.

Phytodegradation   Degradation of contaminants by plants and their associated microbes. Plants have different varieties of enzymes. The soil contaminants are taken up into the cell system by the plants through their roots and shoots. The plant enzymes then metabolize them into harmless products.

Rhizofilteration   Absorption by plant roots of metals from contaminated water.

Phytostabilization   Immobilization and reduction in the mobility and bioavailability of contaminants by plant roots and associated microbes.

Phytovolatilization   Volatilization of contaminants by plants from the soil into the atmosphere. Volatile contaminants or relatively less volatile contaminants get transformed into more volatile contaminants, which are then removed by plants through the process of transpiration.

Microbial responses to metals

Metal speciation and transport is strongly influenced by the microbial activity in the environment. A number of micro-organisms have been reported to produce *metallothionines* (MTs), which them resistance to heavy metals. Most eukaryotes detoxifiy metals by forming complexes. MTs are low molecular weight (6–7 kDa),

Phytoremediation plants in a bioreactor

cysteine-rich proteins found in animals, higher plants, eukaryotic micro-organisms, and some prokaryotes. Functionally homologous bacterial metallothionines have also been reported. Prokaryotic metallothionines conferring resistance to Zn^{2+} and Cd^{2+} have been studied in the cyanobacterium *Synechococcus*. Sulphate-reducing bacteria display a certain degree of metal tolerance by producing sulphides and immobilizing toxic ions as metal sulphides. However, the possibility of enzymatic metal reduction in these bacteria cannot be ruled out. Chemolithotrophs such as iron- and sulphur-oxidizing bacteria, including thiobacilli and thermophilic archaea, can grow in acidic *p*H, in which metal solubility is quite high, so as to maximize their growth at the highest metal ion concentration.

Different groups of aerobic, anaerobic, and facultative groups of bacteria thus show a variety of responses towards different metal ions at different levels of concentration present in the environment. They have the following applications.

(a) Biomining (recovery of metals from the aqueous phase by specific micro-organisms).
(b) Designing metal-tolerant strains that are better adapted to biodegrade organic pollutants.
(c) Metal bioremediation or mitigation through the breeding of natural or engineered strains.

Biosorption

The term biosorption is used to describe the uptake or binding of heavy metals or radionuclides to cellular components. Plant, algal, or microbial biomass can be actively used for metal recovery (sometimes by treating with a strong base) by enhancing their metal-binding ability, i.e., the ability to remove the metal species from aqueous solutions. At present, bacteria and biopolymers are being used for biosorption. Biosorbents may be viewed as natural ion-exchange materials that primarily contain weakly acidic and basic groups, where the chelation process is unspecific. Metals can be stripped from the matrix after loading by using sulphuric or hydrochloric acid, sodium hydroxide or complexing agents, whether dead biomass or live bacteria are used. The system may then be regenerated by employing alkali treatment for reuse in further sorption–desorption cycles.

Biosorption has several advantages over other chemical techniques.

1. Biosorption methods seem to be more effective than their physico-chemical counterparts in removing dissolved metals at low concentrations (below 2–10 mg L^{-1}).
2. The higher specificity of biosorbents never allows them to be overloaded with alkaline earth metals, a very common problem with chemical techniques such as ion-exchange resins.

3. Genetic modifications can result in strain improvement, which would enable increased bioaccumulation, production of new metal chelating peptides, or even more specificity towards certain metals.

Metal precipitation

Apart from biosorption, bacteria can efficiently precipitate heavy metals by reducing them to a lower redox state, as a result of which less bioactive metal species are produced. Microbiological metal precipitation is a common occurrence that is either the result of a dissimilatory reduction or the secondary consequence of some metabolic processes, unrelated to the transformed metals.

Precipitation is effected in bacteria by the formation of metal sulphides and phosphates (indirect reduction). Presently this mechanism is attracting huge interest in biotechnology (optimization of the process and genetically modified organisms are being tried out) owing to its huge potential in bioremediation.

Enzymatic transformation of metals and metalloids

Enzymatic transformation is one of the specific and most effective technologies for bioremediation. Certain metal species are transformed by a number of enzymatic activities, namely, oxidation, reduction, methylation, and alkylation.

Though there are several techniques of bioremediation (bioaccumulation, precipitation, immobilization, etc.), several other biological reactions generate less poisonous metal species. For example, resistance to mercury is considered a paradigm of metal detoxification by enzymatic transformation to a less noxious species. The mechanism of bacterial resistance to Hg^{2+} is its reduction by mercuric reductase (the product of the *merA* gene) to the less toxic and volatile Hg^0 species. Natural isolates often show broad-spectrum mercury determinants, encoding the capacity to degrade organomercurials such as the highly poisonous methylmercury to Hg^{2+}, which is subsequently transformed into Hg^0. The mobilization of mercury could solve a local problem, but there is public concern that this might eventually contribute to global atmospheric pollution.

Cyanide detoxification

Cyanide detoxification poses a major problem, as many micro-organisms are unable to survive in an environment that contains cyanide, which makes the bioremediation of this metal difficult. Hydrogen cyanide (HCN) is produced on a large scale worldwide to fulfil the demand of major sectors such as steel, electroplating, mining, and chemical industries. Potassium and sodium cyanide are massively employed in industrial operations, particularly for the recovery of precious metals (e.g., gold and silver) from their mineral ores.

Cyanide causes toxicity by binding to cytochrome oxidase and thereby inhibits the process of respiration in aerobic organisms. A dose of cyanide in the range 0.5–3.5 mg kg^{-1} body weight in mammals proves to be lethal. Therefore, cyanide released from industries represents a serious threat to the environment. Wastes containing cyanides should be appropriately treated before being discharged into the environment.

Some common methods for treatment of cyanide-polluted wastewater include the expensive alkaline chlorination procedure, ozonation, wet-air oxidation, and sulphur-based technologies. This also poses a great challenge to the environment, as it involves the release of chemical agents (one or many) which may potentially cause secondary pollution.

Despite these problems encountered in cyanide bioremediation, it is not totally foreign to the biological world. Cyanide and compounds related chemically to it are known to be synthesized, excreted, and decomposed in nature by several species of bacteria, algae, fungi, plants, and insects. Cyanogenic micro-organisms represent a large part (up to 50%) of the microbial community in certain ecosystems such as soil and rhizosphere ecosystems. Consequently, micro-organisms residing in these ecosystems have evolved resistance and/or assimilation capacity to survive the toxicity of either endogenous or exogenous cyanide.

Cyanide bioremediation is affected by rhodaneses (thiosulphate:cyanide sulphurtransferases; EC 2.8.1.1), which are highly conserved and widespread enzymes. Currently, these constitute one of the mechanisms evolved for cyanide detoxification. In vitro rhodaneses catalyse the irreversible transfer of a sulphane sulphur atom from a suitable donor (i.e., thiosulphate) to cyanide, leading to the formation of less toxic sulphite and thiocyanate. The role of rhodanese in cyanide detoxification is also supported by the high concentration of this enzyme in mammalian tissues and organs (e.g., the liver) that have been exposed to cyanide.

Bioremediation of polycyclic aromatic hydrocarbons

The structural similarity of polycyclic aromatic hydrocarbons with lignin enables their degradation by a host of lignin-degrading enzymes such as laccase, lignin peroxidase, manganese peroxidase, versatile peroxidase, and cellobiose dehydrogenase. All these enzymes can be potentially used for bioremediation of these persistent aromatic pollutants. A large number of researchers are involved in devising mechanisms and processes for the application of these enzymes for the biodegradation of xenobiotics.

Laccases have broad substrate specificity with respect to electron donors. They catalyse the removal of hydrogen atom from the hydroxyl group of ortho- and para-substituted mono and polyphenolic substrates and form aromatic amines by one electron abstraction to form free radicals. These radicals are capable of undergoing further depolymerization, repolymerization, demethylation, or quinone formation.

Lignin peroxidase, on the other hand, has no substrate specificity. It reacts with a wide variety of lignin and related compounds. Lignin peroxidase can oxidize both phenolic and non-phenolic lignin-related compounds resulting in the cleavage of the $C\alpha$–$C\beta$ bond, the aryl $C\alpha$ bond, the aromatic ring opening, phenolic oxidation, and demethoxylation. It can be assayed by the oxidation of VA to veratraldehyde, the formation of which is monitored at 310 nm.

The mode of action of manganese peroxidase (MnP) is based on an initial one electron oxidation of the substrate by enzyme-generated Mn(III), which produces a phenoxy radical intermediate. This radical is further oxidized by Mn(III) to form a carbon-centred cation. The subsequent loss of proton yields the ketone dimer, whereas an attack by water on the cation, which is followed by alkyl–phenyl cleavage of the aryl glycerol-β-aryl structure, produces other products. Various monomeric and dimeric phenols, including phenolic lignin compounds, can be oxidized by MnP.

Hazardous Wastes

Hazardous wastes pose substantial hazard or potential risk to humans or other living organisms because they are non-degradable, lethal, can be biologically magnified, or may otherwise have some detrimental cumulative effects. Hazardous wastes can be characterized by certain common properties such as ignitability, corrosivity, reactivity, toxicity, carcinogenicity, radioactivity, and mutagenicity. These wastes are released from various sources.

Hazardous wastes from residential sources   These include household cleaners, automotive products, paint products, garden products, etc.

Hazardous wastes from commercial sources   Some examples are inks from print shops, solvents from dry cleaning establishments, cleaning solvents from auto-repair shops, and paints and thinners from painting contractors.

Hazardous wastes from hospitals   These include all materials used and discarded from hospitals such as cotton, bandage materials, syringes, date-expired medicines, plastics, paper, insecticides, radioactive elements, and so on.

Management of hazardous wastes

The best way to manage hazardous wastes is to collect them separately at the point of generation itself. The quantity of hazardous wastes is generally not very large. Therefore, these should not be mixed with large volumes of non-hazardous wastes.

The various methods of solid waste treatment, discussed earlier in this chapter, can be selectively used for the treatment of hazardous solid wastes also. Incineration and sanitary landfilling methods can be used selectively for treating hazardous organic solid wastes. Non-organic hazardous solid wastes, except radioactive elements, should be sterilized before disposal.

Biological Treatment: Bioreactors and other Ex Situ Bioremediation Methods

Biological treatment in bioreactors offers the benefits of degradation under controlled conditions with a continuous monitoring system and is known as the ex situ method of bioremediation. Several types of bioreactors such as batch, continuous, sequential batch biofilm, membrane, fluidized bed, biofilm, and airlift bioreactors are available nowadays. They are employed for treating a wide array of organic wastes. Despite the advantage of a controlled environment for treatment, bioreactors suffer from the limitations of high capital requirement and operation costs, and also excavation of contaminated sites.

Other ex situ bioremediation methods include land farming, which has of late been considered as a disposal alternative; composting; and biopiles engineered systems, a combination of land farming and composting. However, these methods suffer various disadvantages such as large space requirements, extended treatment time, mass transfer problems, and restricted bioavailability of contaminants. Further, bioremediation till date has not been effective for all chemicals and hence a combined effort involving chemical, biological, and physical techniques can be attempted for complete degradation of the persistent chemicals. Despite the prominent strides of researchers in this arena, bioremediation of xenobiotics is essentially an 'end-of-the-pipe' approach meant for *damage control and recovery*, once the damage has been done. Hence, it is always advisable and beneficial to adapt and encourage the 'front-of-the-pipe' approach, which includes adoption of safe processes and technologies that would minimize the release of these substances into the environment.

Summary

Chapter 7 begins with the definition of solid wastes and a brief outline of the various activities involved in solid waste management. Following this, the two methods of solid waste disposal—thermal and biological—have been discussed. Incineration, pyrolysis, and gasification have been covered under thermal methods and aerobic and anaerobic systems have been covered under biological conversion processes.

The various aerobic composting methods of MSW have been presented in detail including the design parameters. The important process of vermicomposting of solid wastes has been detailed, with special emphasis on its advantages. For large-scale treatment of solid wastes, the landfill bioreactor is most suitable and has been described in this chapter. Finally, the various aspects of bioremediation including bioventing, in situ and ex situ bioremediation, bioremediation of xenobiotics, and bioremediation of heavy metals have been presented.

Up to this chapter, we have discussed the scientific and technological aspects of environmental pollution. The final chapter of this book, i.e., Chapter 8 is devoted to the socio-economic aspects of environmental pollution.

Review Questions

1. Define solid wastes and solid waste management.
2. What are the functional elements of a solid waste management programme?
3. State the two basic methods of solid waste disposal.
4. What are the advantages and disadvantages of incineration?
5. Compare the processed solid wastes incineration system with the unprocessed solid waste incineration system.
6. With a neat sketch, describe the operational aspects of the vertical fixed bed gasifier system.
7. What is pyrolysis? State its advantages.
8. What is the purpose of a waste heat boiler? How does it operate?
9. Briefly describe the aerobic composting of solid waste.
10. Before composting aerobically, the empirical formula for a particular solid waste was determined as $(C_6H_{10}O_5)_7$. If 60% of the solid waste is decomposed, calculate the number of moles of oxygen that are required per mole of waste decomposed.
11. Enumerate the parameters that influence the efficiency of an aerobic composting system.
12. Name five methods by which the degree of decomposition of wastes can be measured.
13. Highlight the differences between aerobic and anaerobic composting systems.
14. Classify the anaerobic digestion methods and explain the basic steps involved in the anaerobic digestion/fermentation process.
15. Describe the vermicomposting method for biological treatment and mention its advantages.
16. Is vermicompost an ideal fertilizer? Justify.
17. How would you prepare a worm pit for vermicomposting?
18. State the methods that can be adopted to maintain temperature and moisture content in a vermicomposting bin.
19. What kind of problems may be faced during vermicomposting?
20. Describe the construction and operation of a landfill bioreactor.
21. What is bioremediation? What are the three types of bioremediation strategies?
22. Is there any difference between bioventing and biosparing? How is bioventing applied for the bioremediation of solid wastes?
23. What are the different methods of ex situ bioremediation? How are they performed?
24. What are the characteristics of xenobiotics? Why are xenobiotics so difficult to degrade biologically?
25. Discuss the methods of phytoremediation and bio-augmentation.
26. Clearly define and explain the following: in situ, ex situ, and engineered bioremediation.
27. What are heavy metals? Why is bioremediation of heavy metals a major concern to public health? What is the origin of heavy metals in the environment?
28. What is biomineralization?
29. What are the different methods of bioremediation of heavy metals?
30. Explain in detail the process of biosorption.
31. Why does cyanide detoxification cause pollution?
32. Why is the bioremediation of polycyclic aromatic hydrocarbons (PAHs) so difficult? Which method is adapted to degrade PAHs?
33. What are hazardous wastes and what are their sources?

34. Determine the volume of gaseous fuels at NTP that can be produced by the pyrolysis of 1 t of solid wastes. The relevant data are the following:

(a) Solid waste contains 30% moisture.

(b) The chemical formula of solid waste on a dry weight basis ($C_6H_{10}O_8$).

(c) 3 mols of dry solid waste give 8 mols of H_2O, 2 mols of CO, 2 mols of CO_2, 1 mol of CH_2, and 1 mol of H_2.

How many kilograms of liquid fuel (C_6H_8O) and pure carbon (C) will be obtained from the same quantity of solid waste if 1 mol of liquid fuel and 7 atoms of C are produced from 3 mols of solid waste?

35. If a municipal corporation generates 1000 t of garbage containing 30% moisture and 30% BVS on dry basis and wants to stabilize this aerobically, how many tonnes of air will be required for this purpose theoretically if the empirical formula of the solid waste is $(C_4H_6O_3)_{10}$?

References

Benefield, L.D., C.W. Ran-lall 1980, *Biological Process Design for Wastewater Treatment*, Engelwood Cliffs, NJ: Prentice Hall.

Franzius, V. 1981, 'Dangers of landfill gas', *Müll und Abfall*, vol. 62, no. 9.

Jowitt, Z.L., S.M. Donald, and G.K. Harlan 2003, 'Pilot-scale operation of thermophilic aerobic digestion for volatile fatty acid production and distribution', *J. Environ. Eng. Sci./Rev. Gen. Sci. Env.*, vol. 2, no. 3, pp. 187–97.

Metcalf and Eddy, Inc. 1991, *Waste Water Engineering: Treatment, Disposal and Reuse*, 3rd edn, McGraw-Hill, New York.

Pind, P.F., I. Angelidaki, and B.K. Ahring 2003, 'Dynamics of the anaerobic process: effects of volatile fatty acids', *Biotechnol. Bioengin.*, vol. 82, no. 7, pp. 791–801.

Ramalho, R.S. 1977, *Introduction to Wastewater Treatment Processes*, Academic Press, London.

Schmidt, I., O. Sliekers, M. Schmid, I. Cirpus, M. Strous, E. Bock, J.G. Kuenen, and M.S.M. Jetten 2002, 'Aerobic and anaerobic ammonia oxidizing bacteria: Competitors or natural partners?', *FEMS Microbiol. Ecol.*, vol. 39, pp. 175–81.

Senior, E., I.A. Watson-Craik, and G.B. Kasali 1990, 'Control/promotion of the refuse methanogenic fermentation', *Critical Reviews in Biotechnology*, vol. 10, pp. 93–118.

Economical and Social Aspects of Waste Treatment

Introduction

The economic feasibility of using anaerobic treatment process for the production of biogas from wastes may differ widely. It is dependent on various factors such as availability of a domestic source of energy, cost of imported fuel, uses and actual benefits from biogas production, the public and private costs associated with the development and utilization of biogas, and the technology used to generate biogas. Moreover, if some energy is required to process the biogas further, then that must also be considered when evaluating the amount of net usable energy produced. All these parameters have to be taken into account while analysing the cost-effectiveness of such a project along with an assessment of the skills available for successful completion. An important point to be borne in mind is that the cost of biogas production from wastes is not predicted to increase to the extent that the price of conventional fuels is expected to rise in the future. Of course, conventional fuels additionally suffer from the limitation of availability. An interesting cost–benefit analysis and comparison has been reported (Bhattacharyya 1979) on the production of 230,000 t of nitrogen fertilizer annually using chemical technology and alternate biogas technology. This comparative study is presented in Table 8.1, which is self-explanatory. This analysis has been done for biogas produced using cattle dung as raw material. The main advantage of cattle dung is that it has uniform characteristics, unlike other organic wastes whose characteristics vary depending on their source. According to livestock

Table 8.1 Comparison of chemical technology with biogas technology (based on 1975 cost data)

Item	Chemical technology	Biogas technology
No. of plants	1 (coal based)	26,150 (8 t/plant/year)
Capital cost	Rs 1120 million	Rs 100 million (Rs 3800/plant)
Foreign exchange	70 million dollars (approx.)	Nil
Capital–sale ratio (at 510 dollars/t N_2)	1.2	1.07
Employment	1000	130,750 (5 persons/plant)
Energy (approx.)	0.1 million MWh/year consumed	6.35 million MWh/year generated
Effect on rural development	Centralized	Highly decentralized
Potable water	Required	Not required

census figures, cattle population in India is about 240 million. If we consider the dung collection efficiency to be about 75%, then the quantity of dry dung available in the country that can be used for biogas production is more than 200 million t per year. If processed anaerobically in thousands of biogas plants throughout the country, then this quantity of dung can produce more than 100 million m^3 biogas per day, having an energy equivalent of 560 million kWh or kerosene equivalent of 30,000 t. Each tonne of dry dung produces about 1.2 t of organic fertilizer. Thus, annual production of organic manure will be more than 240 million t, containing about 3.7 million t of N_2, which can further save about 1.75 million t of naphtha per year that can be put to better use such as in petrochemical production.

All this highlights the potentiality of cattle dung as one of the best feedstocks for the production of biogas and biofertilizer in India. Further, the large-scale utilization of dung in biogas plants would simultaneously produce gaseous fuel and organic manure, which, in turn, will reduce the country's dependence on fossil fuels as the main source of energy and on naphtha as a fertilizer.

The main hindrance to the use of cattle dung for large-scale production of biogas is the fact that the collection of feedstock from scattered sources is problematic. However, an integrated biogas plant attached to a dairy farm can solve the problem to a large extent. The author has carried out a detailed case study with the help of his student at the Government dairy farm at Haringhata, West Bengal, to examine the economical feasibility of a biogas plant attached to a dairy farm. The results of the findings are quite interesting and ensure that if properly executed, such a project would be economically viable. See Case Study 1 at the end of the chapter.

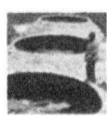

Community Biogas Plant

The quality of life or standard of living is directly related to the per capita consumption of energy, which, on the other hand, depends on the availability of energy resources. With the rapid depletion of fossil fuels, the prices of non-renewable fuels such as coal and petroleum products are gradually increasing and becoming beyond the reach of the common man. Furthermore, due to the lack of proper road facilities in villages in a country like India, these fuels are not easily available to the local populace. As a result, people in the rural sector depend mostly on wood, straw, leaves, dung cakes, etc. as their main source of energy. These fuel materials are mostly burnt in primitive ovens, with minimum fuel efficiency. That means, per unit of useful energy, fuel consumption is very large. Other disadvantages are that trees are cut to be used as fuel and thereby cause deforestation, which leads to ecological imbalance. Burning of cattle dung, an extremely good source of fertilizer-cum-improver of soil structure, as fuel increases the dependency of farmers on costly chemical fertilizers.

In recent years attention has rightly been drawn to these problems and efforts are being directed at seeking alternative sources of energy for the rural population. A permanent solution to this problem is only possible if the source of energy is made available locally. Such a source of energy should ideally have high thermodynamic availability (quality of energy source) and should be renewable, ecologically inoffensive, based on technology suitable for the rural sector, and, finally, operable on a large scale. The alternative energy sources for the rural sector can be categorized as directly solar and indirectly solar.

Every part of India is blessed with abundant sunshine, which can be utilized directly as a source of energy for various purposes such as cooking, drying, and generation of electricity. Biomass, both primary and secondary, is also a product of solar energy, directly or indirectly. This includes all cellulosic materials such as vegetation and cattle, human, and agricultural wastes. Although biomass can be burnt directly as fuel, it is not advisable to do so for two reasons. First, all types of biomass have good fertilizer value, which the village people can directly use for their lands. Therefore, these materials should not be wasted by burning them as fuel. Second, biomass is a low-grade fuel. If burnt directly, only about 10% of its energy content is utilized, and the remaining gets wasted. An appropriate technology for the rural community will be one will extract high-grade fuel value from biomass and also ensure that it does not lose its fertilizer value. The process should also be easy enough to be handled by the local populace. If biomass is fermented in the absence of air in a closed chamber, gas enriched with methane can be produced, which will not only have a high fuel value but will also produce sludge rich in N, P, and K.

Concept

Biomass is abundantly available in villages, although all families may not have the requisite quantity on an individual basis. In such cases family-size biogas plants will not be able to solve the energy problem of the entire village. Some families may have surplus amount of biomass, while some other families may not have a sufficient enough source of biomass. If all kinds of cellulosic waste materials (or biomass) available in a village are collected together and treated in a common fermentor, the gas produced from such a plant may be sufficient to cater to the energy needs of the entire village. There is another reason in favour of community biogas plants. Most of the time there is shortage of land that can be spared for individual household biogas plants and it is difficult to ensure that they are conveniently located near houses. The third point is the lack of private toilets in rural areas, which forces people in the villages to develop the unhealthy habit of going to the field to relieve themselves.

A suitably designed and conveniently located community biogas plant can solve the aforementioned problems and will have the following advantages.

1. *Land economy* A community biogas plant occupies a much smaller area of land than the total land area put together needed for individual gas plants for a village.
2. *Toilet facility* Community toilets can be constructed and attached to a biogas plant. This will improve the hygiene condition of the village on one hand and allow for the utilization of the waste material for production of gas and fertilizer on the other hand.
3. *Maintenance problem* The smooth operation of a gas plant depends on its proper maintenance. It will be less troublesome and more economical to maintain a single large unit than a large number of small units.

The main problem faced by a community plant will be in terms of its product distribution, which means sharing of fuel gas and fertilizer among the participating families of the village.

Size

The size of a biogas plant depends on various factors, which include the following.

(a) Nature and quantity of biomass available in a village, purpose of utilization of biogas
(b) Necessity of a single unit or more than one in a village.

The quality and quantity of biomass available in a village depends on the population, cattle heads, agricultural waste, household garbage, poultry litter, etc.

The biogas thus produced in a community plant can have applications such as cooking, lighting, operation of irrigation pumps and other agriculture machines, generation of electricity, and so on. Depending on the size of the village, its population and layout, the number of biogas units may be selected after considering the problems related to the collection of biomass and distribution of gas to families.

Another important factor that influences the size of a community gas plant is the retention time, which is again dependent on various factors. Since a community biogas plant is normally connected with community toilets and the digested sludge is used as fertilizer, the destruction of the pathogenic bacteria present in the biomass fed into the digester should be one of the prime concerns in deciding the retention time. The volume of gas that can be stored further determines the size of the plant. See Case Study 2 at the end of the chapter.

Economic Analysis of Family-size Biogas Plant

The economic analysis of any venture involves expenditure and return statements. To evaluate the cost–benefit viability of any project, the total investment incurred in the venture and the annual earnings thereof are to be necessarily considered. However, it should be noted that benefits from all investments do not, as a rule, always translate into cash in hand. A good example is the construction of a house, where investment is a tangible factor but annual income from such investment is only notional.

The same is true in the case of a family-size biogas plant. Here, the construction cost of the plant can be estimated with sufficient accuracy, but the products are not sold in the market to show cash earnings against investment. This does not mean that the products do not have any value. The two important products, biogas and biomanure, are used by the owners themselves and, therefore, indirectly save the expenditure the farmer has to bear in order to meet these two demands from an outside source. If an equivalent cost (amount required to be spent if fuel and manure are purchased from market) is attributed to the products, then the annual return on investment can be reasonably arrived at.

Cost–benefit Analysis of 6-m³ Family-size Biogas Plant

Table 8.2 gives the estimated costs of family-size biogas plants for different popular models. The cost figures may vary with time and place. However, for the purpose of cost–benefit analysis, these figures may be considered sufficiently accurate. The size of the plant given in the table indicates the volume of gas produced daily. It may further be noted from the same table that with increase in plant size, the capital cost per unit volume of gas produced decreases.

Table 8.2 Capacity and cost of various models of family-size biogas plant

No. of cattle	Size of plant (m³)	Estimated cost as per April 1991 (Rs)			Capital cost per unit volume of gas production/day (Rs/m³)		
		KVIC	Janta	Deen bandhu	KVIC	Janta	Deen bandhu
2–3	1	–	4200	3100	–	4200	3100
3–4	2	9400	7200	4500	4700	3600	2250
5–6	3	11,500	8000	5500	3840	2670	1830
7–8	4	13,000	9900	6800	3250	2480	1700
9–10	6	15,800	12,100	9100	2630	2020	1520

Let us now perform the cost and income analysis for a 6-m³ biogas plant. As the KVIC (Khadi and Village Industries Commission) model has the highest cost, this model has been considered for economic analysis. This is because if the KVIC model is found economically viable, then the other two models, which cost much less than the KVIC model, will definitely be economically more attractive. The analysis is based on a comparison with previous use of cattle dung as a farmyard manure or fuel. It is also based on the assumption that gas production is about 0.04 m³/kg of wet dung with 40 days retention period. As mentioned, the gas plant is in conventional KVIC design with a floating gas holder. The analysis is based on the estimated cost as in April 1991.

The following two alternative approaches have been adopted in this analysis to arrive at the profit–loss balance.

1. The total capital invested belongs completely to the owner. In such a case, depreciation only, and not interest, will be charged on the capital during cost estimation. Also the per cent return on capital investment will be calculated. If any financial grant is available from the government, then that may be considered as an extra benefit.
2. In the second case the total investment is considered as on loan at the rate of 15% interest. Here no depreciation will be charged as cost, but loan repayment with interest will be shown as cost for the first five years, by which time the loan money will be paid back totally. Thereafter, there will only be maintenance and raw material expenditure. Any kind of subsidy or financial assistance obtained will be considered as an extra incentive.

Basic cost of gas holder	Rs 6300
Pipeline and appliance	Rs 1700
Civil construction	Rs 7800
Total cost of plant	Rs 15,800

Case 1: Capital investment by owner

Table 8.3 gives details of the capital investment by the owner.

Table 8.3 Capital investment by owner

Annual working costs	Previous use of cattle dung	
	Farmyard manure (Rs)	Fuel (Rs)
Gas holder (10-yr life) 10%	630	630
Pipeline and appliances (15-yr life) 6.7%	114	114
Civil work (25-yr life) 4%	312	312
Painting gas holder	600	600
Maintenance	300	300
Cost of dung as pit manure (27 t)	1620	-
Cost of dung as fuel with 20% solid (in terms of kerosene equivalent @ Rs 4/L)	-	2178
Total cost	Rs 3576	Rs 4134
Annual income — Manure (40 t @ Rs 60/t)	Rs 2400	
Annual income — Biogas (2160 m³ @ Rs 4/L kerosene equivalent)	Rs 5357	
Total income	Rs 7757	

Table 8.4 lists the profit details.

Table 8.4 Profitability

Previous use of dung as	Income (Rs)	Costs (Rs)	Net profit (Rs)	% of return on investment
Farmyard manure	7757	3576	4181	26
Fuel	7757	4134	3623	23

As farmers usually use a part of dung as fuel and the rest as manure, the payback period can be calculated by taking the average profit as the basis, i.e., on Rs 3902.

$$\text{Payback period} = \frac{\text{initial investment}}{\text{annual (profit + depreciation)}}$$

$$= \frac{15,800}{3902 + 1056}$$

$$= \frac{15,800}{4958}$$
$$= 3.2 \text{ years}$$

That is, the capital investment will be recovered in about 3 years.

Case 2: Capital investment on loan

Table 8.5 gives the details of the situation in which a loan has been taken for the capital.

Table 8.5 Investment on loan

Annual working cost (for first five years of loan repayment)	Previous use of cattle dung	
	Farmyard manure (Rs)	Fuel (Rs)
Interest on capital @ 15% (five equal payments of interest on declining balance)	1420	1420
Annual instalment of loan repayment (Rs 15,800 ÷ 5)	3160	3160
Painting gas holder	600	600
Maintenance	300	300
Cost of dung as manure	1620	-
Cost of dung as fuel	-	2178
Total costs	Rs 7100	Rs 7658
Annual income	Manure	Rs 2400
	Biogas	Rs 5357
Total income	Rs 7757	

The profit details are provided in Table 8.6.

Table 8.6 Profitability for the first five years

Previous use of dung as	Income (Rs)	Cost (Rs)	Net income (Rs)
Farmyard manure	7757	7100	657
Fuel	7757	7658	99

Profitability from the sixth year onwards is shown in Table 8.7. After the first five years there will be no burden of loan repayment or payment of interest and, therefore, the annual working cost will be reduced by that amount.

Table 8.7 Profit from the sixth year onwards

Previous use of dung as	Income (Rs)	Cost (Rs)	Net income (Rs)
Manure	7757	2520	5237
Fuel	7757	3078	4679

From this cost–income analysis it is evident that the biogas plant is an attractive proposition and can be implemented by rural communities. It may once again be noted here that this cost analysis exercise has been done considering the KVIC design. The capital investment will be reduced considerably if the Janta or Deenbandhu model is considered.

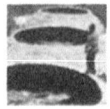

Biogas Scheme—Scope for Rural Employment

Biogas plants can be considered as the starting point of advancing rural development and thereby creating a situation that would discourage the rural population from migrating to overpopulated towns. In this respect the integrated farming system forms an especially attractive possibility, which allows countryside inhabitants a self-dependent humane lifestyle along with the use of technology. Biogas plants can bring many positive changes both at the family and society levels. A few considerations will highlight their importance in the present scenario.

1. Waste and dung pits, which are breeding grounds for flies and insects, can be eliminated.
2. Hygiene conditions can be improved by building toilets, which may be connected to the biogas plants.
3. Immediate ready-to-use hearth for cooking can be provided, doing away with the troublesome smoky kitchens.
4. Source of energy for street lighting can be provided, which encourages activities such as reading and thereby result in an increase in the standard of education.
5. From the fermented decomposed sludge, organic manure of very high grade is obtained, which can be used for enhancing the fertility of the soil.

In Table 8.2, it is seen that the plant cost decreases per unit of gas produced if the plant size is large. It is also known that in villages, most families do not have sufficient cattle heads to operate their own biogas plants. In this context community biogas plants will prove to be beneficial, especially to the weaker sections of the rural population. Also, through the joint production and utilization of biogas, village communities will develop a cooperative attitude and the much needed sense of self-assurance. Educational projects concerning biogas should, therefore, be promoted and supported in schools, social and adult education centres, etc. In this respect the involvement of *Gram* (village) *Panchayats* can bring about significant improvement.

The success of the *biogas scheme*, a useful tool for rural upliftment, is very much dependent on public acceptance. For this reason people need to be educated about the usefulness of a biogas plant and its impact on rural economy and social changes. It is also necessary to construct demonstration plants to show, by giving live examples, the utility and functioning of such plants. The *Gram Panchayats*, which still have a stronghold in villages, can do a successful job of convincing people as to the ways by which biogas scheme can help provide better living conditions. Under the *youth rural reconstruction* programme, the *Gram Panchayat* should construct demonstration plants in selected villages to educate the villagers about its operation. It will not be irrelevant to mention here that the success of schemes such as *afforestation and tree plantation* is very much related to the success of the biogas scheme, which provides fuel for the village community and thus nullifies the need for cutting trees for this purpose.

Another facet of the biogas scheme is its ability to create employment opportunities for rural youth. Such community ventures will also aid the personality development of those involved, as the scheme will open up (a) a high rate of cooperation possibilities for those engaged and (b) motivation to establish trade workshops and ancillary apprenticeship facilities in villages, and thus more new jobs will be available to the educated youth in the rural areas. A brief plan for such a venture is outlined here.

1. A Biogas Cooperative Service Centre (BCSC) comprising a few closely placed villages needs to be formulated. The required number of youths from the participating villages can become the working partners of the centre.

2. A plot needs to be procured in a suitable central position for BCSC activities.

3. The activities of the BCSC will comprise all work related to the biogas plant, such as motivating the potential families to have a gas plant, finding the scope for the construction of a community gas plant, acting as a liaison between financial agencies and prospective parties for biogas plant construction material, helping the owners in the day-to-day plant operations, looking after maintenance, repair, and replacement of parts.

4. All the tasks must be undertaken against suitable remuneration, which will be utilized to run the centre including payment to partners, who need not necessarily be full-time workers.

5. All the partners will undergo the training necessary to run the centre. The Regional Centre for Biogas Development and Training established

by the Department of Non-conventional Energy Sources, Government of India, can provide the necessary training to the members of the BCSC. One such centre has been established by the author at Indian Institute of Technology Kharagpur, West Bengal.

6. The BCSC should have the necessary facilities for undertaking major projects. For complicated tasks, if necessary, such as welding, they can seek help from suitable organizations.

Again, the active involvement of *Gram Panchayats* is very much needed to transform the idea of a BCSC into actual practice. *Gram Panchayats* need to act as the nuclei for BCSCs throughout the country:

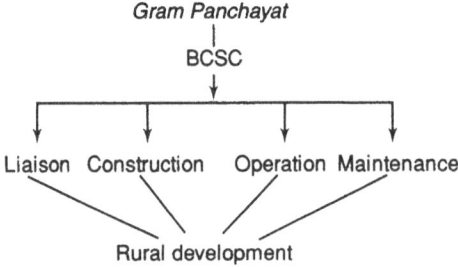

Ethical Issues Involved in Environmental Biotechnology

We must have a common understanding of the terms ethics, morals, and values before we relate them to our discipline of study, i.e., environmental applications of biotechnology. To put it simply, ethics is the science of morals as found in human conduct or behaviour and is related to the study of moral principles. Here, the term 'moral' is concerned with the accepted rules and practices of human behaviour in a society. It encompasses the concepts of right or wrong and goodness or badness of human character. On the other hand, the term 'value' is basically the worth associated with the outcome of the work being considered. Ethics are, therefore, evaluative of the decisions one makes and the actions an individual takes. Morality depends on values to determine the goodness or badness of an action. However, law, religion, and customs should not be related to morality. Laws framed by law-makers define only what is right or wrong, and accordingly a suitable punishment is formulated by the legislature for law-breakers, whereas right and wrong according to religion may be different and are decided by the scriptural authority.

Many individuals feel that nature and environment exist for the benefit of human beings alone. Therefore they indiscriminately use plants, animals, and other environmental resources to meet their selfish purposes. However, there is also a parallel attitude, which demands that nature be considered a sacred creation and that it must be respected and not tampered with. Does this respect mean that human beings should not manipulate nature? In general, some of the ethical issues involved are the impact of biotechnology on human health and safety and on the environment, the extent of undesirable intrusion into the natural order, issues of rights and justice, and the economics of biotechnological applications. It is important that both the benefits and the risks of biotechnology are considered while making ethical decisions about the discipline.

The specific issues of concern about the development and application of bio-technology include damage to the environment, injury to human health, food safety, and socio-economics. Societies are also concerned about the resultant contamination from bioengineered organisms or products, which could pose a threat to human health and damage natural resources such as forests, water bodies, and land. There is a growing concern about the destruction of the aesthetic value of nature and the survival of wild animals and plants being put at stake by destroying their natural habitat. Hence, it becomes our moral obligation to take care of and preserve nature.

Intellectual property right (IPR) is an ethical issue that a society observes to ensure that an inventor is protected from unfair use of his/her invention by any unauthorized person. This ethical principle is supported by law. Copyrights, trademarks, patents, etc. come under IPR. Music, paintings, software, literary works, customary protection of published data in the area of bio-technology, etc. are covered under copyright. Trademarks are important for all businesses including biotechnology companies. Patents are the most important among various rights. Copyrights, trademarks, etc. have somewhat limited relevance in biotechnology, but patents have a very significant role in bio-technology. The three basic types of patents are utility patent, design patent, and plant patent.

Environmental Laws and Policies in India

Environmental legislation provides a legal tool for regulating the undesirable activities affecting the environment. Currently, three approaches are being used. The first method is limited in scope, as it deals with only one aspect of environmental protection through a particular legislation. For example, water-pollution control, air-pollution control, noise-pollution control come under this category. The laws for prevention and control of water pollution and air pollution were enacted in 1974 and 1981, respectively, by the Indian Parliament. This approach will ultimately help to form a comprehensive environmental pollution control policy.

The second approach for environmental pollution control is more comprehensive in nature. This policy covers with all types of pollution, namely, water, air, land, noise, under a single legislation.

The third approach is the best. It envisages the integration of environmental protection with national development planning. In this approach, environmental protection is integrated with the national plans for economic development.

It is pertinent to note that irrespective of the approach being used, the various environmental legislations must be supported by an appropriate scientific and technical knowledge base, so that the planned objectives are achieved without overly taxing the society. Pollution control laws cannot be implemented legally. They need to be viewed as techno-legal aspects.

CASE STUDY 1

Design and Economic Evaluation of an Integrated Biogas Plant for a Dairy Farm

The heart of a biogas production system is the biological digester. The success of the design depends entirely on the proper functioning of this unit. Hence, extensive research has been done on the performance characteristics of the digester. This research has helped to establish certain facts that need to be taken into consideration for the successful execution of the design of a digester. (These points are a recapitulation of what has already been discussed in detail in Chapter 3.)

1. Primarily, there are two categories of bacteria responsible for the biological degradation of biomass resulting in the production of biogas. These are the acid-forming facultative type and methane-forming obligatory type of bacteria. Both the categories function under anaerobic conditions. The obligatory type is, however, strictly anaerobic and therefore the biogas digester has to be necessarily operated under anaerobic conditions.

2. Further, a two-stage digestion system converts biomass into biogas more efficiently than a one-stage unit. During the first-stage process, predominantly volatile acids are produced, while in the second-stage, a methane-rich biogas is produced. A retention period of 4–5 days in the first stage and 8–10 days in the second stage is considered to be ideal.

3. Both the types of bacteria, especially the obligatory types, are very sensitive to any change in pH. Therefore, to maximize the output, the pH must be maintained between 6.8 and 7.2.

4. Temperature is another important parameter that affects the operation of the digester. Efficient performance has been observed both in the mesophilic range of 30°C–40°C (optimum is 35°C) and in the thermophilic range of 50°C–60°C (optimum is 55°C). In the thermophilic temperature range, although the bacteria are much more active, the energy required to constantly maintain that condition would also be quite high. If conservation of energy is a primary concern during the design of the treatment plant, then the mesophilic temperature range would be more ideal and economical, as is the case in India. In order to maintain this temperature range during the winter months, a solar heating device can be incorporated.

5. Intermittent or continuous stirring of the sludge in large-sized digesters improves the digestion rate. This arrangement can be introduced by various means. For example, circulation of sludge-slurry in the digester through the solar heater during the daytime alone may be adopted for purpose of mixing.

6. Healthy growth of the acid-forming and methane-forming bacteria is necessary for the digestion system to function properly. Apart from factors such as optimum range of pH and temperature, certain other conditions such as carbon to nitrogen (C/N) ratio in the feed also need to be maintained. Cattle dung slurry provides the necessary nutrients and optimum C/N ratio for the proliferation of bacterial population. This is another reason why cattle waste is considered to be the ideal feed material for biogas production.

7. Biogas consists of primarily CO_2 and CH_4 in the ratio of 2:3. Due to the presence of CO_2 in large quantities in the gas, the calorific value of the raw biogas is only about 6000 kcal/m^3. If CO_2 is removed from the gas, then the purified gas will have a calorific value of approximately 9500 kcal/m^3. Accordingly, purified biogas shows higher thermal efficiency and lower heat loss when used as fuel gas. The cost of storage will also be reduced. It is, therefore, recommended that a large-sized biogas plant should also have a gas purification system.

8. A biogas plant produces a large quantity of digested sludge everyday, having excellent fertilizer value, which needs to be disposed of. This biofertilizer can be sold profitably if stored properly. For this purpose a sludge filtering unit should also be installed.

From the aforementioned considerations, it can be concluded that a reliable and efficient biogas plant system should consist of the following units.

1. Slurry proportioning tank
2. (a) Primary digester, (b) secondary digester
3. Solar heater
4. Raw gas scrubber
5. Purified gas storage system
6. Sludge filter

The system flow diagram of a biogas plant is presented here.

Plant Capacity

The capacity of a biogas plant depends on the quantity of cattle waste available from a diary farm. For the design under discussion, the West Bengal Government Dairy Farm at Haringhata has been chosen. Cattle dung available from this farm is about 185 t/day.

Raw Gas Yield

If the digestion temperature is set at 35°C with the help of solar heating, then gas production is expected to be around 100 m^3/t of dung per day. This will give a daily gas yield of 18,500 m^3 (std), containing 65% methane and 35% carbon dioxide. Accordingly, the yield can be calculated as follows:

$$CH_4 \text{ produced/day} = 12{,}000 \text{ m}^3$$
$$= 8.6 \text{ t}$$

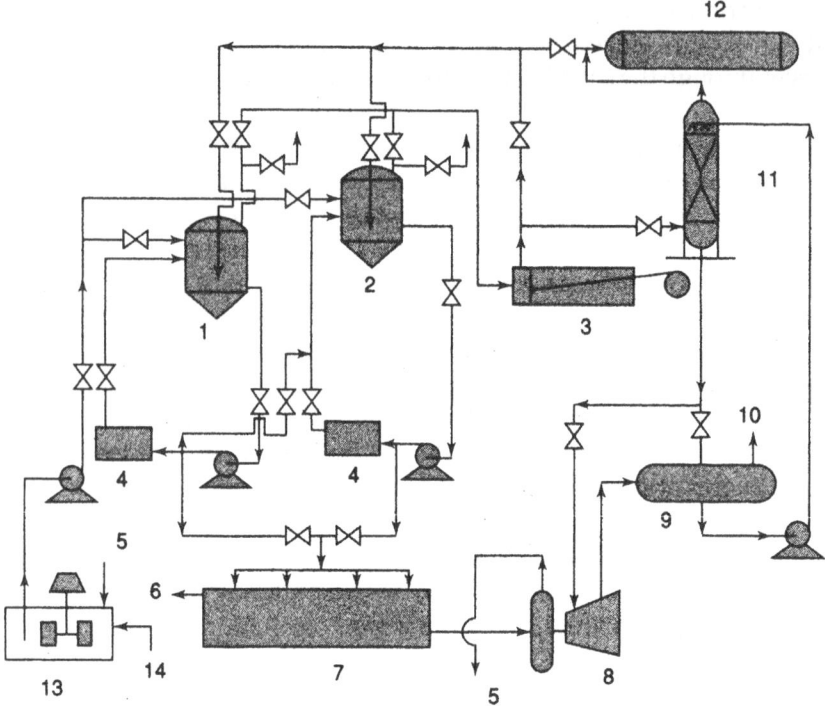

System flow diagram of an integrated biogas plant:
(1) Primary digester, (2) secondary digester, (3) compressor, (4) solar heater, (5) water, (6) manure, (7) sludge filter, (8) turbine pump, (9) flash tank, (10) CO_2, (11) absorber, (12) CH_4 storage, (13) proportioning tank, (14) dung from cattle shed

$$CO_2 \text{ produced} = 6470 \text{ m}^3$$
$$= 12.7 \text{ t}$$

Biofertilizer Production

Each tonne of wet dung produces 730 kg of biomanure. Therefore, the daily yield of fertilizer amounts to 135 t, which contains 2% N_2, 1% P_2O_5, and 0.8% K_2O.

Digestion Optimization

All the components of a biogas plant are responsible to different degrees for ensuring that the digestion process is optimized. Next we will discuss the role of each of these major components.

Feed proportioning tank

The function of this unit is to mix raw dung with water in the ratio of 1:1, by weight. The total weight of the slurry for our design plant is 370 t/day or 336 m^3/day.

Slurry can be prepared in six batches of 4 hours each. For this purpose an open top rectangular tank with a capacity of 60 m^3 and provided with a paddle agitator will be suitable. The optimum dimensions of this tank are calculated on the basis of the minimum walling area. The final dimensions selected are length 5.0 m, width 5.0 m, and height 2.5 m.

Primary digester

Based on the design plan, the appropriate slurry retention time in the primary digester is found to be 5 days. If the feed rate is taken as 336 m^3/day, then the volume of slurry in the digester will be 1680 m^3. After providing 20% extra space above the slurry level for gas disengagement and accumulation, the total volume of the primary digester can be fixed at 2000 m^3.

The shapes chosen for different parts of the digester are domed top, cylindrical shell, and conical bottom. The slope of the conical base is 1 in 5 and the raise of the dome is one-fifth the shell diameter. A prestressed concrete construction would be suitable for this type of tank. Taking into consideration the cost of construction and heat loss through the various surfaces, the dimensions of the ideal digester as per the design requirements can be calculated on the basis of the minimum surface area. Thus, the dimensions selected are diameter 15.0 m, cylindrical height 10.0 m, dome height 3.0 m, cone depth 1.5 m, and surface area 855 m^2.

Secondary digester

Here, the slurry retention time is 10 days. Therefore, the capacity of the secondary digester needs to be 4000 m^3, i.e., double the capacity of the primary digester. The shape will be same as that of the primary digester. On the basis of the same argument that applies to the primary digester, dimensions that would be most favourable are computed. The final dimensions selected are diameter 18.0 m, cylindrical height 12.0 m, dome height 3.6 m, cone depth 1.8 m, and surface area 1235 m^2.

Solar heater

Solar heat energy is used for the following purposes in a biogas plant.

1. Preheating the feed slurry to the digester temperature.
2. Compensating loss of heat from the surface, both during daytime and in the night.

The solar heater consists of a black plate solar collector covered with a glass pane. On a day when the sky is clear, such a collector can extract solar energy from 1000 to 3000 kcal/m^2(day). In India, during winter days an average value of 2000 kcal/m^2 can be safely assumed. For the calculation of heat load on the solar collector, an ambient temperature of 20°C and an average heat loss factor of 0.7 kcal/hr(m^2)°C for the surface of the digester can be reasonably assumed. With this information as the basis, the heat loss from the digester is estimated to be 0.532 × 10^6 kcal/day. However, for preheating the feed slurry from 20°C to 35°C during winter, the amount of heat needed would be 5.55 × 10^6 kcal/day. However, it will not be economical to design a solar heater keeping in view such a huge requirement of heat because this is the case only for a certain period of the year. In other months of the year the ambient temperature will not be so low. Hence, it would be appropriate to install a solar heater with a capacity of about 2.6 × 10^6 kcal/day or 300 kW, which can be used for a major part of the year. The excess heat requirement of the short period of winter can be met with the use of additional heating devices.

Raw gas scrubber

It has already been indicated that purified biogas has many advantages over the raw biogas, as the latter contains about 30–40% of CO_2. For the removal of CO_2 several processes such as *chemisorption* or *physical absorption* can be employed. Raw gas scrubbing is primarily a two-step process. The first step involves the absorption of CO_2 in a solvent and the second stage involves its desorption to regenerate the same. Selection of the solvent and the process conditions is the main criteria to be borne in mind while designing a biogas purification system. Some commonly used solvents are water, aqueous ammonia, alkaline salt solutions, and ethanolamines such as monoethanolamine (MEA).

Each solvent comes with its own advantages and limitations. For example, water is considered the most inexpensive solvent for CO_2 absorption and requires a simple system design, but a limitation is that the system needs a very high pumping load because of poor CO_2 removal efficiency. To arrive at a fair selection of solvent, a relative comparison is made among the different solvent systems based on various design and cost parameters. The results are shown in Table 8.8.

Table 8.8 will help us to conclude that mono ethanolamine solution would be the best solvent for bulk removal of CO_2, as it has the best overall cost performance advantage over the other three solvents. However, it is to be noted that the extra energy requirement for the recovery of solvents is also quite substantial. Therefore, if the source of energy is already the main problem, then for such systems solvents requiring extra energy should not be chosen. If we were to consider this argument, then it can be said that water is one of the best solvents. To improve the absorption efficiency of water as a solvent, the system may be operated at a higher pressure, say, about 15 bar.

One more interesting point may be discussed in connection with the use of water as a solvent. After regeneration the discharged water is found to be saturated with CO_2. This water will be extremely

Table 8.8 Comparative cost factor of different solvents for CO_2 removal

Item	MEA	K_2CO_3	NH_3	Water
Chemical cost	10	7	6	1
Pumping cost	1	5	5	10
Steam requirement	4	3	5	0
Heat exchange requirement	5	4	1	0
Cooling water requirement	5	4	0	0
Efficiency of absorption	0	10	15	25
Base of regeneration	0	5	5	10
Corrosiveness	7	5	6	10
Stability	5	5	5	0
Chemical loss	5	2	3	0
Poison	8	5	8	1

useful for the production of algae by photochemical reaction, which can then be used as fodder for the farm cattle.

For the removal of CO_2 from raw gas containing 35% CO_2 and 65% CH_4 under 15 atm pressure using water as the solvent, a packed tower of 0.9 m diameter and 10 m height is found to be suitable for the capacity given for the said dairy farm. The purified gas will contain 99% CH_4.

Sludge filter

From economic considerations, a sand filter has a definite advantage over other filtering devices. A sand bed of size 2 m by 6 m will be suitable for the said capacity. Along with the sand filter, a glass-covered space will be required for solar drying the manure so that it can be stored properly.

Energy Potentiality of Biogas Plant

The following analysis shows the energy potentiality of a biogas plant, which can produce up to 18,500 m^3 of raw gas or 12,000 m^3 of pure CH_4/day. The heat of combustion of CH_4 is 9500 kcal/m^3.

Heat generation capacity

If the entire quantity of gas is burnt, then the amount of heat that will be available on a daily basis is given as

$$12,000 \times 9500 = 114 \times 10^6 \text{ kcal/day}$$

Generation of electricity

If the thermal efficiency of a generator engine is taken as 25% and the generator efficiency as 80%, then the overall conversion efficiency of heat to electricity becomes 20%.

The total heat energy available is 114×10^6 kcal/day. The heat equivalent of electricity is one-fifth of this value, i.e., 22.8×10^6 kcal/day or 0.95×10^6 kcal/hr.

$$1 \text{ kW} = 858 \text{ kcal/hr}$$

Therefore, with the given amount of gas generated daily, a 1.11-MW generator can be operated.

Waste heat utilization

From the calculations done so far it can be seen that only 20% of the available heat is converted into electricity. Therefore, the quantity of heat getting wasted is $(114 - 22.8) \times 10^6$ or 91.2×10^6 kcal/day. 50% of this waste heat can be utilized for the production of low-pressure process heating steam. That is, 45.6×10^6 kcal/day will be available for steam generation or heating purposes. Steam can be produced at the rate of 10,000 kg/hr with the help of a waste heat boiler. However, this much steam will not be required by the dairy farm. Hence, the excess quantity can also be utilized for preheating the feed slurry to the digester.

Power requirement for biogas plant

The total power requirement for this plant is about 300 h.p., which is needed to operate compressors, pumps, etc. To this value, if the power required for lighting, etc., are also added, then the total power requirement for the biogas plant will add up to much less than 300 kW.

Power supply to dairy farm

From the analysis of all the data and the various design parameters it can be established clearly that a dairy farm can become self-sufficient in terms of its energy requirement if an integrated biogas plant is installed and attached to it.

Plant Economics

To test the feasibility of a project, an analysis of its economic considerations is essential. This will reveal whether or not the project is economically viable. Furthermore, economic evaluation would also provide an estimate of the total capital investment required, the annual fixed and operating costs, returns from the sale of products, profitability of the project, and other such factors of immense importance while undertaking any industrial investment.

Fixed capital investment

FCI or fixed capital investment refers to the following aspects.

(i) Installed cost of process equipment
(ii) Expenditure including cost of land, site development, and buildings
(iii) Investment towards laboratory equipment and facilities
(iv) Consultancy, technical know-how, engineering, and supervision
(v) Instrumentation and electrical costs
(vi) Contingency funds

Table 8.9 provides the list of process equipment and other installed costs.

Working capital

This includes the cost of raw materials, at least one month's salary of workers, credit sale, and inventory. The main raw material is cattle dung, which is available practically free of cost. The working capital requirements would, therefore, be less. Let us take the working capital to be Rs 5,000,000.

Total capital investment

TCI = FCI + working capital
= Rs 136,835,000, say, Rs 137,000,000

Manufacturing costs (on yearly basis)

Table 8.10 provides data regarding the costs involved in manufacturing on an annual basis.

Earnings (on yearly basis)

There are two main products that are potential sources of revenue for the investor. These are methane gas and biofertilizer. A large quantity of pure CO_2 $(2.45 \times 10^6 \text{ m}^3)$ will also be available for various industrial applications. In the present profitability analysis, however, CO_2 has not been considered as a revenue earning product. The price of methane gas (9000 kcal/m^3) is fixed at Rs $15/\text{m}^3$ on the basis of the price of fuel oil (8200 kcal/L). Biofertilizer is very cheap and the price can be set at Rs $100/\text{t}$. Table 8.11 shows the earnings on an annual basis.

Table 8.9 Process equipment and other costs

Item	Cost (Rs)
Component cost	
Two digesters (6000 m^3; prestressed and RCC construction; @ Rs 3500/m^3)	21,000,000
CO_2 absorption column (0.9 m diameter 3 12 m height; total weight without packing 5 t; fabricated @ Rs 1,000,000/t; cost of packing Rs 1,000,000; installation charge 15%)	6,900,000
Pressure laid down tank (1 m diameter 3 3 m length; total weight 0.9 t; fabricated @ Rs 1,000,000/t; installation charge 15%)	1,035,000
Two gas storage tanks (2 m diameter 3 10 m length; weight 11 t each; fabricated @ Rs 1,000,000/t, saddle supports Rs 200,000/tank; installation charge 15%)	25,760,000
Dung mixing tank (RCC construction with agitator)	500,000
Three solar heaters (heat duty 300 kW; fabricated @ Rs 15,000/kW)	4,500,000
Pumps with motors (a) Two feed pumps (b) Four recirculation pumps (c) Two absorber water pumps (d) Three process water pumps (e) One filtrate pump with water turbine	500,000 4,000,000 2,000,000 480,000 160,000
Two compressors (50 h.p., 400 m^3/hr capacity methane compressor; installation charge Rs 100,000 total)	9,000,000
Pipeline (total weight 40 t, @ Rs 500,000/t; installation charge 25%)	25,000,000
	Total = Rs 100,835,000
Cost of land, site development, buildings, etc.	10,000,000
Laboratory equipment and facilities	1,000,000
Consultancy, technical know-how, engineering, and supervision (5% of equipment cost)	5,000,000
Instrumentation and electrical costs (5% of equipment cost)	5,000,000
Contingency funds (10% of equipment cost)	10,000,000
	Total FCI = Rs 131,835,000

Table 8.10   Manufacturing costs

Item	(₹) lac	Rs
Raw material		Nil
Energy requirement (1,620,000 kWh @ Rs 5/unit)		8,100,000
Salary and wages		4,000,000
Depreciation charges on:		
Building and digester (5%)		1,550,000
Other equipment (10%)		858,000
Maintenance and repair (4% of equipment and buildings)		4,600,000
Insurance (2%)		2,000,000
Interest on borrowed capital (@ 12% on Rs 100,000,000)		12,000,000
Total		Rs 33,108,000

Table 8.11   Earnings (on yearly basis)

Item	Rs
Annual earnings from 4.32 $\times$ 10^6 m^3 methane gas @ Rs 15/m^3	64,800,000
Annual income from biofertilizer (48,600 t) @ Rs 100/t	4,860,000
Total	Rs 69,660,000

Profitability

1. Annual gross profit = Revenue − costs
$$= 69,660,000 - 33,108,000$$
$$= \text{Rs } 36,552,000$$

2. Unit profit before tax = Rs 540/t of wet dung

3. Gross profit ratio
$$= \frac{\text{gross profit}}{\text{total sale}} \times 100$$

$$= \frac{36,552,000}{69,660,000} \times 100 = 52\%$$

4. Return on investment (before tax)

$$= \frac{\text{gross profit}}{\text{total investment}} \times 100$$

$$= \frac{36{,}552{,}000}{137{,}000{,}000} \times 100$$

$$= 27\%$$

5. Payback period

$$= \frac{\text{initial investment}}{\text{annual (profit + depreciation)}}$$

$$= \frac{131{,}835{,}000}{33{,}108{,}000 + 2{,}408{,}000}$$

$$= 3.7 \text{ years}$$

Benefits

By investing a little over Rs 140 million for the installation of a biogas plant adjacent to a farm, multiple benefits can be enjoyed as shown in this case study.

The main benefit is that the plant will provide an avenue for better utilization of 66,000 t/year of dairy farm waste, mainly cattle dung. This will help in solving the pollution problem and also the problems associated with the daily disposal of huge quantities of solid and liquid biodegradable waste, and that too without incurring any extra expenditure. It will also produce 48,000 t/year of high-grade biofertilizer containing 960 t of N_2, 480 t of P_2O_5, and 380 t of K_2O.

The analysis pertaining to our design plant, though based on old cost data, clearly indicates that a biogas plant attached to a dairy farm is economically feasible. Capital investment can be realized in about 2 years time. The pollution problem can be solved free of cost. On the other hand, it also produces high-grade biofertilizer as well as pollution-free fuel gas at a very low price. The supply of methane gas and carbon dioxide could prove to be a boon to newly set up small industries in the surrounding areas, thereby creating more employment opportunities.

CASE STUDY 2

Design Analysis of a Community Biogas Plant

A community biogas plant has been constructed in a village near Kharagpur and is funded by a DST project of the Rural Development Centre of IIT Kharagpur, in which the author was involved during the design and construction of the plant.

The quantity of biomass that may be available daily from various sources was estimated as shown here.

Raw cattle dung	750 kg
Human excreta	50 kg

Agricultural waste and   200 kg
    household garbage

Hence, this amount of material was to be treated, mixed with an equivalent volume of water in the community biogas plant, to which eight toilets were attached. The major dimensions of the plant were decided upon as follows.

Rectangular entry tank   3.2 m length × 3.2 m breadth × 1.7 m depth
Gas generator diameter   5.0 m
Cylindrical height   4.0 m and conical top
Rectangular exit tank   3.2 m × 3.2 m × 1.7 m
Sludge collector   3.2 m length × 5 m depth × 1.5 m depth

Gas production capacity

Under normal conditions this plant will produce 60 m^3 of biogas daily. This amount of gas can be utilized for the following purposes.

1. If all the gas is used for cooking, then it will be possible to cook food for 200–250 heads per day.
2. By running a small generator with 60 m^3/day of biogas, it will be possible to produce 60–70 kWh of electricity.
3. A part of the gas can be used for cooking and the rest for street lighting.
4. This gas can be used directly as fuel to run irrigation pumps.

Cost–benefit Analysis

The community biogas plant will provide both direct and indirect benefits to the village. Indirect benefits are those that cannot be estimated quantitatively. The most important indirect benefits are listed here.

1. The system will improve the health conditions of the village community. Many of the diseases prevailing in the rural areas are due to the lack of proper sanitary toilets. The community biogas plant will provide this facility and thus improve the hygiene standard of the village.
2. It will improve cleanliness in the village, as all kinds of degradable waste material can be fed to the biogas plant to obtain useful products.

The direct benefits can be evaluated by their money equivalent. Such direct benefits accrue from the production of gas and good quality fertilizer. It can be mentioned here that the normal practice of producing dung manure in the pit yields only about 500 kg of manure per tonne of raw dung. On the other hand, each tonne of raw dung when treated in the gas plant will give more than 700 kg of a better quality manure. If it is assumed that the plant maintains its normal gas production capacity for, say, 300 days in a year, then the quantity of gas available per year is = 60 × 300 = 18,000 m^3. With a minimum price of Rs 9/m^3, the return from the gas/year will be Rs 162,000. If the cost of dung is estimated to be equivalent to the cost of the same amount of pit manure, then the direct profit available from the excess amount of fertilizer produced in the gas plant can be estimated as follows:

Quantity of excess manure

$$= \frac{700 - 500}{1000} \times 350 = 70 \text{ t}$$

The yearly net return from manure will be around Rs 7000. The total return (minimum) from the proposed biogas plant will be equal to Rs (162,000 + 7000) = Rs 169,000. The cost of the plant including a shallow tube-well for water supply will be about Rs 300,000. The payback period for such a plant is, therefore, 1.8 years.

Management Policy

It is to be remembered that a community biogas plant is not an individual's property but belongs to the particular locality or entire village as the case may be. This brings into focus the management problem for such a plant. As such maintenance is not a major problem because autogenously operated biogas plants are relatively easy to care for. The only major problem that is foreseen is how to arrange for the collection and feeding of biomass on a regular basis. Biomass can be fed more than once in a day or even once in two days. The amount of gas produced will accordingly depend on the quantity fed.

Once the major problem with regard to the collection of biomass from various sources and mixing it with water and feeding it into the feed tank is solved, other problems can be tackled more readily, which include (i) distribution of gas for cooking and (ii) keeping an account of the dung collected from different sources.

To tackle all such problems, a *users group* can be formed to ensure successful running of the plant. While formulating management policies, one should consider the biogas plant as a processing machinery. By feeding raw biomass (such as cattle dung), we are obtaining a processed product (i.e., biomanure) and for this some token amount needs to be paid in the form of money or labour. The users group's main responsibilities would be to

1. check the cleanliness of the toilets,
2. keep an account of the dung received from the various families (so that an equivalent amount of manure can be allocated to them when needed),
3. draw labour from different families as per the plan chalked out to collect biomass such as dung, garbage, agricultural waste, and so on,
4. grow water hyacinth or algal plants in the entry and exit tanks of the biogas plant to increase the quantity of biomass for plant feed, and
5. distribute gas for cooking purposes.

Biogas Distribution

Biogas is the main by-product of the treatment plant and hence every family of the village or locality must get the benefit of using it. A biogas distribution system needs to be developed keeping this in mind. Gas distribution through pipelines is a relatively expensive affair. Alternatively, two methods can be explored.

1. Distribution of gas through portable gas holders
2. Construction of community kitchen rooms near the plant

This case study clearly provides evidence that projects such as community biogas plants are not only viable but also profitable. These plants indirectly aid in improving the hygiene condition of the village and provide higher efficiency fuel for cooking and manure for cultivation. The quantity of manure obtained can also be increased by this process. However, the area of management of a community biogas plant still remains a concern. Proper operation of the plant needs the development of a cooperative attitude among the users. To achieve this end, people need to be educated and inspired to set an example. The proposed plant considered under this case study has been constructed to test the possibility of achieving this purpose.

Summary

This chapter deals with the socio-economic aspects of biogas production by anaerobic digestion of biodegradable waste material, which otherwise may cause environmental pollution. It lays stress on the application of the anaerobic waste treatment process for improving the hygiene and economic condition of rural India.

In this chapter we have discussed the various aspects of small- and large-scale biogas plants including design and cost benefit analyses. The most interesting aspect of this chapter is the inclusion of two case studies—one on community biogas plants and the other on the concept of an integrated biogas plant for a dairy farm. The chapter also outlines a plan to generate employment by establishing community biogas plants.

Finally, the chapter discusses the ethical issues involved in the applications of environmental biotechnology and provides an overview of the environmental laws and policies prevalent in India.

Review Questions

1. Discuss the factors that influence the economic viability of a waste treatment process.
2. Compare biogas and chemical technologies, in the context of production of nitrogen fertilizer.
3. Enumerate the concept of an integrated biogas plant and state its role in pollution abatement.
4. Without the use of conventional fuels, how can the operating temperature of a biogas plant be maintained?
5. How do community biogas plants help in the maintenance of hygienic environment in rural areas?
6. Discuss the advantages of using purified biogas containing mainly CH_4.
7. Why are biofertilizers preferred over chemical fertilizers?
8. If a biogas plant produces 100,000 m^3 of CH_4 gas per day and it is decided to utilize the entire gas produced for providing electricity, then what will be the capacity of the power generator? Will it be possible to run one low-pressure waste heat boiler simultaneously? If yes, what will be its maximum steam generation capacity at the maximum permissible pressure level?
9. List the main advantages of a community biogas plant.
10. Discuss the various factors that need to be considered while deciding the size of a community biogas plant.
11. Examine the possible means by which biogas can be distributed to rural families.
12. Compare family-size and community-size biogas plants.
13. How can the biogas scheme contribute towards rural employment?

14. A family consists of six adults and six children. This family is interested in constructing a biogas plant to meet their cooking requirements. (a) Which type of plant size would you suggest? (b) Determine the minimum number of cattle heads the family should possess to run a 3-m³ biogas plant. (c) If the same family does not possess the required number of cattle, can they resort to any alternative means to produce the same volume of gas? (d) Calculate the volume of gas needed by the same family if they want to use the gas for the following purposes:
   (i) cook for the entire family,
   (ii) light three- 100-W power lamps for 4 hrs at night, and
   (iii) runn one 5-h.p. irrigation pump for 2 hours daily.

15. A small dairy farm in a village possesses 100 cattle heads with an average dung collection of 15 kg per cattle. The farm owner is interested in utilizing the entire amount of dung for the generation of electricity. If the dung is to be treated in a biogas plant for this purpose, what should be the plant dimensions? How much electricity (kWh) can be generated daily? How much waste heat will be available from the engine generator for heating the gas plant?

16. Design a solar heater for a 3-m³ biogas plant to maintain the temperature at 30°C during winter months, when the ambient temperature is around 10°C.

17. Design a system to work as a batch or fed-batch system for solid waste digestion, to produce 3 m³ biogas/day containing approximately 65% CH_4 at 30°C. Determine the requirement of feed material and the size of the digester. The following data may be assumed: digestion rate constant $k = 0.02$ d^{-1}, biodegradability of feed = 53% of TS, feed biomass to biogas conversion = 0.69 kg BVS/m³, initial TS concentration in the digester = 70 kg/m³ or less, and final TS concentration in the digester = 150 kg/m³ or less (after 6 months of operation).

18. If a two-stage biomethanation system is to be designed based on solid waste digestion to produce 3 m³ biogas (70%) at 35°C per day, calculate the quantity of feed material required and determine the sizes of the first and second-stage digesters? The following data may be used: moisture content of feed solid = 5%, biodegradability of TS (dry) = 50%, β = COD/VS = 1.1 kg/kg, $K_s = 0.15$ kg/m³, $Y_{mo} = 0.33$ m³ (CH_4 at STP)/kg (acetic acid), μ_m = 0.3/d, k (rate constant in the first stage) = 0.05/d, initial TS (dry) concentration in the first-stage solid-state digester = 100 kg/m³, acetic acid concentration in the influent leachate of second-stage digester = 2 kg per m³, acetic acid conversion efficiency in the second-stage digester = 95%.

19. If a large-scale biogas plant is designed for the generation of electricity, what major units the plant will have?

20. What are the social and economic advantages of using biogas plants for waste treatment?

21. What is the use of a solar heater in a waste treatment plant and how can it be designed?

22. Is it necessary to scrub biogas before using it in an internal combustion engine?

23. How much electricity can be produced from 20,000 m³ of biogas per day?

24. If a biogas plant with a production of 3 m³/day costs Rs 12,000, how will the pay back period be decided?

25. What is the concept of a community biogas plant?
26. How can a community biogas plant improve the quality of life in rural areas?
27. How is the appropriate size of a community biogas plant determined?
28. Why can biogas not be distributed like Indane gas?

References

Bhattacharyya, B.C. 1979, 'A low cost gobar gas plant', *NSS Bulletin*, vol. VI, IIT Kharagpur.

Bhattacharyya, B.C. 1979, 'Design and economy of gobar (bio) gas plant for rural use', Monograph, *Chemical Age of India*.

Bhattacharyya, B.C. 1978, 'Biogas as an alternative source of energy,' *Chemical Engineering world*, vol. XIII, no 12, pp. 49–52.

Bhattacharyya, B.C. 1978, 'Economic design of gobar (or bio) gas plant for rural community', *Chemical Age of India*.

Chawla, O.P. 'Significance of biogas technology in Indian economy', *Proceedings of Bio-energy Society*, Third Convention and symposium, 1986, 219.

Mitra, T. 1978, 'Design of biogas plant attached to a dairy farm and its economic evaluation', BTech Home Project, Department of Chemical Engineering, IIT Kharagpur.

Glossary

Adenosine diphosphate (ADP) Ribonucleoside diphosphate serving as a phosphate-group acceptor in the energy cycle of a cell.

Adenosine triphosphate (ATP) Ribonucleoside triphosphate serving as a phosphate-group donor in the energy cycle of a cell. It is the major carrier of chemical energy in living cells.

Aerobes Group of organisms that require air or oxygen for their survival and growth.

Aerobic respiration Respiration that occurs in an oxygen-rich environment.

Agar Complex mixture of polysaccharides obtained from marine red algae.

Amino acid Basic unit or building block of proteins comprising a free amino (NH_2) end, a free carboxyl (COOH) end, and a side group (*R*).

Anaerobes Group of organisms that cannot tolerate the presence of air or oxygen, or survive in the absence of air or oxygen.

Anaerobic Without air or oxygen.

Anthropogenic Originating in human activity.

Archaebacteria Most primitive type of micro-organism among prokaryotes.

Asexual reproduction Non-sexual (vegetative) reproduction.

Biochemical pathway Various steps involved in any bioconversion process.

Biofuel Fuel, e.g., hydrogen, ethanol, and methane, produced from a biological source by the action of micro-organisms.

Biogas Gas produced when organic wastes are digested by micro-organisms under anaerobic conditions (in the absence of air or oxygen). The major constituents of biogas are CH_4 and CO_2.

Bioleaching Process of using micro-organisms to recover metals from their ores.

BOD Biological oxygen demand. BOD quantifies the amount of oxygen required by micro-organisms to oxidize organic wastes to CO_2 and H_2O.

Biomarker Biochemical that quantitatively measures the effects of exposure to xenobiotic substances on a biological system.

Biomass Total dry matter of all organisms in a particular sample, population, or area.

Biomining Process of recovering metals from ores using micro-organisms. Biomining is also used for extraction of metallic pollutants from solid or liquid wastes.

Bioreactor Apparatus used to carry out biological reactions or processes, especially on an industrial scale.

Bioremediation Process of using organisms to consume or otherwise help remove pollutants from the environment.

Biosensor Analytical device used to determine the concentration of substances and/or other predefined parameters by converting a biological response into an electrical signal.

Biotechnology Exploitation of biological processes for industrial and other purposes.

Carbohydrates Class of carbon-hydrogen-oxygen compounds usually represented chemically by the formula $(CH_2O)_n$, where n 3.

Carcinogen Cancer-causing agent.

Catabolism Breakdown of complex biological molecules into simpler ones, usually accompanied with the release of energy in the form of ATP.

Clone Exact genetic replica of a specific gene, cell, or an entire organism.

Cloning Technique involving mitotic division of a progenitor cell to give rise to a population of identical daughter cells.

Co-enzyme (co-factor) Organic molecule that binds to an enzyme and is required for its catalytic activity.

COD Chemical oxygen demand. COD quantifies the amount of oxygen required to convert all the oxidizable chemicals in wastes to stable end products such as CO_2, H_2O, and NO_3.

Corrosion Destruction of metal by chemical, electrochemical, or biological action.

CRT Cell retention time. CRT indicates the duration for which active cell mass remains inside a bioreactor before it is decomposed.

Electrophoresis Technique for separating molecules based on the differential mobility in an electric field.

Enzyme Organic, protein-based substance produced by living cells to catalyse biochemical reactions.

Eukaryote Organism whose cells have a nucleus and other membrane-bound organelles.

Fermentation Enzyme-catalysed reaction in which molecules are broken down under anaerobic conditions.

FOTE Field oxygen transfer efficiency. FOTE is the practical value of the amount of oxygen transferred to wastewater per unit power consumption. This is lower than SOTE (see SOTE).

Gas chromatography Analytical technique employed to separate compounds from a mixture, based primarily on their volatilities. It can also quantify methane in biogas.

Gene Sequence of nucleotides in the genome of an organism to which a specific function can be attributed.

Gene cloning Process of synthesizing multiple copies of a particular DNA sequence using another organism as a host.

Genetic engineering Genetic manipulation of an organism directly involving DNA at the molecular level.

Glycolysis Metabolic process in which sugars are broken down into smaller compounds along with the release of energy.

Green technology Pollution-free technology, or technology in which pollution is controlled at source, used for manufacturing useful products.

HPLC High-pressure liquid chromatography. Based on their chemical or molecular properties, chemical compounds can be separated through specific stationary and mobile phases of HPLC. HPLC is mainly used as an analytical technique in biotechnological, biomedical, and biochemical research and in pharmaceutical industries.

HRT Hydraulic retention time. HRT is the ratio of the effective volume of a reactor to the volumetric flow rate of feed into the reactor under steady-state conditions.

In situ In the natural or original position.

In vitro In a test tube (in glass) or an artificial environment.

In vivo In a living organism.

Isozymes (isoenzymes) Multiple forms of an enzyme that differ in properties such as substrate specificity and maximum activity.

Ligation Formation of a phosphodiester bond to link two adjacent bases separated by a nick in one strand of the double helix of DNA.

Lysis Process of cell disintegration or membrane rupturing.

Macromolecules Large molecules, contained within a cell, with molecular weights ranging from a few thousand to hundreds of millions.

Messenger RNA (mRNA) Class of RNA molecules that copies genetic information from DNA in the nucleus and carries it to ribosomes for translation into proteins.

Monosaccharides Chemical building blocks of carbohydrates with the empirical formula $(CH_2O)_n$.

Mutagen Chemical or physical agent capable of producing a genetic mutation in a living organism.

Mutant Product of a mutation or heritable genetic change.

Mutation Any change that alters the sequence of nucleotide bases in the DNA of the cell of an organism.

Nucleic acid Macromolecule containing a phosphate group, a sugar group, and nitrogenous bases.

Nucleoside Building block of DNA and RNA, consisting of a nitrogenous base linked to a five-carbon sugar.

Nucleotide Building block of DNA and RNA, consisting of a nitrogenous base, a five-carbon sugar, and a phosphate group.

Oligomer Relatively short molecular chain consisting of repeating units.

Oligonucleotide DNA polymer consisting of a few nucleotides.

Oligosaccharide Relatively short molecular chain consisting of 10 to 100 simple sugar units.

Organelles Membrane-bound structures with distinct functions contained in eukaryotic cells.

OUR Oxygen uptake rate. OUR is the amount of oxygen consumed per unit time to stabilize waste.

Peptide Two or more amino acids covalently joined by peptide bonds.

Peptide bond Covalent bond between the ± amino group of one amino acid and the ± carboxyl group of another amino acid.

Photosynthesis Process in green plants of converting carbon dioxide and water into sugar using light as the source of energy.

Phytoremediation Use of certain plants to remove contaminants or pollutants either from soil or from the environment.

Polyacrylamide gel electrophoresis (PAGE) Technique of electrophoresis in which molecules are separated on the basis of size and charge in a polyacrylamide gel.

Polymerase chain reaction (PCR) Reaction that uses DNA polymerase to catalyse the amplification of a DNA strand through repeated cycles of DNA synthesis.

Polypeptide (protein) Polymer composed of multiple amino acid units linked with peptide bonds.

Polysaccharide Long chain molecule composed of multiple units of monosaccharides.

PSW Processed solid waste. After removing the inorganic constituents of solid waste, the organic constituents are shredded for size reduction and finally compacted into pellets or cubes to facilitate handling and use in incinerators or other thermal systems.

Protoplast Part of the cell that includes the cell membrane and all intracellular components, except the cell wall.

Prokaryotes Primitive type of micro-organisms.

Recombinant Cell formed by a recombination of genes.

Recombinant DNA Technology of cutting and recombining DNA fragments from different sources.

Recombination Formation of new gene combinations by joining different genes, sets of genes, or parts of genes.

Restriction endonuclease (enzyme) Class of endonucleases that cleaves DNA after recognizing a specific sequence.

Sequencing (of DNA molecules) Process of obtaining the sequential order of nucleotides in the DNA backbone.

SOTE Standard oxygen transfer efficiency. SOTE is the theoretical value of the amount of oxygen transferred to wastewater per unit power consumption.

Stoichiometry Science of balancing the material and energy involved in a chemical or biological conversion process.

Transformation Process of transferring a foreign piece of DNA into a cell.

USW Unprocessed solid waste. USW is solid waste that has not been processed to isolate organic constituents from inorganic constituents or subjected to any size reduction before being dumped into an incinerator or any other treatment system.

Vector Agent used in recombinant DNA research to carry new genes into cells.

Waste stabilization Process of reducing the BOD or COD of organic wastes to render them harmless.

Wild type Natural form of an organism.

Xenobiotic Group of chemicals unfamiliar or foreign to micro-organisms and thus not easily degradable by them.

Index